Insect–Plant Interactions

Volume V

Insect–Plant Interactions

Volume V

Edited by

Elizabeth A. Bernays, Ph.D.
Regents Professor and Head
Department of Entomology
University of Arizona
Tucson, Arizona

CRC Press
Taylor & Francis Group
Boca Raton London New York

CRC Press is an imprint of the
Taylor & Francis Group, an **informa** business

Preface

This is the fifth volume of a unique book series concentrating on in-depth reviews of topics that are particularly important at the present time or are in research areas that are currently breaking new ground. The research areas generally lack recent reviews of any kind elsewhere. For those wishing to be up-to-date on some of the new and controversial elements of insect–plant interactions this series is a must. For those needing access to the literature, surveys and informed value judgments by leaders in the field, or evaluation of research directions, Insect–Plant Interactions will be extremely useful. Volume V contains seven chapters covering six diverse areas that are each of interest and importance to a wide readership.

Ian Baldwin gives a very clear review of chemical changes in plants as a result of insects feeding on their leaves. He provides a well rounded study of the induction of high nicotine levels in tobacco, showing the factors that influence the plant and how the plant invests energy in different chemical products.

Dan Papaj presents the dynamic elements of the use and avoidance of host fruit by tephritid flies, as a function of the presence of other flies. Chemicals produced or damage done by occupants can influence flies that come later in very different ways depending on many aspects of the host, the flies, and a variety of environmental parameters.

Heidi Dobson's review of floral volatiles in insect biology opens up a new area for many researchers interested in insect–plant interactions. She presents a thorough literature review for this exciting research that has expanded considerably in the last few years.

Alison J. Popay and Daryl D. Rowan provide a review of endophytic fungi as mediators of insect–plant interactions. They focus on grasses and demonstrate that the endophytes are widespread and extremely important in mediating host acceptability and host suitability in many cases.

Jonathan Gershenzon takes a new look at the cost of chemical defense against herbivory. He shows that many of the studies previously available in the literature have some problems in terms of biochemical reality. Quoting widely from an expanding literature and using a sound biochemical perspective, Jonathan provides a particularly useful and different review of this topic.

Simon Leather discusses life-history traits of insect herbivores in relation to host quality. What is host quality? How does host quality impact migration? How does insect phenology interact with host quality? What is the best measure of fitness in relation to host quality?

Heidi Appel provides the first available review on physicochemical conditions of the gut lumen from an ecological perspective. Clearly such factors as pH, redox potential, and ionic content have profound effects on digestion, and on how different insect species deal with plant allelochemicals, but, in addition, there are differential effects on the pathogens of the insects. Relatively little is understood so far, but Heidi opens up the field and demonstrates the need for new projects.

Elizabeth A. Bernays, Ph.D.

The Editor

Elizabeth A. Bernays, Ph.D., is Regents Professor of Entomology and Head of the Department of Entomology, and Adjunct Professor of Ecology and Evolutionary Biology, at the University of Arizona, Tucson.

Dr. Bernays graduated in 1962 from the University of Queensland, Australia, with a B.Sc. with honors in zoology and entomology. After a period of high school teaching, she obtained an M.Sc. in 1967 and then a Ph.D. in 1970 from the University of London. The same university awarded her a D.Sc. in 1991 for her contributions to biology. After receiving her Ph.D. she became a scientist on the British government service, and in 1983 was appointed Professor of Entomology and Adjunct Professor of Zoology at the University of California, Berkeley. She took her present position at the University of Arizona in 1989.

Dr. Bernays is a member of several societies, including the Royal Entomological Society, the American Zoological Society, and the Entomological Society of America. Among several awards, she won the 1987 gold medal of the Pontifical Academy of Science and has published 130 scientific papers. Her research is funded by the National Science Foundation. She regularly presents research and review papers at national and international meetings, as well as at universities in various countries. She is assistant editor of three research journals, two books, and the review series Insect–Plant Interactions, of which this is Volume V.

Dr. Bernays is best known for her work on the physiological mechanisms underlying insect–plant relationships. This area of research began with her studies on the regulation of food intake in locusts. Many of the novel laboratory experiments on plant–insect interactions are based on field situations, so that they combine physiological understanding with appropriate behavior or ecology. Through this work she has brought a new understanding of the process of host-plant selection by insect herbivores, ranging from the analysis of chemical stimuli, through acceptance for feeding, to the nutritional implications of the resultant diet. In particular, she has demonstrated that many plant secondary substances, while critical as behavioral cues, are often surprisingly inactive metabolically, and many have subtle benefits—shedding new light on such general questions as diet breadth in insects, and the costs and benefits of different host-plant ranges. Most recently her elegant experiments on predators have highlighted the significance of higher tropic levels in the maintenance of specialized feeding habits.

Advisory Board

Contributors

Heidi M. Appel, Ph.D.
Research Associate
Department of Entomology
Pennsylvania State University
University Park, Pennsylvania

Ian T. Baldwin, Ph.D.
Assistant Professor of Biology
Department of Biological Sciences
State University of New York at Buffalo
Buffalo, New York

Heidi E. M. Dobson, Ph.D.
Assistant Professor
Department of Biology
Whitman College
Walla Walla, Washington

Jonathan Gershenzon, Ph.D.
Assistant Scientist
Institute of Biological Chemistry
Washington State University
Pullman, Washington

Simon R. Leather, Ph.D.
Lecturer in Pest Management
Department of Biology
Imperial College at Silwood Park
Ascot, Berkshire
United Kingdom

Daniel R. Papaj, Ph.D.
Assistant Professor
Department of Ecology and Evolutionary
    Biology
University of Arizona
Tucson, Arizona

Alison J. Popay
Scientist
Pastoral Research Institute of New
    Zealand, Ltd.
Palmerston North, New Zealand

Daryl D. Rowan, Ph.D.
Scientist
Horticulture and Food Research Institute
    of New Zealand, Ltd.
Palmerston North, New Zealand

Contents

Insect–Plant Interactions

Volume V

Chapter 1

CHEMICAL CHANGES RAPIDLY INDUCED BY FOLIVORY

Ian T. Baldwin

TABLE OF CONTENTS

0-8493-4125-6/94/$0.00+$.50

1

I. INTRODUCTION

The chemical characteristics of leaves are remarkably protean. Concentrations of primary and secondary metabolites change throughout ontogeny as leaves develop photosynthetic competence, mature, and senesce.[42,120,141,142] These ontogenetic chemical changes can be influenced by biotic and abiotic environmental factors affecting plant growth. The observation that the concentrations of many secondary metabolites increase after folivory has captured the imagination of the ecological community; at least superficially, the fact of their inducibility appears to support the notion that these metabolites function defensively. Moreover, the inducibility of these metabolites is consistent with the economic framework proposed to explain the patterns of secondary metabolites: inducibility may minimize the costs of producing a chemical defense.[11,73,89,94,168]

Herbivory-induced responses have been the subject of many excellent reviews,[18,27,36,63,70,91,93,109,133,166-168,176,179,203] including one in this series[217] and a recent book.[194a] Induced responses can be studied on three time scales: *preformed-induced responses,* which occur immediately upon damage, are restricted to the damaged tissues, and usually result from the mixing of previously isolated enzymes and substrates (e.g., cyanogenesis); *rapidly-induced responses*, which occur within hours or days of the injury and can be systemic or localized to the damaged leaf; and *delayed-induced responses,* which occur in the next season's foliage or later. The delayed-induced responses have received the most ecological attention because of their potential importance in regulating herbivore populations.[95]

This review takes a phytocentric view of induced responses,[42,96] recognizing that folivory results not only in defense-related changes in secondary metabolism but also in changes in the "civilian" chemistry of a leaf. These changes in primary metabolites result from a suite of physiological responses that contribute to a plant's resilience to herbivory. This review will be restricted to the rapidly-induced responses, because they occur when a growing plant reconfigures its methods of acquiring and partitioning resources in response to folivory.

Folivory is, in many ways, a tolerable form of herbivory. Leaves are engineered for obsolescence, and if an herbivore removes productive leaves prematurely, plants have at their disposal a battery of physiological responses that help them regrow lost tissues. One of these physiological responses — the changes in gas exchange after herbivory — has been thoroughly reviewed in this series.[218] These physiological responses are not limited to the resource-acquiring leaves and roots; they are part of a whole-plant response to damage that influences the patterns of resource allocation and partitioning. Considering induced "defense" responses in the context of the functional reorganization of a plant that occurs after herbivory is important for both ecological and mechanistic reasons. The physiological responses to herbivory may provide mechanisms for the induced changes in secondary metabolism: the "civilian" physiological responses may alter the pools of metabolites that supply "defensive" metabolic pathways. Moreover, since the food quality of a leaf for an herbivore is determined by the interaction of both its primary and secondary metabolites,[48,101,207] understanding folivory-induced changes in plant function is essential for understanding herbivore resistance induced by chemical changes.

II. CHEMICAL DEFENSE THEORIES AND THEIR APPLICATION TO HERBIVORY-INDUCED RESPONSES IN PLANTS

A. PHILOSOPHICAL ISSUES

Two classes of theories address the intraspecific patterns in secondary chemistry: 1) the carbon-nutrient theory (C/N) and the growth-differentiation theory, which emphasize respectively the physiological and ontogenetic constraints imposed by resource availability on the allocation of resources to growth, reproduction, and defense;[13,24,43,127,201,202] and 2) the optimal defense theory, which emphasizes the adaptive value of secondary metabolites based on considerations of tissue value,[132,133] apparency to herbivores,[61,169] or the optimization of defense allocation.[18,168] Although these two classes of hypotheses both make predictions about the interspecific, intraspecific, and within-individual patterns of secondary metabolites, they are not alternative hypotheses, for they are posed at different levels of analysis (see Sherman, Reference 185). Alternatives to the C/N and growth-differentiation theories and the optimal defense theories need to be posed at mechanistic and functional levels, respectively. While the C/N and growth-differentiation theories are not strictly functional-level hypotheses, they have been used as a null model for the defensive, functional-level hypotheses by proposing that damage results in "incidental" changes in secondary metabolism.[26] However, as Tuomi et al. and Haslam have pointed out, the two mechanistic-level hypotheses do not preclude the defensive tailoring of "incidentally" increased "overflow" metabolites.[92,203] This "defensive" view of a damage-induced chemical response derived from the C/N theory differs from the "defensive" view derived from the optimal defense theory on one important point: no tradeoffs are predicted to occur between the allocation to defensively-functioning metabolites and other plant functions, and therefore, the economic argument that inducible defenses evolved as a cost-savings measure would not apply. Other models for the evolution of inducibility that do not invoke the trade-off argument have been proposed.[67,109] For example, an inducible plant presents herbivores and pathogens with a protean, heterogeneous target, to which adaptation might be more difficult than to a static, constitutively defended target.[109]

B. C/N THEORY

This theory argues that allocation of a plant's resources to growth is its highest priority, and allocation to secondary metabolites increases in response to an imbalance in resources needed for growth. Hence plants with an amount of carbon in excess of growth requirements (which, by definition, have a high C to N ratio) would be predicted to produce more carbon-intensive secondary metabolites — such as tannins, phenols, and terpenes — than plants with a low C to N ratio. This theory has successfully predicted the shifts in constitutive levels of carbon-intensive metabolites after nitrogen fertilization, shading, and watering regimes;[27,139,159,202,217] however, other manipulations such as exposure to different CO_2 concentrations have not been consistent with the mechanism set forth by the C/N theory.[105] In addition, many of the patterns of delayed-induced alterations in the phenolic contents of trees are consistent with the C/N theory.[27,200,203] Most of the studies that have examined this model have defined the C to N ratio as the availability of these resources in the plant's external environment. However, it is the internal levels of these resources that determine their availability to a plant,[143] and these internal levels are in a state of flux as plants grow.[216] Since the internal C to N ratio will change as a plant grows, predictable ontogenetic changes in the production of secondary metabolites should occur in concert with the changes in the internal C to N ratio.

The C/N theory was originally formulated to explain changes in carbon-intensive metabolites. Much less has been done with this theory for plants producing nitrogen-intensive defenses, such as alkaloids, cyanogenic glycosides, and proteinase inhibitors.[13] The patterns of constitutive (as opposed to induced) nitrogen-intensive secondary metabolite production appear to be consistent with the C/N theory. High nitrogen supply rates increase the concentration of some but not all of these metabolites.[3,70,103,150] Low nitrogen supply rates, by lowering both the substrate availability and the concentrations of alkaloid biosynthetic enzymes, consistently lower the constitutive concentrations of nitrogen-intensive secondary metabolites[70,217] but not necessarily the proportion of total nitrogen in a secondary metabolite.[8]

Although the general predictions of the C/N theory regarding changes in secondary metabolites induced by defoliation are clear, specific predictions for particular plants are murky. Three considerations make specific predictions difficult. First, specific predictions depend on the type of nutrient imbalance, if any, that is hypothesized to result from the loss of leaf, and we lack both good estimates and methods for measuring such an imbalance. A defoliation-induced nutrient imbalance could result not only from the nutrients lost to herbivores but also from defoliation-induced remobilization of stored reserves and changes in resource acquisition that can alter the C to N ratio of plant tissues. Second, the nutrient imbalances are generally considered to be operating on a whole-plant level, whereas secondary metabolites are frequently synthesized in specific tissues. In order to make specific predictions regarding the effect of different nutrient imbalances on secondary metabolite production, one must consider nutrient imbalances occurring within different parts of the plant where the specific secondary metabolites are synthesized. Thus, an adequate understanding of both the location of the defoliation-induced imbalance and the sites of synthesis must precede specific predictions. Third, the C/N theory makes predictions about the *production* of secondary metabolites; most researchers only measure metabolite pool sizes and do not explicitly measure rates of production. The turnover rate of the metabolite determines whether the metabolite accumulates in response to a particular nutrient imbalance.[165] Because metabolism rates of secondary metabolites have been found to differ dramatically between detached and intact plant parts[46,138] and in response to damage of intact plants,[229] predictions regarding metabolite accumulation are further complicated.

C. OPTIMAL DEFENSE THEORY

This theory predicts that defenses should be allocated to plant parts that have a high probability of attack and contribute significantly to a plant's fitness. The theory thus predicts that the constitutive allocation of defenses to young leaves with high photosynthetic rates and long photosynthetic lifetimes as well as to reproductive parts should be most generous.[133] The predictions of optimal defense theory are less clear with regard to defoliation-induced responses, except when defoliation increases the value of the tissue to the plant or indicates further herbivory. Harper has argued that the value of a leaf is determined by the fate of its exported assimilates.[84] If folivory alters the rate of assimilate acquisition and partitioning, the value of the leaf and its defensive allocation would be predicted to change in concert. Clearly, the ability of the optimal defense theory to predict the patterns of induced defenses rests in part on an understanding of the damage-induced "civilian" responses that alter a leaf's fitness value.

III. PATTERNS OF RAPIDLY INDUCED CHEMICAL CHANGES

Dynamic processes are inherently difficult to study, and induced chemical changes are no exception. As Welter concluded in his review of the effects of herbivory on plant gas exchange, the choice of experimental methods can determine the conclusion drawn.[218] Almost every component of experimental design has been found to influence important parameters of the induced response.[56] How a leaf is damaged can be more important than the quantity of leaf removed.[6] Type of damage has been found to be important in responses as diverse as increases in alkaloids,[4] phenylalanine ammonia lyase (PAL) activity,[87] ethylene production,[112] and photosynthetic capacity.[218] The frequency of damage can influence the magnitude of the response.[5,146] The amount of leaf damage can determine whether the concentration of metabolites increases or decreases. Defoliations of greater than 50% can result in decreases in indole alkaloids[65] and phenolics,[56,211] while less severe defoliations may increase concentrations. The choice of control leaf tissues is similarly important; concluding whether an induced response occurred depends on whether the chemical changes are measured in damaged, adjacent undamaged, or distal undamaged but vascularly connected leaves.[106] The age of the tissues and the growing conditions can influence the strength of the response.[42] Nutrient supply rate can profoundly influence the magnitude of certain induced responses. The absolute magnitude of damage-induced alkaloid accumulation is generally enhanced by fertilization,[7,14,103] while

TABLE 1
Survey of Damage-Induced
"Secondary" Metabolites

Phenolics
    "Total phenolics" numerous studies[10,50,51,56,86,131,145,180,206,211]
    Phenolic glycosides[40]
    Flavonoid resins — decreases[102]
    Isoflavonoids[115,151]
    Pterocarpans[115]
    Coumestans[115]
    Furanocoumarins[231,233]
    Hydrolyzable tannins[55,57,174,180]
    Condensed tannins[56,174,210,211]
Terpenes[71]
    Monoterpenes[129]
    Diterpene acids[28,52,76,214]
    Triterpenes[194]
    Sesquiterpenes[78]
Alkaloids[88]
    Tobacco[3-8]
    Tropane[114]
    Quinolizidine[15,103,104,222,225,227] — no effect[164]
    Pyrrolizidine — no effect[209]
    Indole[65,144]
Silicates[136] — no effect[39]
Volatile hydrocarbons[193,204,220]
Defense-related proteins[21]
    HRGP
    GRP
    PRP45
    Peroxidases[48,62]
    CAD
    Callosesynthase
    Proteinase inhibitors[23,176]
    Amylase inhibitors
    Endohydrolases
    Thaumatin-like proteins
    PAL[37,85-87,111]
    Monoterpene cyclase[72]
    Chalcone synthase[45]
Defense-related hormones[99,112]
Defense-related gene products[21,36,45]

many induced phenolic responses are attenuated under high nutrient supply rates.[27,201,203] The induced responses of tobacco and cotton are also affected by rooting volume;[3,108] plants grown in small pots are not inducible. Given the range of experimental factors which are known to influence the responses to damage, it becomes increasingly difficult to distinguish between a truly unresponsive plant and an experiment conducted under noninducing conditions.

A. SECONDARY METABOLITES: BIOSYNTHETIC PATTERNS

Table 1 lists reports of damage-induced secondary metabolites. As a quick glance underscores, most of the major classes of secondary metabolites are inducible. Thus, few generalizations can be made about intrinsic proclivities toward inducibility in any class of metabolites. The phenolics, as a class of compounds, have received by far the most attention. The multitude of examples of induced phenolics is due in part to a historical phenomenon—phenolics played a seminal role in early plant-herbivore theory,[61,169] in part to technical reasons—the Folin-Denis general phenolic reagent was one of the first second-ary metabolite reagents to enter the ecologist's toolbox, and in part to the ubiquity and responsiveness of the key enzyme of one of the phenolic biosynthetic pathways, PAL.[86]

Many classes of compounds would not be expected to be induced by damage, such as the compounds that make up the preformed-inducible metabolites. Consistent with this expecta-tion, the glucosinolates have yet to be reported as being inducible, but we know of no explicit examinations. Contrary to this expectation, one report gives circumstantial evidence that cyanogenesis in *Cynodon plectostachyus* is induced.[69]

Many metabolites are synthesized in specialized tissue types that are active only during particular periods during leaf ontogeny. Their inducibility may thus be constrained by mor-phological responses requiring new growth for their expression. For example, many terpe-noids are stored in secretory idioblasts, laticifers, and glandular trichomes, cell types active early in leaf development, which may constrain induced responses to the next leaves produced after damage.[71] However, in the case of induced oleoresin production in conifers,[163] the morphological "packaging" of biosynthesis and storage do not constrain inducibility; the quiescent resin acid-producing epithelial cells of fully differentiated resin ducts in *Pinus pinaster* reactivate after damage and increase production.[129,214] Plant development and orga-nization are known for their plasticity,[197] so it should not be surprising to find biochemical plasticity in differentiated tissues and to find that some tissues are more plastic than others. Nitao found no evidence for induction in total or specific furanocoumarins in the fruits of wild parsnip,[148] but Zangerl found dramatic evidence for induction in leaf tissues in the same genotypes of the same species.[231,233]

The overwhelming pattern that emerges from this survey is that plants have a diverse array of secondary metabolites able to be induced upon damage and that the complexity of these responses is only beginning to be perceived. *Nicotiana* leaves, for example, are known to respond to damage with increased concentrations of alkaloids, phenolics, proteinase inhibi-tors, PR proteins, and sesquiterpenoid phytoalexins.[3-5,7,8,11,21,36,78,80,176,177] Even this list is probably incomplete.

B. SECONDARY METABOLITES: TEMPORAL PATTERNS

Temporal variation is the hallmark of an induced response, yet only rudimentary descrip-tions of the phenology of most induced responses are available, probably due to the onerous sample sizes that such a description requires. One obvious pattern emerges from these few studies; the closer the site of damage is to the site of synthesis, the faster the induced response will be. Leaf-produced furanocourmarins, phenolics, quinolizidine alkaloids, and proteinase inhibitors can reach maximum concentrations within 2 d, while root-produced tropane and pyridine alkaloids require 9 to 10 d to reach maximum leaf titers (see References in Table 1). Induced responses that require *de novo* synthesis usually have a lag time of 10 to 24 h before induced accumulation is detected. Responses that occur in a matter of minutes (e.g., induced cucurbitacins) are likely to be synthesized previously and transported to the site of damage.[195] The duration of the response has been studied only for quinolizidine,[103,104] tropane,[114] and pyridine alkaloids,[5] proteinase inhibitors,[176] and PAL activity.[37] For many of these responses, the accretion of the metabolite lasts about as long as the waning of the response, provided the plant is damaged only once. The duration of the accretion phase can be prolonged if more than one episode of damage is administered.[5] Induced immune responses in animals are character-ized by long-term memory of previous damage events,[91] but nothing is known about this possibility in plants.

The temporal pattern of chemical change must be interpreted in light of the changes that occur throughout the ontogeny of a leaf.[56] The ontogenetic changes in the pool sizes of many secondary compounds follow the overall patterns predicted by the C/N theory:[24,43] nitrogen-intensive metabolites tend to be higher in young leaves (e.g., alkaloids[213]) and decrease in concentration with leaf age, while carbon-intensive metabolites exhibit the opposite pattern (e.g., tannins[56,60]). Some of these associations of secondary metabolite pools with leaf internal C to N ratio may, however, be more apparent than real. For example, the monomeric indole alkaloids, vindoline and catharanthine, decrease in concentration as a leaf ages,[65,144] but this decrease is likely due to their conjugation to form the dimer, 3'-4'-anhydrovinblastine, which increases in concentration with leaf age.[144] Clearly, more information is needed on how rates of production and accumulation change through ontogeny.

C. SECONDARY METABOLITES: MECHANISMS

Five partially overlapping mechanistic models have been proposed to explain how secondary metabolites change after damage. The first three models argue that secondary metabolites accumulate in response to imbalances existing between growth-related processes and metabolite production. The three hypotheses differ in how this imbalance is perceived.

1. C/N theory (see Section II) views the carbon and nutrient requirements of growth and the availability of these resources to a plant as being imbalanced. Resources existing in excess of growth requirements are shunted into secondary metabolism;
2. Substrate/enzyme imbalance theory[92,217] argues that secondary metabolites accumulate as a result of "overflow" metabolism and emphasizes differential enzyme compartmentalization and regulation;
3. Growth/differentiation balance theory recognizes that all secondary metabolites have an ontogenetically determined phenology, and their synthesis is emphasized during periods of differentiation.

Presumably, folivory-induced changes in metabolite production are a result of induced imbalances in one of these three processes. The remaining two theories argue that damage results in signals which directly activate secondary metabolism, but the two theories differ in the specificity of the signalling system:

4. Generalized Stress Response (GSR) Theory;
5. Active Defense Response (ADR) Theory.[18,36]

Chapin has proposed that plants have a hormonally mediated (e.g., abscisic acid, cytokinin) centralized system of physiological responses for coping with water and nutrient stress.[33] Stress-related plant hormones are some of the signals that activate some damage-induced secondary metabolites (abscisic acid,[156,158] auxin,[113] and jasmonic acid[58] are signals for proteinase inhibitor production) and strongly influence the induced response in others (auxin on pyridine[5] alkaloids). Since wounding is known to alter the concentrations of many plant hormones,[45,59,99] it seems reasonable that this GSR might also be responsible for some of the observed induced secondary metabolite accumulations. The ADR theory[18] postulates that damage cues or cues specific to the intruding organism activate specific defense responses. The ADR theory appears to describe many phytoalexin responses[2,44] and many of the induced responses found in colonial invertebrates as well.[89,91] The revolutionary discovery from Ryan's laboratory of the first polypeptide plant hormone,[155] which at femtomolar concentrations transcriptionally activates the proteinase inhibitor genes in tomato, lends substantial support to the ADR theory.

The debate over what controls secondary metabolism is analogous to the debate regarding positive and negative controls on plant growth.[197,198] Although this oversimplification does a disservice to some of the above theories, the five theories can be codified into two groups, according to their respective views on the regulators of secondary metabolism. The three balance theories see changes in secondary metabolism as being under negative control, mediated by the use of cues or substrates by competing sinks; the latter two theories emphasize enzyme regulation by direct control with damage cues. The regulation of metabolites derived from the phenyl propanoid pathway is perhaps the best-studied in this regard. Strong evidence exists for both types of regulation. The evidence for substrate regulation comes principally from whole-plant manipulations of external resources and from delayed-induced responses (see Section II.B). We know of no studies that explicitly examine a rapidly-induced response in the context of substrate control over secondary metabolism, although a case could be made for its consideration (see Section III.E). The evidence for direct enzymatic control over induced phenylpropanoid metabolism is overwhelming and recently reviewed.[21,81] The key enzyme in this pathway, PAL, exists in multiple isomeric forms, which are differentially transcriptionally induced by distinct stimuli (wounding, UV light, polysaccharide elicitors). The induction of PAL is in some cases known to be coordinated with the transcriptional activation of other enzymes (for example Co-A ligase) later in the biosynthetic sequence for the production of the appropriate metabolite by the appropriate stimuli. The view that has emerged from molecular biological studies of phenylpropanoid metabolism as being highly choreographed and precisely regulated by external stimuli appears difficult to reconcile with the view that many of its products are the "flotsam and jetsam on the metabolic beach"[92] derived from an "overflow" pathway; however, it is possible that pathways are regulated differently depending on their cellular localization, as has been proposed for the cytosolic and plastid-localized components of phenolic metabolism (see references in Reference 217). Only more detailed mechanistic studies will determine the relative importance of these two types of regulation.

D. PRIMARY METABOLITES

Once an herbivore alters the distribution of resources in a plant, the damaged plant responds by altering the distribution of resources in the remaining, uneaten tissues to regain balanced growth. Resource allocation theory predicts that resources should be allocated so that all resources should equally limit growth.[19] Because folivores remove photosynthetic tissues, the whole-plant responses include decreases in root growth,[31,54] movement of reserves from root to shoot,[20,29,192] increases in resource acquisition in roots and shoots,[29,32,34,175,218] activating new meristems,[29,130,219] and altering of patterns of leaf senescence.[149] These are mediated responses that result in the reconfiguration of plant function,[68] and they are likely to be every bit as complicated as changes in secondary metabolism. These are not universal responses. The types and magnitudes of these physiological responses depend on the ecological and evolutionary history of the plant[77] and have been found to differ even among genotypes of the same species that differ in their grazing history.[49] Thus, whether a plant increases its allocation of stored or newly acquired assimilates to below-ground organs, new meristems, or remaining leaves is as likely to be an evolutionary response to folivory as is the production of toxic metabolites.

All of these physiological processes influence the distribution of primary metabolites in leaves after damage (see Table 2 for examples). The ecological literature is replete with reports of changes in leaf nitrogen content after herbivory, and all possible responses are reported. In woody plants, growth rate, evergreenness, and whether the herbivory removes meristems influence the nitrogen dynamics after folivory.[25,27] The nitrogen mobilized from root to shoot after folivory can come from previously stored reserves or from supplies acquired after damage. The postfolivory increases in leaf nitrogen accumulation may result from a damage-induced increase in root-specific uptake[175] and from alterations in soil nitrogen cycling processes that increase the availability of nitrogen in the soil solution.[98] In perennial ryegrass, movement of nitrogen into new growth occurs primarily from organic nitrogen remobilized from roots and stubble during the first 6 d after damage and from newly acquired and assimilated nitrogen for the remainder of the regrowth process.[153] Carbohydrate pools are in

TABLE 2
Survey of Damage-Induced
"Primary" Metabolites

Proteins
    "Total proteins"[57,210,211]
    Photosynthetic proteins[117,118,215] and pigments[230]
    Storage proteins[188,189]
Carbohydrates[49,74,135,192,205]
Nutrients
    NO_3[153,154]
    Ca[205]
    K[131,145,175,205]
    Na[145]
    P[32,34,145]
    Cu[131]
    Na[131]

a similar state of flux after damage, though plants appear to mobilize a smaller proportion of their stored carbohydrate reserves than of their nitrogen reserves.[33,97] Nitrate frequently accumulates in high concentrations after damage; it is thought to play an osmoregulatory role, helping to drive damage-induced increases in the rates of leaf elongation.[154,190,228] Once reduced, the remobilized nitrogen likely plays important roles in supplying the nitrogen demands of leaf growth, increases in carboxylation capacity, and defense.

The best-documented physiological response to leaf damage is the increased rate of photosynthesis in the undamaged leaves of damaged plants as compared to that of same-aged leaves or new leaves of undamaged plants.[12,29,116,117,215,218,230] Increased rates of photosynthesis are frequently associated with increases in photosynthetic pigments and RuBPCase (ribulose bisphosphate carboxylase/oxygenase), the primary carboxylating enzyme of C3 plants (and the most abundant protein, representing approximately 10% of total leaf protein or approximately 50% of soluble protein[53]). The coordination of changes in photosynthetic capacity and protein content of leaves after damage has been used to argue that the carboxylating capacity of leaves, as determined by RuBPCase concentration, is one of the principal rate-limiting steps in photosynthesis.[215] However, recent experiments with plants transformed with antisense genes for the small subunit of RuBPCase indicate that tobacco plants typically function with 15 to 40% more leaf protein in RuBPCase than is needed to avoid carboxylation capacity limitations on photosynthesis.[161,162] Since this enzyme plays a role in storage,[140] the damage-induced accumulation of this protein may reflect increases in leaf nitrogen storage. In this regard, it is interesting to note that leaf damage is also known to regulate transcriptionally the rapid accumulation of vegetative storage proteins in soybean leaves.[188]

E. INTERACTIONS BETWEEN INDUCED PRIMARY AND SECONDARY METABOLITES

Wagner was one of the first to report induced increases in total phenolics and condensed tannins co-occurring with increases in protein contents.[211] Few studies have examined the coordination of these responses in detail. Most of the studies examining the delayed induced phenolic responses have monitored changes in leaf nitrogen contents in order to test the predictions of the C/N theory. Few studies of rapidly-induced responses have even monitored this parameter. Rapidly-induced furanocoumarins occur without any change in leaf nitrogen content,[231] while increases in tobacco alkaloids are coordinated with increases in photosynthetic capacity, protein, nitrate, and total nitrogen contents.[8] A number of studies have found regrowth foliage to be higher in both primary (protein content) and secondary (tannins[57] or alkaloids[207]) metabolites as compared to mature foliage, but since the ontogenetic changes in undamaged plants are not well understood, we cannot ascertain whether the chemistry of regrowth foliage differs from that of primary-growth leaves at the same stage in ontogeny. Given that senescence is a reversible process in plants,[149] many of the rapidly-induced chemical changes that occur in plants will probably resemble the process of rejuvenilization. By examining the chemical dynamics induced by folivory in the context of the changes that occur during leaf ontogeny, we will be able to understand just how tightly coupled the folivory-induced suites of changes are.

Linkages between the induced changes in primary and secondary metabolism are relevant for both classes of mechanistic hypotheses proposed to explain changes in secondary metabolism. The increases in secondary metabolites may be a consequence of altered C to N ratio in plant tissues, resulting from a physiological response to damage. For example, Clausen et al. have proposed that the rapidly induced increases in the phenolic glycosides, salicortin and tremulacin, may be a consequence of increases in leaf photosynthetic capacity after damage.[41] This increased carbon assimilation rate may result in the shunting of assimilates into phenolic biosynthetic pathways rather than in new growth. Similarly, the damage-induced increase in root-produced tropane and pyridine alkaloids may be a result of lowered root C to N ratio, which, in turn, could result from a damage-induced stimulation of root nitrogen uptake. Even if these induced responses in secondary metabolites were in fact a consequence of an induced physiological response (such as increased photosynthetic capacity or root nitrate uptake), the question of why the induced physiological response would shunt the newly acquired resources into secondary metabolic pathways rather than into growth would remain. Clearly, it is very difficult to ascribe "defensive" or "incidental" labels to responses that result from alterations in a plant's resource balance without clearly understanding the controls over the allocation and partitioning of these resources.

Direct evidence of coordination among induced responses comes from the discovery that apparently damage-specific cues appear also to affect processes that are clearly part of primary metabolism. Treatment of potato leaves with polysaccharide fungal elicitors results in the rapid transcriptional activation of a series of defense-related genes both locally in the treated leaf (PAL; 4-coumarate : CoA ligase; chitinases; PR proteins[21]) and systemically in adjacent leaves (1,3-glucanases[119]). Kombrink and Hahlbrock have found that these same elicitors also result in the rapid systemic transcriptional repression of the small subunit of RuBPCase.[118] Thus, some induced defense responses may also involve rapid regulation of primary metabolism. This coordination may be specific to responses elicited by fungal attack, for the systemic induction of proteinase inhibitors by wounding in potato did not influence the genes encoding for the small subunit of RuBPCase or the tuber storage protein, patatin.[157] Farmer and Ryan have presented a model for the damage-induced activation of proteinase inhibitor genes.[59] They propose that both long- and short-distance damage cues (systemin and PIIF, respectively) interact with plasma membrane receptors, which, in turn, activate a secondary damage messenger signal (jasmonic acid) and the genes coding for proteinase inhibitors. Interestingly, jasmonic acid is also known to transcriptionally induce vegetative storage proteins in soybean leaves.[189] These proteins are involved in temporary nitrogen storage and may be the "civilian" analogue in soybean of the defense proteins, proteinase inhibitors. Alternatively, the role of jasmonic acid in mediating damage responses may be an example of a generalized stress response, where stress-related hormones influence changes in both primary and secondary metabolism. Defense and regrowth may indeed be alternative evolutionary strategies for coping with herbivory,[160,208] but it is more likely that most plants employ both strategies.

IV. FUNCTIONAL ROLES FOR RAPIDLY INDUCED CHEMICAL CHANGES

Many of the previous reviews of induced responses have wrestled with the difficult question of whether these changes are evolved "defense" responses or incidental changes resulting from herbivory. The perceived intractability of this question[147] stems in part from the lack of alternative functional-level hypotheses to the defensive hypothesis for induced responses. This lack of alternative hypotheses stems in part from the assumption that "secondary" metabolites are truly secondary, that they are not playing any particular role in the primary metabolism of plants. Although a number of authors have stressed the possible primary roles of secondary metabolites and argued that we need to know more about the biochemistry and physiology of secondary metabolism before we can understand the evolution of chemical defense,[133,181-183,213] such a view is only beginning to be incorporated into current research efforts.[35,165,194] This section updates Seigler's review of alternative functional roles for secondary metabolites that could provide alternatives to the purely defensive hypothesis for induced secondary metabolites.[181] The evolutionary context for induced responses should recognize the varied functional-level roles that these metabolites may be playing and the constraints that these roles impose on defensive functions. As Tallamy has pointed out, labeling compounds as "defensive" is to imply that they are a chemical adaptation "derived and maintained for the singular purpose of defending their producers from herbivores" is an oversimplification.[195] More importantly, this view fails to recognize that the process of evolution is analogous to the process of tinkering,[100] whereby evolutionary "problems" are solved by modifications of the available materials.

A. ECOLOGICAL ROLES

The antiherbivore and antipathogen functional roles for induced secondary metabolites have been the principal focus of the previous reviews and therefore will be discussed here only in terms of the specificity of the responses. A plant's response to pathogen invasion can be specific both to the attacker and to the elicited response.[2,44,186] The antiherbivore responses appear to be more generalized. Induction of resistance by either spider mites or *Verticillium* fungus has reciprocal effects on the other species.[107] The alkaloids of lupine and tobacco, and the isoflavonoids, pterocarpans, and coumestans of soybean that are induced by leaf damage and herbivore attack have deleterious consequences for a variety of other predators and pathogens.[115,121,222,223,225] Moreover, many insecticidal allelochemicals also have fungicidal and bactericidal properties.[121] Damage-induced PAL activity and total phenolic contents were found to be more pronounced in response to leaf damage caused by insects, nonsterile scissors, and scissors coated with insect regurgitate than from sterile scissors,[86,87] indicating that these phenolic responses may have been activated by microbial contamination of the wound site. Interestingly, the largest responses found in this study occurred in leaves adjacent to the damaged leaves rather than at the wound site.[87]

Although these results suggest that a plant's responses to leaf damage are not specific, most antiherbivore studies, unlike antipathogen studies, have not employed methods that would allow detection of specificity in either the signaling of the damage response or in their consequences for the attacker. Most studies of induced resistance to herbivore attack have not explicitly compared responses to different herbivores or to an appropriate mechanical damage treatment that mimicked the timing, magnitude, and location of herbivore damage.[6] Moreover, most studies have not analyzed the plant's response with more than one bioassay organism or with much chemical resolution. Studies that have analyzed the chemical responses in detail and examined the sensitivity of the herbivore to this chemical complexity have found evidence suggestive of "defensive" tailoring of the chemical response. For example, feeding by *Trichoplusia ni* larvae resulted in approximately two-fold increases in the concentrations of five furanocoumarins within 48 h in wild parsnip. The chemical responses induced by insect feeding were greater than those induced by mechanical damage, and of the two induced furanocoumarins that were examined in artificial diets, the compound to which *T. ni* was most sensitive, xanthotoxin, attained the highest concentrations in induced leaves.[231] Tests with more compounds and herbivores will be needed before the case for defensive tailoring of induced responses is robust.

If these induced chemical responses are part of a generalized stress response of plants, then the scope of biotic interactions affected by these responses must be broadened to include competition and facilitation. Many of the compounds that are induced by folivory are potent allelopathic compounds,[212] affecting processes as diverse as seed germination (e.g., alkaloids[38,221,223]) and rates of soil nitrification (e.g., tannins[9]).

B. PHYSIOLOGICAL ROLES

Unfortunately, in the 15 years since Seigler argued that secondary metabolites might have primary roles,[181] this topic has received little attention. Some secondary metabolites clearly have primary roles as growth regulators[172,191] or modulators of plant hormones (e.g., gibberellins[75,79]) or as signals mediating responses as diverse as nitrogen-fixation symbiosis and resistance to Tobacco Mosaic Virus infection (e.g., flavonoids[83] and salicylic acid,[128,137] respectively). However, these functions occur at μM to nM concentrations and are not likely to explain why plants produce the large (mM) increases after damage. Folivory results in wounding; wounds lose water[152] and other volatile compounds[193,204,220] and permit the entry of pathogens, all of which are ecologically and physiologically important. Increased synthesis of phenolics at the wound site and transport to the wound site are likely important in sealing the wound, but this physiological role does little to explain why plants accumulate these compounds systemically without invoking the anticipation of future wounding. Two of the many primary roles Seigler proposed are relevant to the large, damage-induced increases in secondary metabolites and will be considered in greater detail: carbon and nitrogen storage and transport and photosystem protection.

1. Carbon and Nitrogen Storage and Transport

Nowacki et al. proposed that enhanced alkaloid production under high nitrogen supply rates was a means of avoiding amino acid toxicity.[150] We are unaware of any evidence that folivory results in amino acid toxicity or that alkaloids are any less toxic than amino acid transport compounds. Alkaloids deter attacking organisms more effectively than do amino acids, and folivory mobilizes the transport of carbon and nitrogen reserves from root to shoot in order to regrow lost tissues and regain balanced growth. Cyanogenic glycosides,[183,184] purine,[64] and quinolizidine[226] alkaloids and nonprotein amino acids[173] are known to be stored at high concentrations in the endosperm of seeds, metabolized, and used by the developing seedling. Although these specific compounds are not likely to play a long-distance nitrogen or carbon transport role in a mature plant, tropane[114] and tobacco[5] alkaloids are synthesized in large quantities in the roots in response to folivory and transported to the remaining leaves. These damage-induced alkaloids may function as a well-defended transport system for reduced nitrogen from root to shoot after damage, but this hypothesis remains untested. Similarly, if damage-induced phenolic accumulation results from "overflow" metabolism, the accumulated phenolics could be utilized as carbon transport and storage compounds.

The transport and storage hypothesis for induced secondary metabolites depends critically on whether these compounds are catabolized and whether the products are returned to primary metabolism. Turnover rates of secondary metabolites have played a prominent role in recent chemical defense theories,[43] because high rates of turnover increase both the direct and indirect costs of employing a secondary metabolite as a defense.[73] Unfortunately, the turnover of secondary metabolites has not received the attention it deserves, and the available results depend on the techniques employed. Many of the turnover studies have introduced a labeled metabolite into a detached plant and followed the loss of label from the metabolite pool. The loss of label from a pool of metabolites is only the first step of many required for complete catabolism. For example, nicotine has been reported to be metabolically labile in tobacco plants[199] with a half-life of less than 1 d.[82,124,125] However, some of the these studies did not use ring-labeled nicotine and thus measured the demethylation of nicotine to nornicotine.

Nornicotine normally accumulates in most members of the *Nicotiana* genus,[178] and does not necessarily indicate further catabolism.[213] Moreover, the use of detached organs may significantly influence the rates of turnover. In an elegant study, Mihaliak et al. re-examined rates of monoterpene turnover in peppermint and found that the rapid (10 h) turnover reported from studies using detached cuttings was not detectable in intact plants.[138] Similar results have been reported for the turnover of indole alkaloids in *Catharanthus roseus,* where cuttings had higher rates of loss of the monomeric alkaloids than did intact plants.[46] However, damage may result in higher rates of dimerization of the monomers, and thus the apparent loss of monomers may have been due not to alkaloid catabolism[144] but, rather, to alkaloid dimerization. In a study using ring-labeled nicotine, Yoshida found that topping of a tobacco plant dramatically reduced alkaloid degradation in comparison to that measured in undamaged plants.[229] In this study, complete alkaloid degradation occurred, for labeled nitrogen from the alkaloid's two rings was recovered from the plant's protein and amino acid pools. However, treatments were not replicated in this study, and the results need verification.

Thus the available but scanty evidence indicates that the rate of catabolism of mobile secondary metabolites is influenced by prior damage to the plant. A transport and storage role for induced secondary metabolites would predict that mobile carbon- and nitrogen-intensive metabolites, such as monoterpenes and alkaloids, would be catabolized more quickly under conditions of carbon- and nitrogen-limited growth, respectively. Defoliation of a deciduous plant results in relative foliar carbon-stress and foliar nitrogen-surplus (if root reserves are readily mobilized). If the pattern holds that monoterpenes are more quickly degraded while alkaloid catabolism is depressed after damage, the available evidence would be consistent with this prediction. Catabolism may represent both a plant's "cost recovery" program for its investment in chemical defense and a means of mobilizing reserves needed for regrowth processes. If secondary metabolites are both toxic and metabolically unavailable to the aggressor, the storage of reserves in these structures may be favored when the probability of leaf loss is great; thus, the rate of catabolism may have evolved to reflect a plant's "perception" of the probability of leaf loss in the near future. It should be noted, however, that even for labile metabolites such as alkaloids, for which there is strong evidence for complete catabolism, few studies have examined whether alkaloid catabolism can sustain growth under nitrogen-limiting conditions. In the only study of which we are aware that addresses this important question, Wink found that lupine cell suspension cultures grown on agar with the quinolizidine alkaloid, sparteine, as the sole nitrogen source survived longer than cells grown on agar without nitrogen.[224] Thus, it is not clear whether plants can turn their inducible "guns into butter."

2. Protection of the Photosystem

Depending on the severity of the damage, folivory can change the amount and spectral quality of solar radiation arriving at a leaf. Grazing can result in the instantaneous exposure of lower canopy, shade-adapted leaves to full sunlight. Moreover, these changes in light intensity and quality occur concurrently with a reorganization of water, light, and nitrogen use in photosynthetic processes. The amount of inactive RuBPcase (see Section III. D) not used in carboxylation may be reduced as a plant reallocates nitrogen reserves for canopy regrowth. Decreasing RuBPCase pools may increase thylakoid energization, decreasing light-use efficiency.[162] The combination of these internal and external changes may mean that the photosynthetic surface area remaining after folivory may be more vulnerable to photo-oxidative damage.

Secondary metabolism is known to be responsive to the light environment of a plant. The influence of variations in visible light on secondary metabolism has been reviewed in this series.[217] Most thoroughly documented are the UV-induced increases in small phenolics and flavonoids,[16,80,81,170] but increases in alkaloids,[1,110,196] cucurbitacins,[194] and furanocoumarins[232] have also been reported. The observations that these compounds accumulate in response to UV irradiance and that UV light regulates many of the important regulatory enzymes of the phenylpropanoid and flavonoid biosynthetic pathway[81] suggest that they may protect plants from the deleterious consequences of this radiation. Additionally, the latitudinal clines in the frequency of alkaloid-containing taxa[126] suggest that the occurrence of alkaloids may have evolved in response to the latitudinal decrease in UV-B irradiance.[30]

Protection from changes in radiation is thought to consist of two processes that may involve secondary metabolites: avoidance of the effects of radiation by 1) shielding and 2) quenching of the reactive free radicals and excited singlet and triplet species produced when photosensitizers absorb excess visible or UV radiation.[17,47] Shielding is largely achieved by pigments with large molar extinction coefficients in the UV and visible wavelengths, which accumulate in epidermal layers, frequently in response to exposure. These "sunscreens" may decrease the penetration of radiation into the mesophyll tissues, reducing the rates of damage to keep pace with the rates of repair[171] or to reduce the amount of excitation energy in the photochemical apparatus. Some of these pigments are secondary metabolites that are known to be induced by leaf damage, such as flavonoids, phenolics, or alkaloids.

The second component of protection, that of quenching of reactive species before they cause photo-oxidative damage, is accomplished by an enzymatic system utilizing catalases and superoxide dismutases in combination with the antioxidants, reduced ascorbate and glutathione.[47] Interestingly, some plant alkaloids,[17,122,123] including two that are known to be induced by folivory (nicotine and atropine), are potent quenchers of singlet oxygen. *In vitro* measurements of the rate of $^1O_2$ quenching by nicotine are very high[123] at concentrations ($\beta = 0.26$ mM) that are well within the range found in *Nicotiana* leaf tissues.[5] Moreover, nicotine has been found to increase in concentration after exposure to UV-A and -B radiation.[1,110,196] Thus, in so far as folivory alters the amount and spectral quality of solar radiation impinging on a leaf, some inducible secondary metabolites may be playing a physiological role in protecting the photosystem against photo-oxidative damage.

V. SYNTHESIS AND FUTURE RESEARCH DIRECTIONS

The question of whether the rapid changes in seondary metabolism that are induced by herbivory are "incidental" or "evolved" is central to much of the interest in this form of chemical plasticity. Barring the unlikely possibility of actually observing the evolution of this trait in a species, the evidence brought to bear on this question will be indirect and will come from multiple levels of analysis. Understanding the phylogeny, ontogeny, and mechanism of these plastic responses will complement and refine tests of their functional roles in a plant; eventually, the combination of these approaches will resolve the "incidental versus evolved" debate. While framing the study of induced chemical plasticity as a debate about adaptation may not be the most productive research approach, it provides a framework to address the implications of the principal take-home message of this review, the fact that these changes in secondary metabolism are only one and, perhaps, only a small portion of the biochemical plasticity that is induced by folivory. The coordination, or lack thereof, between induced changes in primary and secondary metabolites and among different secondary metabolites induced in a plant will profoundly influence how the "incidental versus evolved" debate is framed at each of the levels of analysis.

At a mechanistic level of analysis, we are confronted with fundamental questions about regulation of both primary and secondary metabolism. Are the damage-induced changes in secondary metabolism the metabolic consequences of induced changes in primary metabolism?

Or are they responses regulated by damage-specific cues, as appears to be the case in damage-induced changes in proteinase-inhibitors? Clearly plants have multiple defense systems, yet little is known about the coordination among the various defense responses. Are there generalized defense responses that involve the coordination of multiple metabolic pathways producing different chemical defenses? How specific are the damage cues that activate the various induced responses? Is there competition among the inducible responses? While mechanisms do not determine functional roles, detailed mechanistic understanding will inform the "evolved versus incidental" debate. Induced responses that are transcriptionally regulated by herbivore-specific damage cues are more likely to be an evolutionary response to herbivory than ones resulting from "overflow" metabolism.

At a functional level of analysis, recognition of the multiplicity of chemical changes that result from herbivory highlights the practical difficulties in conducting tests of the defensive utility of a particular chemical response to damage. Clearly, comparing damage rates or suitability of previously damaged tissues with those of undamaged tissues is likely to be a poor test if a suite of primary metabolites is also induced that increases the suitability of the tissue to herbivores. Examinations of the functional importance of induced chemical plasticity need to recognize the potentially conflicting demands of defense and regrowth; given that some secondary metabolites may be playing a part in both functions, this may not be an easy task.

The question of whether the allocation of resources to chemical defense competes directly with allocations to growth and reproduction is a critical prediction distinguishing the C/N and optimal defense hypotheses; the former theory predicts the absence of resource trade-offs between defense and other functions, while the latter predicts the existence of tradeoffs. Chapin et al. have proposed an economic framework that defines resource allocations to the biochemical processes of growth, storage and defense as being allocations that compete with each other.[35] Whether they do in fact compete is the critical question. Since plants reorganize their patterns of resource acquisition, allocation, and partitioning in response to herbivory, descriptions of resource allocation to growth and defense processes in the newly reconfigured plant provide a unique opportunity to examine a plant's allocation priorities. However, allocations to regrowth and defense may have evolved as coordinated processes. The optimal defense theory would predict that allocations to defense and regrowth should be coordinated when herbivory results in physiological rejuvenilization of older tissues that increase their fitness value to a plant. Comparisons of induced chemical plasticity in leaves with that in other tissues that are less likely to physiologically rejuvenilize (e.g., fruits) might provide insights into coordination among induced responses. The coordination of physiological and defensive responses to damage highlights a need to better understand the ontogeny of secondary metabolite production and accumulation in unattacked plants. The normal ontogenetic plasticity in metabolite production and pool size can serve as a null model against which damage-induced plasticity can be compared.

The future study of induced chemical responses at all levels of analysis will be technically demanding, yet the rewards will undoubtedly be great. Studying the interactions among the multitude of induced responses may provide a unique opportunity to address Stahl's century-old question: How much of what we call a "plant" is generated by its interactions with other organisms?[187]

ACKNOWLEDGMENTS

This research was supported by National Science Foundation grants BSR9157258 and BSR-9118452. Emily Wheeler is thanked for superlative editorial assistance.

REFERENCES

1. **Anderson, R. and Kasperbauer, M. J.**, Chemical composition of tobacco leaves altered by near uv and intensity of visible light, *Plant Physiol.*, 51, 723, 1973.
2. **Ayers, A. R., Goodell, J. J., and DeAngelis, P. L.**, Plant detection of pathogens, *Rec. Adv. Phytochem.*, 19, 1, 1985.
3. **Baldwin, I. T.**, Damaged-induced alkaloids in tobacco: pot-bound plants are not inducible, *J. Chem. Ecol.*, 4, 1113, 1988.
4. **Baldwin, I. T.**, The alkaloidal responses of wild tobacco to real and simulated herbivory, *Oecologia*, 77, 378, 1988.
5. **Baldwin, I. T.**, The mechanism of damage-induced alkaloids in wild tobacco, *J. Chem. Ecol.*, 15, 1661, 1989.
6. **Baldwin, I. T.**, Herbivory simulations in ecological research, *TREE*, 5, 91, 1990.
7. **Baldwin, I. T.**, Damage-induced alkaloids in wild tobacco, in *Phytochemical Induction by Herbivores*, Raupp, M. J. and Tallamy, D. W., Eds., John Wiley & Sons, New York, 1991, chap. 2.
8. **Baldwin, I. T. and Ohnmeiss, T.**, Coordination of photosynthetic and alkaloidal responses to leaf damage in pot-bound and inducible *Nicotiana sylvestris* plants, *Ecology,* submitted.
9. **Baldwin, I. T., Olson, R. K., and Reiners, W. A.**, Protein binding phenolics and the inhibition of nitrification in subalpine balsam fir soils, *Soil Biol. Biochem.*, 15, 419, 1983.
10. **Baldwin, I. T. and Schultz, J. C.**, Damage- and communication-induced changes in yellow birch leaf phenolics, *Proc. N. Am. For. Biol. Workshop*, 25, 1984.
11. **Baldwin, I. T., Sims, C. L., and Kean, S. E.**, Reproductive consequences associated with inducible alkaloidal responses in wild tobacco, *Ecology*, 71, 252, 1990.
12. **Bassman, J. H. and Dickmann, D. I.**, Effect of defoliation in the developing leaf zone on young *Populus X euramericana* plants. I. Photosynthetic physiology, growth, and dry weight partitioning, *For. Sci.*, 28, 599, 1982.
13. **Bazzaz, F. A., Chiariello, N. R., Coley, P. D., and Pitelka, L. F.**, Allocating resources to reproduction and defense, *BioScience,* 37, 58, 1987.
14. **Bentley, B. L. and Johnson, N. D.**, Plants as food for herbivores: the role of nitrogen fixation and carbon dioxide enrichment, in *Plant–Animal Interactions: Evolutionary Ecology in Tropical and Temperate Regions*, Price, P. W., Lewinsohn, T. M., Fernandes, G. W., and Benson, W. W., Eds., John Wiley & Sons, New York, 1991, 257.
15. **Bentley, B. L., Johnson, N. D., and Rigney, L.**, Short-term induction in leaf tissue alkaloids in lupines following experimental defoliation, *Am. J. Bot.*, 74, 646, 1987.
16. **Berenbaum, M.**, Effects of electromagnetic radiation on insect–plant interactions, in *Plant Stress–Insect Interactions*, Heinrichs, E. A., Ed., John Wiley & Sons, New York, 1988, 167.
17. **Berenbaum, M. and Larsen R. A.**, Environmental phototoxicity: solar ultraviolet radiation affects the toxicity of natural and man-made chemicals, *Environ. Sci. Technol.*, 22, 354, 1988.
18. **Berryman, A. A.**, Towards a unified theory of plant defense, in *Mechanisms of Woody Plant Defenses Against Insects*, Mattson, W. J., Levieux, J., and Bernard-Dagan, C., Eds., Springer-Verlag, New York, 1988, 39.
19. **Bloom, A., III, F. S. C. and Mooney, H. A.**, Resource limitation in plants — an economic analogy, *Annu. Rev. Ecol. Syst.*, 16, 363, 1985.
20. **Bokhari, U. G.**, Regrowth of western wheatgrass utilizing ^{14}C-labeled assimilates stored in belowground parts, *Plant Soil*, 48, 115, 1977.
21. **Bowles, D. J.**, Defense-related proteins in higher plants, *Annu. Rev. Biochem.*, 59, 873, 1990.
22. **Bradshaw, H. D. J., Parsons, T. J., and Gordon, M. P.**, Wound-responsive gene expression in poplars, *For. Ecol. Manage.*, 43, 211, 1991.
23. **Broadway, R. M., Duffey, S. S., Pearce, G., and Ryan, C. A.**, Plant proteinase inhibitors: a defence against herbivorous insects?, *Entomol. Exp. Appl.*, 41, 33, 1986.
24. **Bryant, J. P., Chapin, F. S., III, and Klein, D. R.**, Carbon/nutrient balance of boreal plants in relation to vertebrate herbivory, *Oikos*, 40, 357, 1983.
25. **Bryant, J. P., Danell, K., Provenza, F., Reichardt, P. B., Clausen, T. A., and Werner, R. A.**, The effects of mammal browsing on the chemistry of deciduous woody plants, in *Phytochemical Induction by Herbivores*, Tallamy, D. W. and Raupp, M. J., Eds., John Wiley & Sons, New York, 1991, 135.
26. **Bryant, J. P., Heitkonig, I., Kuropat, P., and Owen-Smith, N.**, Effects of severe defoliation on the long-term resistance to insect attack and on leaf chemistry in six woody species of the southern African Savanna, *Am. Nat.*, 137, 50, 1991.
27. **Bryant, J. P., Tuomi, J., and Niemala, P.**, Environmental constraint of constitutive and long-term inducible defenses in woody plants, in *Chemical Mediation of Coevolution*, Spencer, K. C., Ed., Academic Press, New York, 1988, 367.

28. **Buratti, L., Allais, J. P., and Barbier, M.,** The role of resin acids in the relationship between Scots pine and the sawfly, *Diprion pini (Hymenoptera: Diprionidae)*. I-Resin acids in the needles, in *Mechanisms of Woody Plant Defenses Against Insects*, Mattson, W. J., Levieux, J., and Bernard-Dagan, C., Eds., Springer-Verlag, New York, 1988, 171.

29. **Caldwell, M. M., Richards, J. H., Johnson, D. A., Nowak, R. S., and Dzurec, R. S.,** Coping with herbivory: photosynthetic capacity and resource allocation in two semiarid *Agropyron* bunchgrasses, *Oecologia*, 50, 14, 1981.

30. **Caldwell, M. M. and Robberecht, R.,** A steep latitudinal gradient of solar ultraviolet-B radiation in the arctic-alpinelife zone, *Ecology*, 61, 600, 1980.

31. **Caloin, M., Clement, B., and Herrmann, S.,** Regrowth kinetics of *Dactylis glomerata* following defoliation, *Ann. Bot.*, 66, 397, 1990.

32. **Chapin, F. S., III,** Nutrient allocation and responses to defoliation in tundra plants., *Arctic Alpine Res.*, 12, 553, 1980.

33. **Chapin, F. S., III,** Integrated responses of plants to stress, *BioScience*, 47, 29, 1991.

34. **Chapin, F. S. and Slack, M.,** Effects of defoliation upon root growth, phosphate absorption and respiration in nutrient-limited tundra graminoids, *Oecologia*, 42, 67, 1979.

35. **Chapin, R. S., III, Schulze, E. D., and Mooney, H. A.,** The ecology and economics of storage in plants, *Annu. Rev. Ecol. Syst.*, 21, 423, 1990.

36. **Chessin, M. and Zipf, A. E.,** Alarm systems in higher plants, *Bot. Rev.*, 56, 193, 1990.

37. **Chiang, H. S., Norris, D. M., Ciepiela, A., Shapiro, P., and Oosterwyk, A.,** Inducible versus constitutive PI 227687 soybean resistance to mexican beetle, *Epilachna varivestis, J. Chem. Ecol.*, 13, 741, 1987.

38. **Chou, C.-H. and Waller, G. R.,** Possible allelopathic constituents of *Coffea arabica, J. Chem. Ecol.*, 6, 643, 1980.

39. **Cid, M. S., Detling, J. K., Brisuela, M. A. and Whicker, A. D.,** Patterns in grass silicification: response to grazing history and defoliation, *Oecologia*, 80, 268, 1989.

40. **Clausen, T. P., Reichardt, P. B., Bryant, J. P., Werner, R. A., Post, K., and Frisby, K.,** Chemical model for short-term induction in quaking aspen (*Populus tremuloides*) foliage against herbivores, *J. Chem. Ecol.*, 15, 2335, 1989.

41. **Clausen, T. P., Reichart, P. B., Bryant, J. P., and Werner, R. A.,** Long-term and short-term induction in quaking aspen: related phenomena?, in *Phytochemical Induction By Herbivores*, Tallamy, D. and Raupp, M. R., Eds., John Wiley & Sons, New York, 1991, 71.

42. **Coleman, J. S. and Jones, C. G.,** A phytocentric perspective of phytochemical induction by herbivores, in *Phytochemical Induction by Herbivores*, Tallamy, D. W. and Raupp, M. J., Eds., John Wiley & Sons, New York, 1991, 3.

43. **Coley, P. D., Bryant, J. P., and Chapin, F. S., III,** Resource availability and plant antiherbivore defense, *Science*, 230, 895, 1985.

44. **Creasy, L. L.,** Biochemical responses of plants to fungal attack, *Rec. Adv. Phytochem.*, 19, 47, 1985.

45. **Creelman, R. A., Tierney, M. L., and Mullet, J. E.,** Jasmonic acid/methyl jasmonate accumulate in wounded soybean hypocotyls and modulate wound gene expression, *Proc. Natl. Acad. Sci. U.S.A.*, 89, 4938, 1992.

46. **Daddona, P. E., Wright, J. L., and Hutchinson, C. R.,** Alkaloid catabolism and mobilization in *Catharanthus roseus, Phytochemistry*, 15, 941, 1976.

47. **Demming-Adams, B. and Adams, W. W., III,** Photoprotection and other responses of plants to high light stress, *Annu. Rev. Plant Physiol. Plant Mol. Biol.*, 43, 599, 1992

48. **Duffey, S. S., Bloem, K. A., and Campbell, B. C.,** Consequences of sequestration of plant natural products in plant–insect–parasitoid interactions, in *Interactions of Plant Resistance and Parasitoids and Predators of Insects*, Boethel, D. J. and Eikenbarry, R. D., Eds., Ellis Horwood, Chicester, England, 1986, 31.

49. **Dyer, M. I., Acra, M. A., Wang, G. M., Coleman, D. C., Freckman, D. W., McNaughton, S. J., and Strain, B. R.,** Source-sink carbon relations in two *Panicum coloratum* ecotypes in response to herbivory, *Ecology*, 72, 1472, 1991.

50. **Edwards, P. J. and Wratten, S. D.,** Wound-induced defenses in plants and their consequences for patterns of insect grazing, *Oecologia*, 59, 88, 1983.

51. **Edwards, P. J. and Wratten, S. D.,** Induced plant defences against insect grazing: fact or artefact?, *Oikos*, 44, 70, 1985.

52. **Ericsson, A., Gref, R., Hellqvist, C., and Lågstrom, B.,** Wound response of living bark of scots pine seedlings and its influence on feeding by the weevil, *Hylobius abietis*, in *Mechanisms of Woody Plant Defenses Against Insects*, Mattson, W. J., Levieux, J., and Bernard-Dagan, C., Eds., Springer-Verlag, New York, 1988, 227.

53. **Evans, J. R.,** Photosynthesis and nitrogen relationships in leaves of C3 plants, *Oecologia*, 78, 9, 1989.

54. **Evans, P. S.,** Root growth of *Lolium perenne* L. II. Effects of defoliation and shading, *N.Z. J. Agric. Res.*, 14, 552, 1971.

55. **Faeth, S.,** Indirect interactions between temporally separated herbivores mediated by the host plant, *Ecology*, 67, 479, 1986.

56. **Faeth, S.,** Variable induced responses: direct and indirect effects on oak folivores, in *Phytochemical Induction by Herbivores*, Tallamy, D. and Raupp, M., Eds., John Wiley & Sons, New York, 1991, 293.

57. **Faeth, S. H.,** Do defoliation and subsequent phytochemical responses reduce future herbivory on oak trees?, *J. Chem. Ecol.*, 18, 915, 1992.

58. **Farmer, E. and Ryan, C. A.,** Interplant communication: airborne methyljasmonate induces synthesis of proteinase inhibitors in plant leaves, *Proc. Natl. Acad. Sci. U.S.A.*, 87, 7713, 1990.

59. **Farmer, E. and Ryan, C. A.,** Octadecanoid precursors of jasmonic acid activate the synthesis of wound-inducible proteinase inhibitors, *Plant Cell*, 4, 129, 1992.

60. **Feeny, P.,** Seasonal changes in oak leaf tannins and nutrients as a cause of spring feeding by winter moth caterpillar, *Ecology*, 51, 565, 1970.

61. **Feeny, P.,** Plant apparency and chemical defense, *Rec. Adv. Phytochem.*, 10, 1, 1976.

62. **Felton, G. W., Donato, K., Del Vecchio, R. J., and Duffey, S. S.,** Activation of plant foliar oxidases by insect feeding reduces the nutritive quality of foliage for noctuid herbivores, *J. Chem. Ecol.*, 15, 2667, 1989.

63. **Fowler, S. V. and Lawton, J. H.,** Rapidly induced defenses and talking trees; the Devil's advocate position, *Am. Nat.*, 126, 181, 1985.

64. **Friedman, J. and Waller, G. R.,** Caffeine hazards and their prevention in germinating seeds of coffee (*Coffea arabica* L.), *J. Chem. Ecol.*, 9, 1099, 1983.

65. **Frischknecht, P. M., Battig, M., and Baumann, T. W.,** Effect of drought and wounding stress on indole alkaloid formation in *Catharanthus roseus*, *Phytochemistry*, 26, 707, 1987.

66. **Frischknecht, P. M. and Baumann, T. W.,** Stress induced formation of purine alkaloids in plant tissue culture of *Coffea arabica*, *Phytochemistry*, 24, 2255, 1985.

67. **Fritz, R. S. and Simms, E.,** *Ecology and Evolution of Plant Resistance*, University of Chicago Press, Chicago, in press.

68. **Geiger, D. R. and Servaites, J. C.,** Carbon allocation and response to stress, in *Response of Plants to Multiple Stresses*, Mooney, H. A., Winner, W. E., and Pell, E. J., Eds., Academic Press, New York, 1991, 103.

69. **Georgiadis, N. J. and McNaughton, S. J.,** Interaction between grazers and a cyanogenic grass, *Cynodon plectostachyus*, *Oikos*, 51, 343, 1988.

70. **Gershenzon, J.,** Changes in the levels of plant secondary metabolites under nutrient and water stress, *Rec. Adv. Phytochem.*, 18, 273, 1984.

71. **Gershenzon, J. and Croteau, R.,** Terpenoids, in *Herbivores: Their Interactions with Secondary Plant Metabolites.*, Rosenthal, G. A. and Berenbaum, M. R., Eds., Academic Press, San Diego, 1991, 165.

72. **Gijzen, M., Lewinsohn, E., and Croteau, R.,** Characterization of constitutive and wound-inducible monoterpene cyclases of Grand Fir (*Abies grandis*), *Arch. Biochem. Biophys.*, 289, 267, 1991.

73. **Givnish, T. J.,** Economics of biotic interactions, in *On the Economy of Plant Form and Function*, Givnish, T. J., Ed., Cambridge University Press, Cambridge, MA, 1986, 667.

74. **Gonzalez, B., Boucaud, J., Salette, J., Langlois, J., and Duyme, M.,** Changes in stubble carbohydrate content during regrowth of defoliated perennial ryegrass (*Lolium perenne* L.) on two nitrogen levels, *Grass For. Sci.*, 44, 411, 1989.

75. **Green, F. B. and Corcoran, M. R.,** Inhibitory action of five tannins on growth induced by several gibberellins, *Plant Physiol.*, 56, 801, 1975.

76. **Gref, R. and Ericsson, A.,** Wound-induced changes of resin acid concentrations in living bark of Scots pine seedlings, *Can. J. For. Res.*, 15, 92, 1985.

77. **Grime, J. P. and Campbell, B. D.,** Growth rate, habitat productivity, and plant strategy as predictors of stress response, in *Response of Plants to Multiple Stresses*, Mooney, H. A., Winner, W. A., and Pell, E. J., Eds., Academic Press, New York, 1991, 143.

78. **Guedes, M. E. M., Kuc, J., Hammerschmidt, R., and Bostock, R.,** Accumulation of six sesquiterpenoid phytoalexins in tobacco leaves infiltrated with *Pseudomonas lachrymans*, *Phytochemistry*, 12, 2987, 1982.

79. **Guha, J. and Sen, S. P.,** The cucurbitacins: a review, *Plant Biochem. J.*, 2, 12, 1975.

80. **Hahlbrock, K. and Grisebach, H.,** Enzymatic controls in the biosynthesis of lignin and flavenoids, *Annu. Rev. Plant Physiol.*, 30, 105, 1979.

81. **Hahlbrock, K. and Scheel, D.,** Physiology and molecular biology of phenylpropanoid metabolism, *Annu. Rev. Plant Physiol. Plant Mol. Biol.*, 40, 347, 1989.

82. **Hall, L. W.,** Effects and mode of alkaloid control in four genotypes of *Nicotiana tobaccum*, Master's thesis, University of Kentucky, 1975.

83. **Harborne, J. B.,** Flavonoid pigments, in *Herbivores: Their Interactions with Secondary Plant Metabolites*, Rosenthal, G. A. and Berenbaum, M. R., Eds., Academic Press, San Diego, 1991, 389.

84. **Harper, J. L.,** The value of a leaf, *Oecologia*, 80, 53, 1989.

85. **Hartley, S. E.,** The inhibition of phenolic biosynthesis in damaged and undamaged birch foliage and its effect in insect herbivous, *Oecologia,* 76, 65, 1988.

86. **Hartley, S. E. and Firn, R. D.,** Phenolic biosynthesis, leaf damage, and insect herbivory in birch (*Betula pendula*), *J. Chem. Ecol.,* 15, 275, 1989.

87. **Hartley, S. E. and Lawton, J. H.,** Biochemical aspects and significance of the rapidly induced accumulation of phenolics in birch foliage, in *Phytochemical Induction by Herbivores,* Tallamy, D. W. and Raupp, M. J., Eds., John Wiley & Sons, New York, 1991, 105.

88. **Hartmann, T.,** Alkaloids, in *Herbivores: Their Interactions with Secondary Plant Metabolites,* Rosenthal, G. A. and Berenbaum, M. R., Eds., Academic Press, San Diego, 1991, 79.

89. **Harvell, C. D.,** The ecology and evolution of inducible defenses in a marine bryozoan: cues, costs and consequences, *Am. Nat.,* 128, 810, 1986.

91. **Harvell, C. D.,** The ecology and evolution of inducible defenses, *Q. Rev. Biol.,* 65, 323, 1990.

92. **Haslam, E.,** Secondary metabolism — fact and fiction, *Nat. Prod. Rep.,* 217, 1986.

93. **Haukioja, E.,** Induction of defenses in trees, *Annu. Rev. Entomol.,* 36, 25, 1990.

94. **Haukioja, E. and Hakala, T.,** Herbivore cycles and periodic outbreaks. Formulation of a general hypothesis, *Rep. Kevo Subaretic Res. Stat.,* 12, 1, 1976.

95. **Haukioja, E. and Neuvonen, S.,** Insect population dynamics and induction of plant resistance: the testing of hypotheses, in *Insect Outbreaks,* Barbosa, P. and Schultz, J. C., Eds., Academic Press, New York, 1987, 411.

96. **Haukioja, E., Ruphomaki, K., Senn, J., Supmela, J., and Walls, M.,** Consequences of herbivory in the mountain birch (*Betula pubescens* ssp. *tortuosa*). Importance of the functional organization of the tree, *Oecologia,* 82, 238, 1990.

97. **Heilmeier, H., Schulze, E.-D., and Whale, D. M.,** Carbon and nitrogen partitioning in the biennial monocarp *Arctium tomentosum* Mil., *Oecologia,* 70, 466, 1986.

98. **Holland, E. A. and Detling, J. K.,** Plant response to herbivory and belowground nitrogen cycling, *Ecology,* 71, 1040, 1990.

99. **Imaseki, H.,** Hormonal control of wound-induced response, in *Hormonal Regulation of Development. III. Role of Environmental Factors,* Pharis, R. P. and Reid, D. M., Eds., Springer-Verlag, Berlin, 1985, 485.

100. **Jacob, F.,** Evolution and tinkering, *Science,* 196, 1161, 1977.

101. **Johnson, N. D. and Bentley, B. L.,** Effects of dietary protein and Lupine alkaloids on growth and survivorship of *Spodoptera eridania, J. Chem. Ecol.,* 14, 1391, 1988.

102. **Johnson, N. D. and Brain, S. A.,** The response of leaf resin to artificial herbivory in *Eriodictyon californicum, Biochem. Syst. Ecol.,* 13, 5, 1985.

103. **Johnson, N. D., Liu, B., and Bentley, B. L.,** The effects of nitrogen fixation, soil nitrate, and defoliation on the growth, alkaloids, and nitrogen levels of *Lupinus succulentus* (Fabaceae), *Oecologia,* 74, 425, 1987.

104. **Johnson, N. D., Rigney, L., and Bentley, B. L.,** Short-term changes in alkaloid levels following leaf damage in lupines with and without symbiotic nitrogen fixation, *J. Chem. Ecol.,* 15, 2425, 1988.

105. **Johnson, R. H. and Lincoln, D. E.,** Sagebrush carbon allocation patterns and grasshopper nutrition: the influence of CO_2 enrichment and soil mineral limitation, *Oecologia,* 87, 127, 1991.

106. **Jones, C. G., Hopper, R. F., Coleman, J. S., and Krischik, V. A.,** Plant vasculature controls the distribution of systemically induced defense against an herbivore, *Oecologia,* 93, 452, 1993.

107. **Karban, R., Adamchak, R., and Schnathorst, M. C.,** Induced resistance and interspecific competition between spider mites and a vascular wilt fungus, *Science,* 235, 678, 1987.

108. **Karban, R., Brody, A. K., and Schnathorst, W. C.,** Crowding and a plant's ability to defend itself against herbivores and diseases, *Am. Nat.,* 134, 749, 1989.

109. **Karban, R. and Myers, J. H.,** Induced plant responses to herbivory, *Annu. Rev. Ecol. Syst.,* 20, 331, 1989.

110. **Kartusch, R. and Mittendorfer, B.,** Ultraviolet radiation increases nicotine production in *Nicotiana* callus cultures, *J. Plant Physiol.,* 136, 110, 1990.

111. **Ke, D. and Saltveit, M. E., Jr.,** Wound-induced ethylene production, phenolic metabolism and susceptibility to russet spotting in iceburg lettuce, *Physiol. Plant,* 76, 412, 1989.

112. **Kendall, D. M. and Bjostad, L. B.,** Phytohormone ecology, *J. Chem. Ecol.,* 16, 981, 1990.

113. **Kernan, A. and Thornburg, R. W.,** Auxin levels regulate the expression of a wound-inducible proteinase inhibitor. II. Chloramphenicol acetyl transferase gene fusion in vitro and in vivo, *Plant Physiol.,* 91, 73, 1989.

114. **Khan, M. B. and Harborne, J. B.,** Induced alkaloid defence in *Atropa acuminata* in response to mechanical and herbivore leaf damage, *Chemoecology,* 1, 77, 1990.

115. **Kogan, M. and Fischer, D. C.,** Inducible defenses in soybean against herbivorous insects, in *Phytochemical Induction by Herbivores,* Tallamy, D. and Raupp, M. J., Eds., John Wiley & Sons, New York, 1991, 347.

116. **Kolodny-Hirsch, D. M. and Harrison, F. P.,** Yield loss relationships of tobacco and tomato hornworms (Lepidoptera : Sphingidae) at several growth stages of Maryland tobacco, *J. Econ. Entomol.,* 79, 731, 1986.

117. **Kolodny-Hirsch, D. M., Saunders, J. A., and Harrison, F. P.,** Effects of simulated tobacco hornworm (Lepidoptea : Sphingidae) defoliation in growth dynamics and physiology of tobacco as evidence of plant tolerance to leaf consumption, *Environ. Entomol.,* 15, 1137, 1986.

118. **Kombrink, E. and Hehlbrock, K.,** Rapid, systemic repression of the synthesis of ribulose 1,5-bisphosphate carboxylase small-subunit mRNA in fungus-infected or elicitor treated potato leaves, *Planta,* 181, 216, 1990.

119. **Kombrink, E., Schroder, M., and Hahlbrock, K.,** Several 'pathogenesis-related' proteins in potato are 1,3-B-glucanases and chitinases, *Proc. Natl. Acad. Sci. U.S.A.,* 85, 782, 1988.

120. **Krischik, V. A. and Denno, R. F.,** Individual, population, and geographic patterns in plant defense, in *Variable Plants and Herbivores in Natural and Managed Systems,* Denno, R. F. and McClure, M. S., Eds., Academic Press, New York, 1983, chap. 14.

121. **Krischik, V. A., Goth, R. W., and Barbosa, P.,** Generalized plant defense: effects on multiple species, *Oecologia,* 85, 562, 1991.

122. **Larson, R. A.,** The antioxidants of higher plants, *Phytochemistry,* 27, 967, 1988.

123. **Larson, R. A. and Marley, K. A.,** Quenching of singlet oxygen by alkaloids and related nitrogen heterocyles, *Phytochemistry,* 23, 2351, 1984.

124. **Leete, E.,** Biosynthesis and metabolism of the tobacco alkaloids, *Proc. Am. Chem. Soc. Symp.,* 173, 365, 1977.

125. **Leete, E. and Bell, V. M.,** The biogenesis of *Nicotiana* alkaloids: the metabolism of nicotine in *N. tabacum, J. Am. Chem. Soc.,* 81, 4358, 1959.

126. **Levin, D. A.,** Alkaloid bearing plants: an ecogeographic perspective, *Am. Nat.,* 26, 261, 1976.

127. **Lorio, P. L. J.,** Growth differentiation–balance relationships in pines affect their resistance to bark beetles (*Coleoptera: Scolytidae*), in *Mechanisms of Woody Plant Defenses Against Insects,* Mattson, W. J., Levieux, J., and Bernard-Dagan, C., Eds., Springer-Verlag, New York, 1988, 73.

128. **Malamy, J., Carr, J. P., Klessig, D. F., and Raskin, I.,** Salicylic acid: a likely endogenous signal in the resistance response of tobacco to viral infection, *Science,* 250, 1002, 1990.

129. **Marpeau, A., Walter, J., Launay, J., Charon, J., and Baradat, P.,** Effects of wounds on the terpene content of twigs of maritime pine (*Pinus pinaster Ait.*). II. Changes in the volatile terpene hydrocarbon composition, *Trees,* 4, 220, 1989.

130. **Maschinski, J. and Whitham, T. G.,** The continuum of plant responses to herbivory: the influence of plant association, nutrient availability, and timing, *Am. Nat.,* 134, 1, 1989.

131. **Mattson, W. J. and Palmer, S. R.,** Changes in levels of foliar minerals and phenolics in trembling aspen, *Populus tremuloides,* in response to artifical defoliation, in *Mechanisms of Woody Plant Defenses Against Insects,* Mattson, W. J., Levieux, J., and Bernard-Dagan, C., Eds., Springer-Verlag, New York, 1988, 157.

132. **McKey, D.,** Adaptive patterns in alkaloid physiology, *Am. Nat.,* 108, 305, 1974.

133. **McKey, D.,** The distribution of secondary compounds within plants, in *Herbivores, Their Interaction with Secondary Plant Metabolites,* Janzen, D. H. and Rosenthal, G. A., Eds., Academic Press, New York, 1979, 55.

135. **McNaughton, S. J.,** Interactive regulation of grass yield and chemical properties by defoliation, a salivary chemical, and inorganic nutrition, *Oecologia,* 65, 478, 1985.

136. **McNaughton, S. J. and Tarrants, J. L.,** Grass leaf silicification: natural selection for inducible defense against herbivores, *Proc. Natl. Acad. Sci. U.S.A.,* 80, 790, 1983.

137. **Metraux, J. P., Signer, H., Ryals, J., Ward, E., Wyss-Benz, M., Gaudin, J., Raschdorf, K., Schmid, E., Blum, W., and Inverardi, B.,** Increase in salicylic acid at the onset of systemic acquired resistance in cucumber, *Science,* 250, 1004, 1990.

138. **Mihaliak, C. A., Gershenzon, J., and Croteau, R.,** Lack of rapid monoterpene turnover in rooted plants: implications for theories of plant chemical defense, *Oecologia,* 87, 373, 1991.

139. **Mihaliak, C. A. and Lincoln, D. E.,** Growth pattern and carbon allocation to volatile leaf terpenes under nitrogen-limiting conditions in *Heterotheca subaxillaris* (Asteraceae), *Oecologia,* 66, 423, 1985.

140. **Millard, P.,** The accumulation and storage of nitrogen by herbaceous plants, *Plant Cell Environ.,* 11, 1, 1988.

141. **Mooney, H. A. and Gulmon, S. L.,** Constraints on leaf structure and function in reference to herbivory, *Bioscience,* 32, 198, 1982.

142. **Mooney, H. A., Gulmon, S. L., and Johnson, N. D.,** Physiological constraints on plant chemical defenses, in *Plant Resistance to Insects,* Hedin, P. A., Ed., American Chemical Society, Washington, D. C., 1983, 21.

143. **Mooney, H. A. and Winner, W. A.,** Partitioning response of plants to stress, in *Response of Plants to Multiple Stresses,* Mooney, H. A., Winner, W. A., and Pell, E. J., Eds., Academic Press, New York, 1991, 129.

144. **Naaranlahti, T., Auriola, S., and Lapinjoki, S.,** Growth-related dimerization of vindoline and cantharanthine in *Cantharanthus roseus* and effect of wounding on the process, *Phytochemistry,* 30, 1451, 1991.

145. **Nef, L.,** Interactions between the leaf miner, *Phyllocnistis suffusella,* and poplars, in *Mechanisms of Woody Plant Defenses Against Insects,* Mattson, W. J., Levieux, J., and Bernard-Dagan, C., Eds., Springer-Verlag, New York, 1988, 239.

146. **Neuvonen, S. and Haukioja, E. K.,** The effects of inducible resistance in host foliage on birch-feeding herbivores, in *Phytochemical Induction by Herbivores,* Tallamy, D. and Raupp, M., Eds., John Wiley & Sons, New York, 1991, 277.

147. **Nichols-Orians, C.,** Induction and plant–herbivore interactions, *Ecology,* 73, 711, 1992.
148. **Nitao, J. K.,** Artificial defloration and furanocoumarin induction in *Pastinaca sativa* (Umbelliferae), *J. Chem. Ecol.,* 14, 1515, 1988.
149. **Nooden, L. D. and Leopold, A. C.,** *Senescence and Aging in Plants,* Academic Press, New York, 1988.
150. **Nowacki, E., Jurzysta, M., Gorski, P., Nowacka, D., and Waller, G. R.,** Effect of nitrogen nutrition on alkaloid metabolism in plants, *Biochem. Physiol. Pflanzen.,* 169, 231, 1976.
151. **O'Neill, M. J., Adesanya, S. A., Roberts, M. F., and Pantry, I. R.,** Inducible isoflavenoids from the lima bean, *Phaseolus lunatus, Phytochemistry,* 25, 1315, 1986.
152. **Ostlie, K. R. and Pedigo, L. P.,** Water loss from soybeans after simulated and actual insect defoliation, *Environ. Entomol.,* 13, 1675, 1984.
153. **Ourry, A., Boucaud, J., and Salette, J.,** Nitrogen mobilization from stubble and roots during re-growth of defoliated perennial ryegrass, *J. Exp. Bot.,* 39, 803, 1988.
154. **Ourry, A., Gonzalez, B., and Boucaud, J.,** Osmoregulation and role of nitrate during regrowth after cutting of ryegrass (*Lolium perenne*), *Physiol. Plant.,* 76, 177, 1989.
155. **Pearce, G., Strydom, D., Johnson, S., and Ryan, C. A.,** A polypeptide from tomato leaves induces wound-inducible proteinase inhibitor proteins, *Science,* 253, 895, 1991.
156. **Pena-Cortes, H., Sanchez-Serrano, J. J., Mertens, R., Willmitzer, L., and Prat, S.,** Abscisic acid is involved in the wound-induced expression of the proteinase inhibitor II gene in potato and tomato, *Proc. Natl. Acad. Sci. U.S.A.,* 86, 9851, 1989.
157. **Pena-Cortes, H., Sanchez-Serrano, J., Rocha-Sosa, M., and Willmitzer, L.,** Systemic induction of proteinase-inhibitor-II gene expression in potato plants by wounding, *Planta,* 174, 84, 1988.
158. **Pena-Cortes, H., Willmitzer, L., and Sanchez-Serrano, J. J.,** Abscisic acid mediates wound induction but not developmental-specific expression of the proteinase inhibitor II gene family, *Plant Cell,* 3, 963, 1991.
159. **Price, P. W., Waring, G. L., Julkunen-Tiitto, R., Tahvainen, J., Mooney, H. A., and Craig, T. P.,** Carbon-nutrient balance hypothesis in within-species phytochemical variation of *Salix lasiolepis, J. Chem. Ecol.,* 15, 1117, 1989.
160. **Prins, A. H., Verlaar, H. J., and van den Herik, M.,** Responses of *Cynoglossum officinale* L. and *Senecio jacobaea* L. to various degrees of defoliation, *New Phytol.,* 111, 725, 1989.
161. **Quick, W. P., Schurr, U., Fichtner, K., Schulze, E. D., Rodermel, S. R., Bogorad, L., and Stitt, M.,** The impact of decreased rubisco on photosynthesis, growth, allocation and storage in tobacco plants which have been transformed with antisense rbcS, *Plant J.,* 1, 51, 1991.
162. **Quick, W. P., Schurr, U., Scheibe, R., Schulze, E.-D., Rodermel, S. R., Bogorad, L., and Stitt, M.,** Decreased ribulose-1,5-bisphosphate carboxlase oxygenase in transgenic tobacco transformed with "antisense" rbcS, *Planta,* 183, 542, 1991.
163. **Raffa, K. F.,** Induced defensive reactions in conifer-bark beetle systems, in *Phytochemical Induction By Herbivores,* Tallamy, D. and Raupp, M., Eds., John Wiley & Sons, New York, 1991, 245.
164. **Ralphs, M. H. and Williams, C.,** Alkaloid response to defoliation of Velvet Lupine (*Lupinus leucophyllus*), *Weed Technol.,* 2, 429, 1988.
165. **Reichardt, P. B., Chapin, F. S., III, Bryant, J. P., Mattes, B. R., and Clausen, T. P.,** Carbon/nutrient balance as a predictor of plant defense in Alaskan Balsam Poplar: potential importance of metabolite turnover, *Oecologia,* 88, 401, 1991.
166. **Rhoades, D. F.,** Evolution of plant chemical defense against herbivores, in *Herbivores: Their Interaction with Secondary Plant Metabolites,* Rosenthal, G. A. and. Janzen, D.H., Eds., Acaemic Press, New York, 1979, 3.
167. **Rhoades, D. F.,** Herbivore population dynamics and plant chemistry, in *Variable Plants and Herbivores in Natural and Managed Systems,* Denno, R. F. and McClure, M.S., Eds., Academic Press, New York, 1983, 155.
168. **Rhoades, D. F.,** Offensive-defensive interactions between herbivores and plants: their relevance in herbivore population dynamics and ecological theory, *Am. Nat.,* 125, 205, 1985.
169. **Rhoades, D. F. and Cates, R. G.,** Towards a general theory of plant antiherbivore chemistry, *Rec. Adv. Phytochem.,* 10, 168, 1976.
170. **Robberecht, R.,** Environmental photobiology, in *The Science of Photobiology,* Smith, K. C., Ed., Plenum Press, New York, 1989, 135.
171. **Robberecht, R., Caldwell, M. M., and Billings, W. D.,** Leaf ultraviolet optical properties along a latitudinal gradient in the arctic-alpine life zone, *Ecology,* 61, 612, 1980.
172. **Robinson, T.,** *The Biochemistry of Alkaloids,* Springer-Verlag, New York, 1981.
173. **Rosenthal, G. A.,** Nitrogen allocation for L-canavanine synthesis and its relationship to chemical defense of the seed, *Biochem. Syst. Ecol.,* 5, 219, 1977.
174. **Rossiter, M. C., Schultz, J. C., and Baldwin, I. T.,** Relationships among defoliation, red oak phenolics, and gypsy moth growth and reproduction, *Ecology,* 69, 267, 1988.
175. **Ruess, R. W.,** The interaction of defoliation and nutrient uptake in *Sporobolus kertrophyllus,* a short-grass species from the Serengeti Plains, *Oecologia,* 77, 550, 1988.

176. **Ryan, C. A.,** Insect-induced chemical signals regulating natural plant protection responses, in *Variable Plants and Herbivores in Natural and Managed Systems*, Denno, R. F. and McClure, M. S., Eds., Academic Press, New York, 1983, 43.

177. **Ryan, C. A.,** Insect-induced chemical signals regulating natural plant protection responses, in *Variable Plants and Herbivores in Natural and Managed Systems*, Denno, R. F. and McClure, M.S., Eds., Academic Press, New York, 1983, 43.

178. **Saitoh, F., Noma, M., and Kawashima, N.,** The alkaloid contents of sixty *Nicotiana* species, *Phytochemistry*, 24, 477, 1985.

179. **Schultz, J. C.,** Plant responses induced by herbivores, *TREE*, 3, 45, 1988.

180. **Schultz, J. C. and Baldwin, I. T.,** Oak leaf quality declines in response to defoliation by gypsy moth larvae, *Science*, 217, 149, 1982.

181. **Seigler, D. S.,** Primary roles for secondary compounds, *Biochem. Syst. Ecol.*, 5, 195, 1977.

182. **Seigler, D. S. and Price, P. W.,** Secondary compounds in plants: primary functions, *Am. Nat.*, 110, 101, 1976.

183. **Selmar, D.,** Cyanogenesis and metabolism of cyanogenic compounds: an example of multifunctionality of secondary plant products, *Planta Medica*, 55, 592, 1989.

184. **Selmar, D., Lieberei, R., and Biehl, B.,** Mobilization and utilization of cyanogenic glycosides, *Plant Physiol.*, 86, 711, 1988.

185. **Sherman, P.,** The levels of analysis, *Am. Behav.*, 36, 616, 1988.

186. **Snyder, B. and Nicholson, R.,** Synthesis of phytoalexins in sorghum as a site specific response to fungal ingress, *Science*, 248, 1637, 1990.

187. **Stahl, E.,** Pflanzen und schnecken. Eine biologische studie uber die schutzmittel der pflanzen gegen scheckenfrab, *Jenaische Zeitschr. Naturwiss. und Medizin*, 23, 557, 1888.

188. **Staswick, P. E.,** Novel regulation of vegetative storage protein genes, *Plant Cell*, 2, 1, 1990.

189. **Staswick, P. E., Huang, J. F., and Rhee, Y.,** Nitrogen and methyljasmonate induction of soybean vegetative storage protein genes, *Plant Physiol.*, 96, 130, 1991.

190. **Steingrover, E., Woldendorp, J., and Sijtsma, L.,** Nitrate accumulation and its relation to leaf elongation in spinach leaves, *J. Exp. Bot.*, 37, 1093, 1986.

191. **Stevens, K. L. and Molyneux, R. J.,** Castanospermine — A plant growth regulator, *J. Chem. Ecol.*, 14, 1467, 1988.

192. **Ta, T. C., Macdowall, F. D. H., and Faris, M. A.,** Utilization of carbon and nitrogen reserves of alfalfa roots in supporting N_2-fixation and shoot regrowth, *Plant Soil*, 127, 231, 1990.

193. **Takabayashi, J., Dicke, M., and Posthumus, M.,** Variation in composition of predator-attracting allelochemicals emitted by herbivore-infested plants: relative influence of plant and herbivore, *Chemoecology*, 2, 1, 1991.

194. **Tallamy, D. W. and Krischik, V. A.,** Variation and function of cucurbitacins in *Cucurbita*. An examination of current hypotheses, *Am. Nat.*, 133, 766, 1989.

194a. **Tallamy, D. W. and Raupp, M. J.,** *Phytochemical Induction by Herbivores*, John Wiley & Sons, New York, 1991

195. **Tallamy, D. W. and McCloud, E.S.,** Squash beetles, cucumber beetles, and inducible cucurbit responses, in *Phytochemical Induction by Herbivores*, Tallamy, D. W. and Raupp, M. J, Eds., John Wiley & Sons, New York, 1991, chap. 7.

196. **Tiburcio, A. F., Pinol, M. T., and Serrano, M.,** Effect of UV-C on growth, soluble protein and alkaloids in *Nicotiana rustica* plants, *Environ. Exp. Bot.*, 25, 203, 1985.

197. **Trewavas, A.,** Possible control points in plant development, in *The Molecular Biology of Plant Development*, Smith, H. and Grierson, D., Eds., 1982, 78.

198. **Trewavas, A. J.,** How do plant growth substances work?, *Plant Cell Environ.*, 4, 203, 1981.

199. **Tso, T. C. and Jeffrey, R. N.,** Biochemical studies on tobacco alkaloids. IV. The dynamic state of the nicotine supplied to *N. rustica.*, *Arch. Biochem. Biophys.*, 92, 253, 1961.

200. **Tuomi, J., Fagerstrom, T., and Niemela, P.,** Carbon allocation, phenotypic plasticity and induced defenses, in *Phytochemical Induction by Herbivores*, Tallamy, D. W. and Raupp, M. J., Eds., John Wiley & Sons, New York, 1991, 85.

201. **Tuomi, J., Niemala, P., Haukioja, E., Siren, S., and Neuvonen, S.,** Nutrient stress: an explanation for plant anti-herbivore responses to defoliation, *Oecologia*, 61, 208, 1984.

202. **Tuomi, J., Niemela, P., Chapin, F. S. I., Bryant, J. P., and Siran, S.,** Defense responses in trees in relation to their carbon/nutrient balance, in *Mechanism of Woody Plant Defenses against Insects: Search for Patterns*, Mattson, W. J., Levieux, J., and Bernard-Dagan, C., Eds., Springer-Verlag, New York, 1988, 57.

203. **Tuomi, J., Niemela, P., and Siren, S.,** The panglossian paradigm and delayed inducible accumulation of foliar phenolics in mountain birch, *Oikos*, 59, 399, 1990.

204. **Turlings, T. C. J., Tumlinson, J. H., and Lewis, W. J.,** Exploitation of herbivore-induced plant odors by host-seeking parasitic wasps, *Science*, 250, 1251, 1990.

205. **Valentine, H. T., Wallner, W. E., and Wargo, P. M.,** Nutritional changes in host foliage during and after defoliation, and their relation to the weight of gypsy moth pupae, *Oecologia*, 57, 298, 1983.

206. **van Alstyne, K. L.,** Herbivore grazing increases polyphenolic defenses in the intertidal brown alga *Fucus distichus, Ecology,* 69, 654, 1988.

207. **van der Meijden, E., van Bemmelen, M., Kooi, R., and Post, B. J.,** Nutritional quality and chemical defense in the ragwort–cinnabar moth interaction, *J. Anim. Ecol.,* 53, 443, 1984.

208. **van der Meijden, E., Wijn, M., and Verkaar, H. J.,** Defense and regrowth, alternative plant strategies in the struggle against herbivores, *Oikos,* 51, 355, 1988.

209. **Vrieling, K.,** Costs and benefits of alkaloids of *Senecio jacobaea L.,* Ph.D. thesis, Leiden, 1990.

210. **Wagner, M. R.,** Induced defenses in ponderosa pine against defoliating insects, in *Mechanisms of Woody Plant Defenses Against Insects,* Mattson, W. J., Levieux, J., and Bernard-Dagan, C., Eds., Springer-Verlag, New York, 1988, 141.

211. **Wagner, M. R. and Evans, P. D.,** Defoliation increases nutritional quality and allelochemicals of pine seedlings, *Oecologia,* 67, 235, 1985.

212. **Waller, G. R.,** *Allelochemicals: Role in Agriculture and Forestry,* American Chemical Society, Washington, D.C., 1985.

213. **Waller, G. R. and Nowacki, E. K.,** *Alkaloid Biology and Metabolism in Plants,* Plenum Press, New York, 1978.

214. **Walter, J., Charon, J., Marpeau, A., and Launay, J.,** Effects of wounding on the terpene content of twigs of maritime pine (*Pinus pinaster Ait.*). I. Changes in the concentration of diterpene resin acids and ultrastuctural modifications of the resin duct epithelial cells following injury, *Trees,* 4, 210, 1989.

215. **Wareing, P. F., Khalifa, M. M., and Treharne, K. J.,** Rate limiting processes in photosynthesis at saturating light intensities, *Nature,* 220, 453, 1968.

216. **Waring, R. H., McDonald, A. J. S., Larsson, S., Ericsson, T., Wiren, A., Arwidsson, E., Ericsson, A., and Lohammar, T.,** Differences in chemical composition of plants grown at constant relative growth rates with stable mineral nutrition, *Oecologia,* 66, 157, 1985.

217. **Waterman, P. G. and Mole, S.,** Extrinsic factors influencing production of secondary metabolites in plants, in *Insect–Plant Interactions,* Bernays, E. A., Ed., CRC Press, Boca Raton, FL, 1989, 107.

218. **Welter, S. C.,** Arthropod impact on plant gas exchange, in *Insect–Plant Interactions,* Bernays, E. A., Eds., CRC Press, Boca Raton, FL, 1989, 135.

219. **Whitham, T. G., Maschinski, J., Larson, K. C., and Paige, K. N.,** Plant responses to herbivory: the continuum from negative to positive and underlying physiological mechanisms, in *Plant–Animal Interactions: Evolutionary Ecology in Tropical and Temperate Regions,* Price, P. W., Lewinsohn, T. M., Fernandes, G. W., and Benson, W. W., Eds., John Wiley & Sons, New York, 1991, 227.

220. **Whitman, D. W. and Eller, F. J.,** Parasitic wasps orient to green leaf volatiles, *Chemoecology,* 1, 69, 1990.

221. **Wink, M.,** Inhibition of seed germination by quinolizidine alkaloids, *Planta,* 158, 365, 1983.

222. **Wink, M.,** Chemical defense of *Leguminosae.* Are quinolizidine alkaloids part of the antimicrobial defense system of Lupins, *Z. Naturforsch.,* 39, 548, 1984.

223. **Wink, M.,** Chemical defense of lupins: biological function of quinolizidine alkaloids, *Plant Syst. Evol.,* 150, 65, 1985.

224. **Wink, M.,** Metabolism of quinolizidine alkaloids in plants and cell suspension cultures: induction and degradation, in *Primary and Secondary Metabolism of Plant Cell Cultures,* Neumann et al., Eds., Springer-Verlag, Berlin, 1985, 107.

225. **Wink, M.,** Chemical ecology of quinolizidine alkaloids, in *Allelochemicals: Role in Agriculture and Forestry,* Waller, G. R, Eds., American Chemical Society, Washington D.C., 1987, 326.

226. **Wink, M. and Witte, L.,** Turnover and transport of quinolizidine alkaloids. Diurnal fluctuations of lupanine in the phloem sap, leaves and fruits of *Lupinus albus L., Planta.,* 161, 519, 1984.

227. **Wink, W.,** Wounding-induced increase of quindizidine alkaloid accumulation in Lupin leaves, *Z. Naturforsch.,* 38c, 905, 1983.

228. **Wolf, D. D. and Parrish, D. J.,** Short-term growth responses of tall fescue to changes in soil water potential and to defoliation, *Crop Sci.,* 22, 996, 1982.

229. **Yoshida, D.,** Degradation and translocation of N-15-labeled nicotine injected into intact tobacco leaves, *Plant Cell Physiol.,* 3, 391, 1962.

230. **Yoshinori, W., Watanabe, S., and Kuroda, S.,** Changes in photosynthetic activities and chlorophyll contents of growing tobacco leaves, *Bot. Mag. (Tokyo),* 80, 123, 1967.

231. **Zangerl, A. R.,** Furanocoumarin induction in wild parsnip: evidence for an induced defense against herbivores, *Ecology,* 71, 1926, 1990.

232. **Zangerl, A. R. and Berenbaum, M. R.,** Furanocoumarins in wild parsnip: effects of photosynthetically active radiation, ultraviolet light, and nutrients, *Ecology,* 68, 516, 1987.

233. **Zangerl, A. R. and Berenbaum, M. R.,** Furanocoumarin induction in wild parsnip: genetics and populational variation, *Ecology,* 71, 1933, 1990.

Chapter 2

USE AND AVOIDANCE OF OCCUPIED HOSTS AS A DYNAMIC PROCESS IN TEPHRITID FLIES

Daniel R. Papaj

TABLE OF CONTENTS

I. AVOIDANCE OF OVIPOSITION ON EXPLOITED HOSTS BY PHYTOPHAGOUS INSECTS

Maternal care is a basic family value in many insect species, including many phytophagous forms. Phytophagous insects typically express a simple type of maternal care by choosing where their offspring will develop. A number of species, for example, avoid laying eggs at sites that have been exploited by conspecifics (Table 1),[88,104,110] behavior commonly thought to minimize the level of intraspecific competition incurred by a female's progeny. Intraspecific competition is well-documented in phytophagous species, particularly those that spend part of their life cycle inside stems, buds, fruits, or seeds.[5,20,119-121,125,135] It is therefore not surprising that phytophagous insects commonly avoid use of hosts occupied by conspecifics. Nevertheless, five major qualifications must be made with regard to the notion that avoidance of exploited hosts is a general phenomenon with obvious fitness value in phytophagous species where the host represents a relatively limited resource: 1) only rarely has explicit evidence been provided that avoidance behavior has fitness value;[5,97] 2) theory based on other host-specific insects predicts that level of avoidance should depend strongly on an individual's egg load and level of experience;[57,58,60] 3) in species shown to avoid use of exploited hosts, level of avoidance can be highly variable;[6,48,63,68] 4) a number of species appear not to avoid use of exploited hosts at all;[44,48,88,89,102,112,115] 5) certain species, far from avoiding use of exploited hosts, prefer them as sites of oviposition.[78,80,124] In this review, a phytophagous insect's responses to exploited hosts is viewed as a dynamic process both in a behavioral and an evolutionary sense. With special attention to frugivorous flies in the family Tephritidae, factors that might dispose insects toward either selective avoidance or selective use of exploited hosts are enumerated.

II. OVIPOSITION AVOIDANCE AS A DYNAMIC PROCESS IN TEPHRITID FLIES

Females of the apple maggot fly (*Rhagoletis pomonella*) generally avoid use of occupied hosts. After depositing a single egg in a host fruit, such as a hawthorn berry, a female fly drags her ovipositor over the fruit, thus marking it with a pheromone. This so-called host-marking pheromone (HMP) has been shown to deter the same or other females from subsequently laying eggs in that fruit.[84] Avoidance of occupied fruit reduces competition among larvae[5] and gives rise to uniform distributions of eggs over ripe fruit in hawthorn stands (see Reference 7 and References within, but see Reference 98). Similar behavior and egg dispersion patterns have been observed in other *Rhagoletis* species[1,50,54,99,131] and other tephritids.[36,87,88]

The effect of HMP on apple maggot fly behavior, though generally deterrent, is not absolutely so. As females are deprived of the opportunity to lay eggs, for example, they become progressively less likely to be deterred by HMP.[60,101] This effect of deprivation may reflect an adaptive kind of flexibility in avoidance behavior.[57,58] When fruit are rare and the time between encounters long, it might not pay to avoid occupied fruit once encountered, because the female may not find another fruit before she dies. So long as some progeny survive in an occupied fruit, a female under such conditions would do well to utilize occupied as well as unoccupied fruit.

Response to HMP depends also on an individual's prior experience. Female apple maggot flies are not deterred by HMP without prior experience with marked fruit.[100] Effects of experience, like effects of deprivation, may reflect flexibility in avoidance behavior that is functional.[60] When all fruit in a habitat are infested, for example, it might not pay to discriminate absolutely against occupied hosts. Effects of experience permit females to adjust the degree to which occupied fruit are avoided according to the current level of fruit infestation in the habitat.

TABLE 1
Avoidance of Oviposition on Exploited Hosts by Phytophagous Insects

Group	Host type	References
Coleoptera		
Acanthoscelides obtectus	Legumes	122
Anthonomis grandis	*Gossypium*	118
Ceutorhynchus assimilis	Crucifers	52
Callosobruchus chinensis	Legumes	73, 127
C. maculatus	Legumes	39, 62–65, 67
C. rhodesianus	Legumes	39
Crioceris duodecimpunctata	*Asparagus*	cited in 88
Zabrotes subfasciatus	Legumes	127
Diptera		
Agromyza frontella	*Medicago*	61
Anastrepha suspensa	Numerous	92
A. fraterculus	Numerous	94
Atherigona soccata	*Sorghum*	96
Bactrocera jarvisi	Numerous	36
B. tryoni	Numerous	36
Ceratitis capitata	Numerous	93
Chaetorellia australis	*Centaureaa*	81
Contarinia oregonensis	Firs	66
Dacus oleae	*Oleae*	24
Hylemya spp.	*Polemonium*	134
Orellia ruficauda	*Cirsium*	53
Paraceratitella eurycephala	*Capparis*	34
Rhagoletis basiola	*Rosa*	3
R. cerasi	*Prunus*	13–17, 41
R. cingulata	*Prunus*	91
R. completa	*Juglans*	25
R. cornivora	*Cornus*	91
R. fausta	*Prunus*	85
R. indifferens	*Prunus*	91
R. mendax	*Vaccinium*	91
R. pomonella	Rosaceae	84
R. tabellaria	*Cornus*	91
R. zephyria	*Symphoricarpus*	4
Lepidoptera		
Anthocaris sara	Crucifers	112
A. cardamine	Crucifers	132
Battus philenor	*Aristolochia*	97
Ephestia cautella	Stored products	70
E. elutella	Stored products	70
E. kuehniella	Stored products	26
Hadena bicruris	*Melandrium*	19
Heliothis zea	*Zea*	111
Danaus plexippus	*Asclepias*	33, 128
Heliconius spp.	*Passiflora*	40, 133
Pieris brassicae	Crucifers	109
P. protodice	Crucifers	112
P. rapae	Crucifers	109
P. sisymbrii	Crucifers	51
Plodia interpunctella	Stored products	70
Hymenoptera		
Euura lasiolepis	*Salix*	27
Hoplocampa testudinea	*Malus*	103

Flexible responses to occupied fruit, whether cued by deprivation or experience, should be particularly useful when the host environment is a variable one. Variability in other aspects of a female's external environment, such as weather conditions, presence of conspecifics, and predation risk, may also favor flexibility in the degree to which occupied hosts are avoided. Finally, flexibility in avoidance responses ought not to be a consequence of external factors alone. Any *internal* factor that influences the number of eggs available to be laid by a female insect (i.e., her egg load) over her life should also influence the degree to which occupied hosts are avoided. Such factors are too numerous to discuss fully here but include a female's genetic constitution, mating status, nutritional state as adult or juvenile, and level of parasitism.

Whether a property of external environment or internal milieu, the dynamics of a female's behavior should be consistent with the following general rule. *Any factor that might reduce a female's egg load for a given reproductive period or extend her expected reproductive period for a given egg load should increase the tendency to avoid occupied hosts; any factor that might increase relative egg load or shorten relative reproductive period should reduce that tendency.* This rule appears to account well for the dynamics of egg-laying behavior of species such as the apple maggot fly. However, it would not apply to species in which ovipositing females prefer occupied hosts over unoccupied ones. This kind of preference, reported for an entomophagous parasitoid[123] and described below in detail for several tephritid flies, poses a challenge to theory on the evolution of host selection and host-marking behavior in insects.

III. OCCUPIED HOSTS AS PREFERRED SITES OF EGG LAYING

Flexibility in responses notwithstanding, arguments on the function and evolution of host-marking pheromones generally assume that an exploited host constitutes to a female a relatively low-quality resource.[105] Given a simultaneous choice between an exploited and unexploited host, a female is presumed to prefer always to lay eggs first in the unexploited host. Such appears to be the case for species such as *Rhagoletis pomonella*. However, recent work indicates that this assumption is not valid for other tephritid flies that also use host-marking pheromones. Below, my own studies on three tephritid flies are used as examples in which occupied hosts and, specifically, existing oviposition sites on those hosts are actively exploited by females for egg-laying.

A. MEDITERRANEAN FRUIT FLY

Unlike *Rhagoletis* species which specialize on a narrow range of fruit species, the closely-related Mediterranean fruit fly, *Ceratitis capitata* Wiedemann, attacks hundreds of species of fruits, nuts, and vegetables.[23,55] The host-selection behavior of this fruit parasite (*sensu* Price et al.[83]) is nevertheless similar to that of *Rhagoletis* flies. Upon alighting on a fruit, the female Mediterranean fruit fly (hereafter referred to as medfly) bores into the fruit with her ovipositor. After depositing a clutch of approximately 3 to 4 eggs,[74] the female deposits HMP first locally around the oviposition puncture and eventually over the fruit.[93] Unlike many *Rhagoletis* species, females actively search for and lay eggs into fruit wounds. Papaj and colleagues, for example, showed that navel oranges bearing artificial wounds were visited almost eight times more than paired control oranges in a grove on the Greek island of Chios.[78] Having landed, females were over twice as likely to attempt oviposition into wounded fruit than into control fruit. Females generally bored directly into a wound or, when they did not, in its immediate vicinity. An oviposition puncture is a kind of fruit wound, and one might anticipate that female medflies would deposit eggs into existing punctures. They do. Papaj et al. provided unequivocal evidence that medflies deposit eggs into naturally-formed oviposition punctures, despite the presence of naturally-deposited marking pheromone.[80] Under laboratory conditions, females presented with kumquats (a citrus fruit readily accepted by medfly females) actually preferred to use occupied punctures over establishment of new punctures in the fruit.

TABLE 2
Propensity for Walnut Fly Females to Use Existing Sites: Artificial
Puncture Experiments

		% Boring attempts in punctured fruit		% Boring attempts in punctures		% Clutches in punctures	
Fly species	Field site	%	(N)	%	(N)	%	(N)
R. juglandis	Sonoita Creek	84	(25)	86	(21)	80	(10)
	Carr Canyon	79	(29)	78	(23)	82	(17)
R. boycei	Carr Canyon	94	(31)	93	(29)	78	(9)

Interestingly, the finding that medfly females actively exploit existing oviposition sites was not an original one. As long ago as 1914, Sylvestri noted that female medflies in the field, far from avoiding infested fruit, exploited them actively. In his words, "If, during the exploration, the fly finds any lesion on the surface of the fruit, or the wound produced by the ovipositor of another female which has deposited eggs there, it does not hesitate to sink its ovipositor into the wound and deposit its own eggs there."[113] Since Silvestri's account, there have been repeated anecdotal field observations that females deposit eggs into oviposition punctures.[9,10,23] Yet, over the 75 years that passed before this behavior was first quantified,[80] assertions were made repeatedly that medfly *avoided* (not preferred) laying eggs in previously infested fruit.[93,95] This apparent contradiction is resolved if medfly oviposition behavior is viewed as a dynamic process, as discussed later.

B. WALNUT FLIES

Many *Rhagoletis* species (e.g., *R. pomonella*; Papaj, unpublished data) appear not to use existing oviposition punctures as sites of oviposition. However, we recently discovered that two members of the *R. suavis* species group, *R. juglandis* and *R. boycei*, actively exploit existing oviposition punctures. Like other members of the *suavis* group, these species specialize on walnuts. In southern Arizona, both species are common on Arizona walnut, *Juglans major*, which occurs in montane canyons at elevations between 4,000 and 8,000 ft. *R. juglandis* is distributed across almost all elevations, whereas *R. boycei* appears to be restricted to elevations above 6,000 ft.[18,21]

In one set of field experiments, ripe uninfested walnuts were picked and arranged in pairs according to size. In one member of each pair, we made a puncture with a #0 insect pin and drew a circle 1.5 cm in diameter around the puncture. The same circle was drawn on the control member of the pair, but the fruit was left unpunctured. Fruit pairs were then hung with wire in a host *J. major* tree, and focal observations made. Results are shown for two sites in southern Arizona, Sonoita Creek near Patagonia, where *R. juglandis* is found alone, and Carr Canyon in the Huachuca Mountains, where both species co-occur (Table 2). Both species initiated oviposition more often into punctured fruit than control fruit (Table 2, column 1). Most oviposition attempts on punctured fruit were made directly into artificial punctures (Table 2, column 2), and the majority of all clutches laid by a particular species at a particular site were deposited directly into punctures (Table 2, column 3), despite the prior occurrence of host-marking behavior on such fruit.

Use of occupied fruit takes on added significance in walnut flies because males guard oviposition sites. As in other *Rhagoletis* flies,[18,21,86,90,116,117] males of each species take residence on fruit and defend those fruit from other males. We discovered recently that, unlike other *Rhagoletis* flies, males take residence near punctures (Figure 1A) and defend them from other males (Figure 1B). For the artificial puncture experiment described above, I tabulated the occurrence and activity of males on fruit of each type. For *R. boycei* at Carr Canyon and for *R. juglandis* at Sonoita Creek, males were far more common on punctured fruit than on

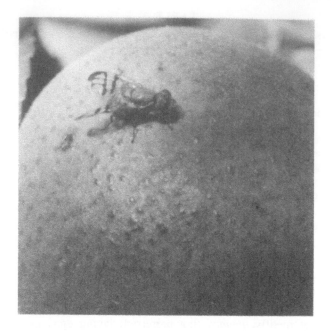

A

B

FIGURE 1. A. *Rhagoletis juglandis* male resident over puncture on host walnut. B. *R. juglandis* males contesting walnut in a boxing display.

control fruit; on punctured fruit, most males spent at least some time resident in the vicinity of punctures (Table 3). At the same time, the frequency of contests among conspecifics was higher on punctured fruit than on control fruit (Table 3).

Results for *R. juglandis* at Carr Canyon are an exception to this pattern. *R. juglandis* at Carr Canyon was sighted only slightly more often on punctured fruit than on control fruit, and

TABLE 3
Walnut Fly Males and Residence at Occupied Sites: Artificial Puncture Experiments

Fly species	Field site	% Males sighted on punctured fruit		% Resident near puncture		% Contests on punctured fruit	
		%	(N)	%	(N)	%	(N)
R. juglandis	Sonoita Creek	71	(241)	61	(170)	90	(157)
	Carr Canyon	61	(218)	13	(134)	83	(40)
R. boycei	Carr Canyon	72	(214)	63	(155)	97	(38)

TABLE 4
Mating in Walnut Flies at Occupied Sites: Artificial Puncture Experiments

Fly species	Field site	% Matings on punctured fruit		Relative mating success (no. matings/male)	
		%	(N)	Punctured fruit	Control fruit
R. juglandis	Sonoita Creek	89	(37)	0.20	0.05
	Carr Canyon	56	(16)	0.10	0.14
R. boycei	Carr Canyon	100	(15)	0.15	0

males on punctured fruit were rarely resident at punctures (Table 3). Because contests between *R. juglandis* males occurred predominantly on punctured fruit at Carr Canyon (as they did at Sonoita Creek), the difference in puncture guarding does not appear to reflect a difference between sites in the value of oviposition punctures. Rather, the difference appears to be a direct consequence of *R. boycei's* occurrence at Carr Canyon. In virtually all contests between males of the two species (96%; N = 89), *R. boycei* males displaced *R. juglandis* males from the fruit.

Dominance by *R. boycei* in heterospecific contests not only precluded puncture guarding by males but apparently influenced where *R. juglandis* matings occurred (Table 4). At Sonoita Creek, *R. juglandis* matings occurred exclusively on punctured fruit. In contrast, only approximately 50% of *R. juglandis* matings at Carr Canyon occurred on punctured fruit. At the same site, *R. boycei* matings occurred exclusively on punctured fruit. At Sonoita Creek, *R. juglandis* males that guarded punctures enjoyed an almost absolute advantage in relative mating success; at Carr Canyon, by contrast, relative *R. juglandis* mating success on punctured fruit was similar to that on control fruit (Table 4).

Under at least some circumstances, puncture guarding constitutes an effective resource-defense mating strategy. Puncture guarding may have additional functions, e.g., parental care. Punctures guarded by males on unmanipulated fruit often contained eggs. Moreover, we sometimes observed males guard a puncture, mate with a female for a brief period of time, and then guard the puncture while the female deposited eggs in it. If males that mate last tend to fertilize the next eggs laid (as seems likely in tephritid flies[72]), such a male would be guarding his own offspring. What benefit would his offspring gain from such guarding? Protection from egg parasites or predators? Perhaps, but we have observed neither egg parasitism nor predation at our field sites to date. Protection from egg deposition by other females? Possibly. Males might conceivably protect offspring by preventing other nonreceptive females from depositing more eggs in the fruit. In support of this idea, males have occasionally been observed to chase conspecific females from fruit. Finally, we have observed flies of an unidentified species depositing eggs in punctures established by *Rhagoletis* females. Males might reduce interspecific competition incurred by their young if they defend punctures from these flies. In support of this idea, males have occasionally been observed to chase these flies from a fruit. Whether in terms of defense from conspecific or heterospecific flies, rigorous support for the parental care hypothesis requires evidence that a male's progeny suffers reduced fitness in his absence. Such evidence is currently unavailable.

Since males sometimes guard punctures while recent mates are depositing eggs in punctures (Papaj, personal observation), puncture guarding may also constitute *de facto* mate guarding. Mate guarding of ovipositing females has been postulated for the walnut husk fly, *R. completa* (S.B. Opp, personal communication). These various functions for puncture guarding are not mutually exclusive and should provide fertile ground for future research. For the purposes of this review, the point of significance to be attached to male puncture-guarding is its strong implication that punctures constitute a resource to egg-laying females.

IV. COSTS AND BENEFITS OF USE OF OCCUPIED SITES

Rigorous evidence of use of occupied sites by tephritid flies can be expected to accumulate rapidly in the near future. Razy, Lalonde, and Mangel (submitted) independently provide strong inferential evidence that another member of the *Rhagoletis suavis* group, *R. completa*, uses existing oviposition punctures as sites of egg-laying. Anecdotal reports exist of use of punctures by a fourth member of the group, *R. suavis* (Brooks 1921, cited in Reference 21; Prokopy et al., submitted). Use of existing punctures has also been noted for *Bactrocera (Dacus)* species[23,43] (Papaj, unpublished data). Selective use of occupied sites is thus more common than previously thought. What are the possible costs and benefits of this behavior?

A. POSSIBLE COSTS
1. Increased Larval Competition
Depositing eggs in occupied sites might increase the level of intraspecific competition incurred by a female's young. That tephritid fly larvae compete within the fruit for food (including larvae of species that oviposit in occupied fruit) has been documented repeatedly.[5,28-31,76]

2. Increased Adult Competition
Another possible cost associated with selective use of existing sites is interference competition among adults for those sites. Assuming that most (if not all) of the fruit surface is potentially suitable for egg-laying, existing sites must be rarer than new ones and potentially a limited resource contested among females. Moreover, any tendency for females to orient to fruit bearing punctures[79] should tend to increase the rate of encounter between females. There is some evidence for medfly that selective use of punctured fruit incurs costs in terms of mutual interference. Medfly females commonly engage in agonistic displays on fruit, typically leading to displacement of one or both females from the fruit. At their field site on Chios, Papaj and colleagues (unpublished data) noted that contests among medfly females were much more common on artificially-punctured fruit than on untreated control fruit (Figure 2). Exactly how such contests reduce female fitness is currently unknown, but it is not unreasonable to suppose that they do.

3. Increased Parasitism of Juvenile Stages
Use of occupied sites might incur a cost in terms of parasitism. Noting that *Bactrocera (Dacus) dorsalis* flies tend to reuse oviposition punctures made by other females, Haramoto ventured that such behavior might increase the foraging effectiveness of an egg parasite of *B. dorsalis*.[43] Increases in parasitism risk may not be restricted to eggs. Where use of occupied sites increases local larval density within a fruit and where larval parasitoids forage more effectively at higher local densities, use of occupied sites by a female might increase her larvae's risk of parasitism.

A. Unmanipulated Oranges

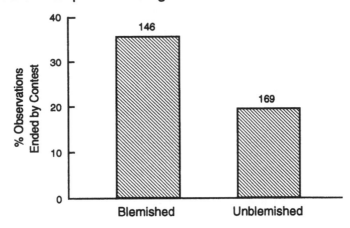

B. Manipulated Oranges

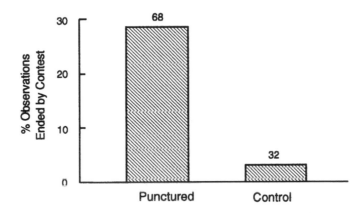

FIGURE 2. Frequency of female-female medfly contests on punctured and unpunctured navel oranges. **A.** Unmanipulated oranges with and without blemishes indicative of oviposition punctures. **B.** Artificially punctured and control oranges.

While plausible, it is not clear that increases in parasitoid foraging effectiveness lead necessarily to an increased risk of parasitism on a per-egg basis. Females risk increased parasitism through use of occupied sites only if eggs or clutches are at higher risk of parasitism when clustered than when deposited alone. Moreover, it seems equally plausible that eggs might suffer lower rates of parasitism when clustered than when distributed randomly or uniformly among sites. If parasitoids are not able to parasitize every egg in a cluster (for example, if interior eggs are shielded by eggs on the outside of a cluster), an egg's risk of parasitism might be reduced by being in the company of others. There is currently no evidence that use of existing sites either increases or decreases parasitism of a female's young.

4. Increased Sexual Harassment

Where males guard oviposition sites, use of those sites may incur a cost in terms of time wasted in interactions with males. For *R. juglandis* and *R. boycei*, for example, mating often takes place when a male mounts a female that has begun to oviposit into an existing puncture. Our field studies indicated that 44% of all *R. juglandis* oviposition attempts (N = 59) and 62%

of *R. boycei* oviposition attempts (N = 34) were interrupted by conspecific males. For females replete with the sperm of high-quality mates, such frequent interruption may represent a significant cost of using existing oviposition sites.

B. POSSIBLE BENEFITS
1. Opportunity for Ovicide

If clutches compete among themselves, use of existing sites might afford females an opportunity to destroy or disable earlier clutches. Price reported that females of an ichneumonid parasitoid, *Pleolophus indistinctus*, thrust their ovipositors into previously established oviposition punctures in their cocooned sawfly hosts, perforating and killing the conspecific or heterospecific egg contained within.[82] Punctures in field-collected cocoons frequently contained both a perforated, deflated egg and an intact one. Price speculated that killing previously laid eggs was a mechanism by which females reduced levels of intraspecific or interspecific competition. Such behavior would be especially functional if females were unlikely to revisit hosts they themselves had parasitized or if females could discriminate between self-parasitized hosts and hosts parasitized by conspecifics or heterospecifics.[32,47,130] At present, we have no evidence to suggest that female tephritid flies kill previously-laid eggs, but the possibility deserves further consideration.

2. Improved Larval Performance

Some investigators believe that tephritid fly larvae, far from competing, cooperate in exploiting host resources. Back and Pemberton, noting high mortality associated with migration of early-instar medfly larvae through the citrus peel and into the pulp, speculated that larvae hatching later might take advantage of migratory routes established by larvae hatching earlier.[9] P. D. Greany (personal communication) suggested that such "social facilitation" might begin a stage earlier if eggs at the periphery of a mass within a puncture shielded eggs on the interior from toxic effects of citrus peel oils. Mourikis claimed (without presenting data) that larvae cooperatively pierce and consume vesicles of citrus pulp.[69] Certain observations hint at the existence of so-called social facilitation among medfly larvae. First, wild medfly females deposit eggs in clutches of approximately 3 to 4.[73] Second, larvae reared on blocks of artificial diet move about in groups over that diet, suggesting a social tendency that could conceivably have a functional basis (E. Boller, personal communication). Cooperation may originate with the females themselves. Tephritid fly females apparently introduce bacteria into the fruit during egg-laying,[37,38,42,46,108,126] which may promote local ripening or phenolic oxidation or have some other effect that improves survival of subsequent clutches deposited at the same site. Social facilitation is of interest here if it causes a larva to be better off as a member of a subsequent clutch in a fruit than as a member of the first. Such facilitation might cause females to exploit egg-infested fruit actively.

3. Reduced Time Expenditures

Use of an existing puncture reduces the time required by females to deposit a clutch in a fruit. Medflies require much less time to deposit eggs in a puncture than at new sites (Table 5; Papaj et al., in preparation). How does time saved in egg deposition translate to increases in female fitness? There are several possibilities. First, a female may be more likely to deposit eggs before interruption at an existing site than at a new one.[80] Interruption has a variety of causes, including harassment by other females and attacks by predators. Observations at the Chios site indicated that about 60% of all medfly oviposition attempts were interrupted by interference or attempted predation. Assuming that females are sometimes unable to relocate the initiated site, such interruptions constitute a cost that can be mitigated by use of existing sites. Second, time saved ovipositing can translate directly into an increase in the number of clutches deposited before reproductive death. If females do not have time left in their lives to

TABLE 5
Median Time Required to Lay Eggs at New Versus Existing Sites

Type of Puncture	N	Median time (sec)	N	Mean clutch size (±1 s.e.)
A. *Laboratory Assay*				
New	17	479a	10	3.4 (±0.27)a
Artificial	29	293b	29	4.2 (±0.50)a
B. *Field Observations*				
New	7	610	—	—
Naturally-formed	3	370	—	—

Note: Values within a column followed by the same letter are not significantly different according to a Mann-Whitney U-test (median time) or a t-test (clutch size).

TABLE 6
Distribution of Kills by Yellowjacket Wasps
According to Female Medfly Activity on Fruit

Female activity	# Attacks	# Kills	% Kills
Oviposition	27	7	26
Other	62	0	0

deposit all eggs available to be laid, use of existing sites might be favored even in the face of costs such as increased larval competition. Third, saving time during oviposition may reduce an adult female's overall risk of predation if oviposition is a relatively risky business. Field observations at the Chios site, for example, indicated that females were much more vulnerable to predation by European yellowjacket wasps (*Vespula germanica*) while laying eggs than while engaged in other activities on the fruit (Table 6; G-test with Yates correction, $p < 0.0001$). On a per-egg basis, females that exploit punctures ought to be less susceptible to predation than females that establish new sites. Whether or not this benefit will be manifested with respect to other important predator species (spiders, robber flies, reduviid bugs, and the like) is not known and may be difficult to determine. Predation rates observed on Chios appeared to be unusually high, and the high vagility of adult tephritid flies makes predator exclusion experiments relatively difficult to conduct.

4. Reduced Ovipositor Wear

Use of existing sites may benefit a female not by reducing the time required to deposit a clutch but by extending the time over which females are capable of laying eggs. Ovipositors are sclerotized structures susceptible to wear, and it has long been known, among those who raise tephritid flies, that females can blunt their ovipositors by attempting to drill into the plastic or wooden frames of their cages. It is conceivable that, through use of existing punctures, females reduce the rate at which their ovipositors degrade and thereby extend their reproductive lifetimes (D. Gerling, personal communication; Razy, Lalonde, and Mangel, submitted). For older females, whose ovipositors are worn to the point where new sites cannot be established, use of existing sites might permit eggs to be laid. In either case, use of existing sites potentially increases a female's reproductive period. There is as yet no clear evidence that such increases are realized in nature.

5. Enhanced Mating Success

In species such as *R. juglandis* and *R. boycei*, where oviposition punctures are sites of mate assembly, the association of males and punctures might constitute not a cost (as specified above), but a benefit to females. Co-occurrence of punctures and males might permit females to forage for both resources more efficiently, a benefit that would be greatest when females were both time- and sperm-limited. Even if females are not sperm-limited, orientation to punctures may improve the quality of a female's mates. In both *R. juglandis* and *R. boycei*, larger males usually win contests (Papaj, unpublished data). Orientation to existing sites in this instance may ensure that females mate with larger and perhaps fitter males. There is currently no evidence that females orient to punctures because males are resident at those sites. Moreover, any advantage in terms of mating implies at least one other advantage to females (unrelated to mating) that gives rise to the co-occurrence of males and punctures in the first place.

V. BALANCING COSTS AND BENEFITS

Are flies that use occupied sites making functional decisions? Our medfly work suggests that use of occupied sites will have costs in terms of larval and adult competition and benefits in terms of time saved during oviposition. Are the benefits likely to exceed the costs? Papaj and Roitberg (in preparation) constructed a dynamic programming model, which weighs these costs and benefits and predicts whether females should use or avoid occupied fruit. The model is dynamic in the sense that a medfly's decisions are characterized with respect to its current egg load and the time of day at which it is foraging. Using parameters estimated from the Chios field site, the model generated several predictions.

First, it predicted that females at the Chios site, from which parameter values were obtained, were severely time-stressed, laying just 3 to 6 eggs out of a total of 30 presumed to be available each day (Figure 3A). This result held true even when females were saving as much time as possible through exclusive use of occupied sites. Recall from the discussion in Section IV.B that time limitation of this type is a key condition underlying several proposed benefits of using occupied sites. Life history studies by Carey and colleagues suggest that a failure to deposit all available eggs over the course of a day is likely to reduce total fecundity over a female's lifetime.[22]

Second, the model predicted that females should exploit occupied sites on fruit even if the fitness of their offspring were reduced by over 50% (Figure 3B). In other words, time savings associated with use of oviposition sites in occupied fruit offset a reduction by more than half in the fitness of their young. Figure 3B also suggests that, where fruit vary with respect to the fitness of a second clutch, female medflies should vary accordingly in their tendency to exploit occupied sites on those fruit. Fruit size is one potential source of such variation. We might suppose reasonably that the fitness of a second clutch in exploited fruit is proportional to the amount of fruit available for larval consumption and, therefore, to fruit size. The smaller the fruit, the lower should be larval fitness when that fruit is already occupied. Does the tendency for female medflies to use exploited fruit (and sites on those fruit) vary according to the size of those fruit?

The answer is yes. In field cages (Papaj, unpublished data), females were permitted to forage in small trees bearing fruit, half of which were untreated and half of which were treated to simulate infestation. Infestation was simulated by puncturing each fruit twice with a small pin and painting the fruit with a methanolic extract of host-marking pheromone. The experiment was run twice in succession, using first kumquats (a relatively large fruit) and then coffee berries (a relatively small fruit). When presented with kumquats, females deposited a majority of clutches in occupied fruit (Figure 4). When presented with coffee berries, in contrast, the same females deposited a majority of clutches in uninfested fruit (Figure 4). These results indicate that individual females can adjust their tendency to use or avoid exploited fruit in a potentially functional way.

A similarly functional pattern was found when size was manipulated within a particular host species. In field cages, Papaj and Messing (in preparation) hung ripe coffee berries in a

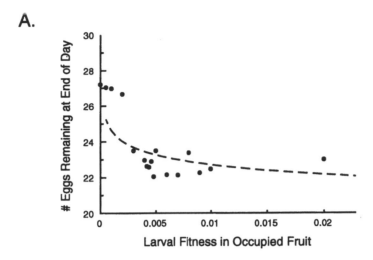

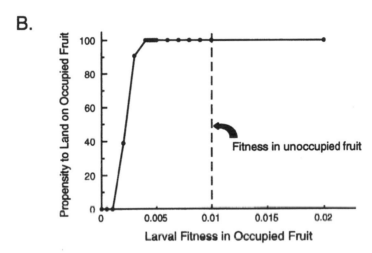

FIGURE 3. Predictions of dynamic programming model of medfly use of occupied sites. **A.** Number of medfly eggs remaining at day's end as a function of clutch fitness at occupied sites. Dashed line indicates line fitted to regression model. **B.** Estimated propensity to deposit eggs in occupied site as a function of clutch fitness at such sites. Each point represents mean propensity of 100 simulated flies. Standard errors around means are negligible and are omitted for clarity of presentation. Dashed line indicates clutch fitness assumed at new site on same fruit.

tree. Half of the berries were twice-infested and twice-marked; half were uninfested. A population of approximately 100 female flies was permitted to forage for fruit while focal observations were made. Over successive presentations, fruit were either very large or very small. Females always deposited more clutches in uninfested berries (Figure 5). However, females avoided use of occupied fruit to a much greater extent on small berries than on large ones. Females thus behave as though costs of larval competition associated with depositing eggs in occupied fruit are smaller when fruit are larger.

In summary, medfly females clearly adjust their behavior to presumed changes in costs associated with larval competition. We might also ask if females can adjust their behavior in response to changes in benefits. Papaj and Messing (in preparation) asked this question in an experiment identical to that described above, except that coffee berry ripeness, rather than coffee size, was manipulated. Although females attempt oviposition readily into unripe coffee berries, they almost never deposit eggs successfully unless oviposition has been initiated into

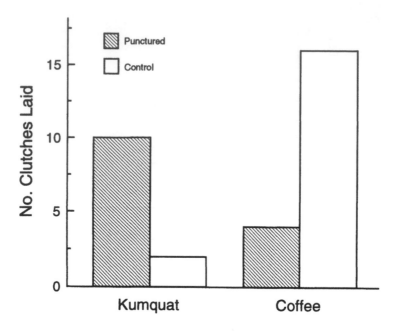

FIGURE 4. Total number of clutches laid by medfly females on focal fruit in kumquat and coffee field-cage presentations. Hatched bars indicate results on fruit bearing artificial punctures and treated with host-marking pheromone extract. Empty bars indicate results on untreated fruit.

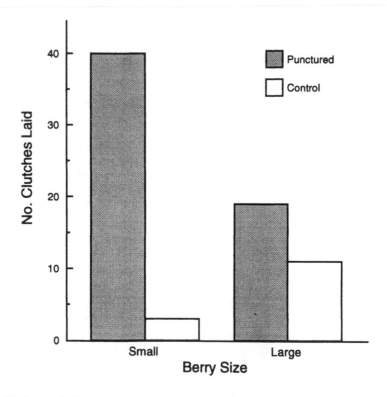

FIGURE 5. Total number of clutches laid by medfly females on focal coffee berries in large berry and small berry field-cage presentations. Shaded bars indicate results on naturally-punctured, naturally-marked fruit. Empty bars indicate results on uninfested fruit.

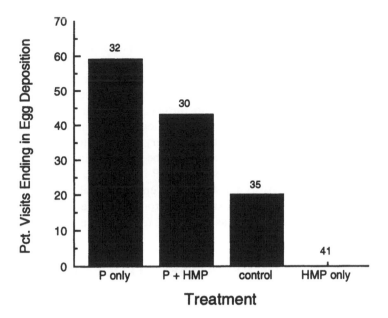

FIGURE 6. Percent visits to a kumquat receiving a medfly clutch according to treatment with host-marking pheromone (HMP) and punctures (P). Number above bar indicates total sample size for that treatment.

an existing puncture. As expected, the same females that avoided oviposition into punctured, marked berries when berries were ripe oviposited almost exclusively into punctured, marked berries when berries were unripe (Papaj and Messing, in preparation). Individual medfly females therefore appear able to adjust their use of occupied fruit according to changes in both the costs and the benefits of using such fruit.

VI. ROLE OF HOST-MARKING IN USE OF OCCUPIED SITES

Host-marking pheromone in tephritids and other insects has often been observed to have deterrent effects on female egg-laying behavior.[8,13-17,53] Yet, for flies that use existing sites, one might expect HMP to have a stimulatory effect. HMP might be a signal to a female that a fruit bears a puncture and might therefore increase her tendency to exploit that puncture. To examine this possibility, Papaj et al.[80] manipulated the occurrence of a puncture independently of the occurrence of HMP on kumquats. Results indicated that a puncture was invariably stimulating and that HMP was invariably deterrent (Figure 6). A kumquat with only a puncture was more likely than any other to receive eggs; a kumquat with only HMP was less likely than any other to receive eggs. Interestingly, a kumquat with both puncture and HMP (i.e., a simulated infested kumquat) was more likely to receive eggs than an uninfested and unmarked kumquat (apparently because the stimulatory effect of the puncture exceeded the deterrent effect of the HMP) (Figure 6). In a separate study, Papaj et al. noted that, even when eggs are laid in a marked fruit, contact with HMP reduces the size of a female's clutch.[79]

A female medfly thus appears to interpret fruit with unoccupied punctures as superior to fruit with occupied punctures. This pattern is fully consistent with the notions that larvae compete within a fruit and that competition increases with egg density;[28-31,77] it is inconsistent with the notion that juveniles in a clutch benefit by being in the company of other clutches. In other words, if eggs or larvae enjoy some benefit by being in larger numbers in a puncture or a fruit, females are not behaving as though they do.

In short, the response evoked by HMP in medfly is not fundamentally different from that evoked in species that do not exploit occupied fruit. The response is consistently a deterrent one. For species like medfly and walnut flies that oviposit multiply in fruit, levels of HMP on the fruit may serve to "inform" females as to the number of clutches present at an occupied site. In this way, a female might assess the level of competition to be suffered by her progeny and respond accordingly. In the presence of only one clutch (and one mark), females might add eggs to a puncture; after a critical number of clutches have been laid in the fruit, females might exploit the fruit no further. This argument has at least three suppositions: 1) females cannot assess the number of eggs in a fruit directly (which would preclude use of HMP); 2) females deposit HMP every time a clutch is deposited in fruit (otherwise the amount of HMP on a fruit would not be a reliable indication of the number of eggs in the fruit); and 3) females respond to HMP in a dosage-dependent fashion. For medfly at least, these suppositions seem reasonable. Females virtually always deposit HMP after eggs are laid,[79] response to HMP extracts is strongly dosage-dependent (Boller and Prokopy, unpublished data), and, as far as is known, females cannot estimate directly the number of eggs in a puncture.

VII. EVOLUTION OF RESPONSES TO OCCUPIED FRUIT

A. RESPONSES TO OVIPOSITION PUNCTURES

Orientation to existing oviposition sites has been observed in four of five members of the *Rhagoletis suavis* group but in no other *Rhagoletis* species. Based on an existing phylogeny,[12] the *suavis* group is derived from *Rhagoletis* groups that evidently do not use punctures. Given use of punctures by the closely-related Mediterranean fruit fly and distantly-related *Bactrocera* (*Dacus*) flies, use of occupied sites appears to have arisen more than once in the Tephritidae.

There are at least two possible routes through which behavior associated with using existing sites might have evolved. The first involves use of fruit wounds that are not of fly origin. Fruit wounds are apparently critical to the use of some hosts by medfly,[71] and an ability to use fruit wounds may have predisposed flies to exploit existing oviposition sites. Alternatively, flies may first have evolved an ability to orient to damage generated during oviposition. Particularly where fruit are difficult to penetrate, female flies frequently do not complete a puncture before they are interrupted by conspecifics or predators. Papaj et al., for example, reported that medfly females initiating oviposition in kumquats completed oviposition less than 60% of the time (see also Section IV.B.2).[80] Similarly, Back and Pemberton noted that as many as 67% of oviposition punctures in citrus fruit infested by medfly were unoccupied.[9] The time wasted initiating and failing to complete a puncture may have favored an ability to resume oviposition in a puncture. In this case, the ability to locate and exploit unoccupied punctures may have disposed flies toward use of occupied ones (as well as toward use of fruit wounds not of fly origin).

B. RESPONSES TO HOST-MARKING PHEROMONE

Some phytophagous insects are able to discriminate between occupied and unoccupied hosts based on cues emitted by the conspecifics themselves[112,113] or cues associated with wounds produced during oviposition.[27,103] Why then has a potentially costly behavior such as host-marking evolved in other phytophagous species? Host-marking systems are postulated to be a means of signalling the presence of conspecifics in circumstances where 1) conspecifics are inconspicuous and 2) other signals associated with oviposition (e.g., damage to the host produced during egg-laying) are weak.[104] In tephritid flies, eggs are laid beneath the fruit surface and are perhaps inconspicuous (condition 1). That other signals associated with oviposition are at least sometimes weak (condition 2) is suggested by the curious behavior of female olive flies (*Dacus oleae*). Following oviposition, these flies use their labellum to spread olive juice over the fruit surface, a behavior that presumably amplifies the intrinsically weak signal associated with tissue damage generated during oviposition. This rudimentary mark subsequently deters egg-laying by female olive flies.[24]

Unfortunately, behavior in some tephritid flies does not satisfy the second condition for the maintenance of host-marking behavior. The behavior of medflies and walnut flies in particular suggests that cues associated with the oviposition puncture are especially strong signals to females. We propose an alternative condition for the maintenance of host-marking in such species: HMP's will be maintained if cues associated with oviposition are an unreliable indication of the presence of otherwise-inconspicuous conspecifics. As noted above (see Section VII.A), female flies often fail to deposit eggs in a puncture that has been initiated. Even in species where females do not exploit existing oviposition sites, it would not be advantageous to reject fruit bearing unoccupied punctures. Host-marking pheromone permits females to discriminate between fruit bearing occupied punctures and fruit bearing yet-to-be-completed ones.

VII. CLOSING REMARKS

Research on egg-laying in tephritid flies is a case study in how thinking about insect behavior has progressed over the last several decades. Early studies were prone to thinking of flies as robots capable of effective but relatively inflexible decisions. In the case of medfly, undue emphasis on the deterrency of host-marking pheromone (and perhaps hope for its application in pest management) caused early observations of selective use of occupied hosts to be ignored for the two decades following the first studies of host-marking in that insect. More generally, the dynamic nature of fly oviposition behavior went essentially unrecognized until Roitberg's groundbreaking studies in the 1980s.[100,101,106]

Future research on the dynamics of oviposition responses to exploited hosts in phytophagous insects should take inspiration (as it has in the past) from studies on entomophagous parasitoids. Recent studies on entomophages have identified a variety of factors that influence use of exploited hosts, including time of season,[107] the presence of other adult females in the habitat,[129] and whether the host is occupied by the female's own young or those of another conspecific.[2,32,47,130] All of these factors ought to be evaluated for phytophages as well.

A more general and equally fascinating issue to be explored in the future is how insects with their comparatively small brains can make decisions about use of occupied hosts that are so remarkably functional, complex, and dynamic. Some of this complex decision-making might be mediated by learning. Over the decade, it has become increasingly clear that phytophagous insects learn and that such learning can play an important role in host use.[75,76,100] However, it would be a mistake to presume that the dynamics of individual behavior outlined here require a great deal of neuronal sophistication. Ethologists have long recognized that relatively complex decision-making could be a consequence of simple fixed responses arranged hierarchically in the nervous system.[56] More recently, Mangel argued persuasively that elementary neural networks may generate relatively complex behavior (specifically, host-selection behavior) that is very nearly optimal.[59] Recent developments in robotics, which use insects as models,[11] would seem to support this point of view.

That fruit flies use relatively simple neural mechanisms to generate complex dynamic behavior was illustrated by a model erected recently to account for medfly behavior on heavily-infested fruit. In that model, no fewer than five potentially functional responses to heavily-infested fruit (e.g., short residence time on fruit, reduced tendency to attempt oviposition, and reduced clutch size if oviposition took place) were dictated by the female fly's adoption of a single rule: when in contact with heavily-infested fruit, increase the rate at which locomotory behavior is initiated.[77] Behavioral sequence data for females on infested and uninfested fruit were wholly consistent with this model. Crudely put, contact with heavily-infested fruit makes medfly females act edgy. This state of "nervous agitation" generates a suite of responses, all of which reduce the number of eggs allocated to heavily-infested fruit. Fruit flies may be robots still, but robots with extraordinary abilities for adapting their behavior to changes in internal state and external environmental conditions.

ACKNOWLEDGMENTS

I thank L. Schoonhoven for useful comments on an earlier draft, R. Lalonde and M. Mangel for access to unpublished manuscripts, and P. Price for providing key references. I am particularly indebted to B. Roitberg for numerous discussions over the years, which have improved greatly my understanding of functional aspects of insect behavior.

REFERENCES

1. **Ali Niazee, M. T.,** The western cherry fruit fly, *Rhagoletis indifferens* (Diptera: Tephritidae). 2. Aggressive behavior, *Can. Entomol.,* 106, 1201, 1974.
2. **van Alphen, J. J. M. and Visser, M. E.,** Superparasitism as an adaptive strategy for insect parasitoids, *Annu. Rev. Entomol.,* 35, 59, 1990.
3. **Averill, A. L. and Prokopy, R. J.,** Oviposition-deterring fruit marking pheromone in *Rhagoletis basiola, Fla. Entomol.,* 17, 315, 1982.
4. **Averill, A. L. and Prokopy, R. J.,** Oviposition-deterring fruit marking pheromone in *Rhagoletis zephyria, J. Ga. Entomol. Soc.,* 17, 315, 1982.
5. **Averill, A. L. and Prokopy, R. J.,** Intraspecific competition in the tephritid fruit fly, *Rhagoletis pomonella, Ecology,* 68, 878, 1987.
6. **Averill, A. L. and Prokopy, R. J.,** Factors influencing release of host-marking pheromone by *Rhagoletis pomonella* flies, *J. Chem. Ecol.,* 14, 95, 1988.
7. **Averill, A. L. and Prokopy, R. J.,** Distribution patterns of *Rhagoletis pomonella* (Diptera: Tephritidae) eggs in hawthorn, *Ann. Entomol. Soc. Am.,* 82, 38, 1989.
8. **Averill, A. L. and Prokopy, R. J.,** Host-marking pheromones, in *World Crop Pests,* Vol. 3A. *Fruit Flies, Their Biology, Natural Enemies and Control,* Robinson, A. S. and Hooper, G., Eds. Elsevier, Amsterdam, 1989, 207.
9. **Back, E. A. and Pemberton, C. E.,** Susceptibility of citrus fruits to the attack of the Mediterranean fruit fly, *J. Agric. Res.,* 3, 311, 1915.
10. **Bateman, M. A.,** The ecology of fruit flies, *Annu. Rev. Entomol.,* 17, 493, 1972.
11. **Beer, R. D., Chiel, H. J., and Sterling, L. S.,** An artificial insect, *Am. Sci.,* 79, 444, 1991.
12. **Berlocher, S. H. and Bush, G. L.,** An electrophoretic analysis of *Rhagoletis* (Diptera: Tephritidae) phylogeny, *Syst. Zool.,* 31, 136, 1982.
13. **Boller, E. F.,** Oviposition-deterring pheromone of the European cherry fruit fly: status of research and potential applications, in *Management of Insect Pests with Semiochemicals,* Mitchell, E. R., Ed., Plenum Press, New York. 1981, 457.
14. **Boller, E. F.,** Information service and new substances, in *Fruit Flies of Economic Importance 84,* R. Cavalloro, Ed., A. A. Balkeman, Rotterdam, 1986, 9.
15. **Boller, E. F. and Hurter, J.,** Oviposition deterring pheromone in *Rhagoletis cerasi*: behavioral laboratory test to measure pheromone activity, *Entomol. Exp. Appl.,* 39, 163, 1985.
16. **Boller, E. F. and Aluja, M.,** Oviposition deterring pheromone in *Rhagoletis cerasi* L.: biological activity of 4 synthetic isomers and HMP discrimination of two host races as measured by an improved laboratory bioassay, *Z. Angew. Entomol.,* in press.
17. **Boller, E. F., Schöni, R., and Bush, G. L.,** Oviposition deterring pheromone in *Rhagoletis cerasi*: biological activity of a pure single compound verified in semi-field test, *Entomol. Exp. Appl.,* 45, 17, 1987.
18. **Boyce, A. M.,** Bionomics of the walnut husk fly, *Rhagoletis completa, Hilgardia,* 8, 363, 1934.
19. **Brantjes, N. B. M.,** Prevention of superparasitism of *Melandrium* flowers (Caryophyllaceae) by *Hadena* (Lepidoptera), *Oecologia,* 24, 1, 1976.
20. **Bultman, T. L. and Faeth, S. H.,** Experimental analysis for intraspecific competition in a lepidopteran leaf miner, *Ecology,* 67, 442, 1986.

21. **Bush, G. L.,** The taxonomy, cytology and evolution of the genus *Rhagoletis* in North America, *Bull. Mus. Comp. Zool. Harv. Univ.,* 134, 431, 1966.
22. **Carey, J. R., Krainacker, D. A., and Vargas, R. I.,** Life history response of female Mediterranean fruit flies, *Ceratitis capitata,* to periods of host deprivation, *Entomol. Exp. Appl.,* 42, 159, 1984.
23. **Christenson, L. D. and Foote, R. H.,** Biology of fruit flies, *Annu. Rev. Entomol.,* 171-192, 1960.
24. **Cirio, U.,** Reperi sul meccanismo stimolo-riposta nell 'ovideposizione della *Dacus oleae* Gmelin (Diptera: Trypetidae), *Redia,* 52, 577, 1971.
25. **Cirio, U.,** Observazioni sul comportameno di ovideposizione della *Rhagoletis completa* in laboratorio, *Proc. 9th Congr. Ital. Entomol. Soc.,* 99, 1972.
26. **Corbet, S. A.** Oviposition pheromone in larval mandibular glands of *Ephestia kuehniella, Nature,* 243, 537, 1973.
27. **Craig, T. P., Itami, J. K., and Price, P. W.,** Plant wound compounds from oviposition scars used in host discrimination by a stem-galling sawfly, *J. Insect Behav.,* 4, 343, 1988.
28. **Debouzie, D.,** Etude de la competition larvaire chez *Ceratitis capitata* (Diptere, Trypetidae), *Arch. Zool. Exp. Gen.,* 118, 315, 1977.
29. **Debouzie, D.,** Effect of initial population size on *Ceratitis* productivity under limited food conditions, *Ann. Zool. Ecol. Anim.,* 9, 367, 1977.
30. **Debouzie, D.,** Variabilite a l'interieur d'une population de *Ceratita capitata* elevee sur un milieu naturel non renouvele (Diptera:Trypetidae), *Ann. Zool. Ecol. Anim.,* 10, 515, 1978.
31. **Debouzie, D.,** Analyse experimentale de l'utilisation des resources dans un systeme simplife forme d'une banane attaquee par la mouche mediterraneenne des fruits *Cerititis capitata, Acta Oecol. Gener.,* 2, 371, 1981.
32. **van Dijken, M. J. and Waage, J. K.,** 1987. Self and conspecific superparasitism by the egg parasitoid *Trichogramma evanescens, Entomol. Exp. Appl.* 43:183.
33. **Dixon, C. A., Erickson, J. M., Kellett, D. N., and Rothschild, M.,** Some adaptations between *Danaus plexippus* and its food plants, with notes on *Danaus chrysippus* and *Euploea core* (Insecta: Lepidoptera), *J. Zool. London,* 185, 437, 1978.
34. **Fitt, G. P.,** Observations on the biology and behavior of *Paraceratitella eurycephala* (Diptera: Tephritidae) in northern Australia, *J. Aust. Entomol. Soc.,* 20, 1, 1981.
35. **Fitt, G. P.,** Factors limiting the host range of Tephritid fruit flies, with particular emphasis on the influence of *Dacus tryoni* on the distribution and abundance of *Dacus jarvisi,* Ph.D. thesis, 1983.
36. **Fitt, G. P.,** Oviposition behavior of two tephritid fruit flies, *Dacus tryoni* and *Dacus jarvisi,* as influenced by the presence of larvae in the host fruit, *Oecologia,* 62, 37, 1984.
37. **Fitt, G. P. and O'Brien, R. G.,** Bacteria associated with four species of *Dacus* (Diptera; Tephritidae) and their role in the nutrition of larvae, *Oecologia,* 67, 447, 1985.
38. **Fytizase, A. B. and Tzanakakis, C. D.,** Some effects of streptomycin when added to the adult food on the adults of *Dacus oleae* and their progeny, *Ann. Entomol. Soc. Am.,* 59, 269, 1966.
39. **Giga, D. P. and Smith, R. H.,** Oviposition markers in *Callosobruchus maculatus* F. and *Callosobruchus rhodesianus* Pic. (Coleoptera, Bruchidae): asymmetry of interspecific responses, *Agric. Ecosyst. Environ.,* 12, 229, 1985.
40. **Gilbert, L. E.,** Ecological consequences of a coevolved mutualism between butterflies and plants, in *Coevolution of Animals and Plants,* Gilbert, L. E. and Raven, P. H., Eds., Univ. of Texas Press, Austin, 1975, 210.
41. **Haefliger, E.,** Das Auswahlvermoegen der Kirschenfliege bei der Eiablage (Eine statistische Studie), *Mitt. Schweiz. Entomol. Ges.,* 26, 258, 1953.
42. **Hagen, K. S.,** Dependence of the olive fruit fly, *Dacus oleae,* larvae on the symbiosis with *Pseudomonas savastanoi* for the utilization of the olive, *Nature,* 204, 423, 1966.
43. **Haramoto, F. H.,** Unpublished M.Sc. dissertation, University of Hawaii, 1953.
44. **Hayes, J. L.,** Egg distribution and survivorship in the pierid butterfly, *Colias alexandra, Oecologia,* 66, 495, 1985.
45. **Hendrichs, J. P., Katsoyannos, B. I., Papaj, D. R., and Prokopy, R. J.,** Mediterranean fruit fly (Diptera:Tephritidae) activities in nature, including feeding and fecundity on natural food sources, *Oecologia,* 86, 223, 1991.
46. **Howard, D. J., Bush, G. L., and Breznak, J. A.,** The evolutionary significance of bacteria associated with *Rhagoletis, Evolution,* 39, 405, 1985.
47. **Hubbard, S. F., Marris, G., Reynolds, A., and Rowe, G. W.,** Adaptive patterns in the avoidance of superparasitism by solitary parasitic wasps, *J. Anim. Ecol.,* 56, 387, 1987.
48. **Ives, P. M.,** How discriminating are cabbage butterflies?, *Aust. J. Ecol.,* 3, 261, 1978.
49. **Jonasson, T.,** Frit fly (*Oscinella frit*) oviposition on oat seedlings: evidence for contagious distribution of eggs, *Entomol. Exp. Appl.,* 32, 98, 1982.
50. **Katsoyannos, B. I.,** Oviposition-deterring, male-arresting, fruitmarking pheromone in *Rhagoletis cerasi, Environ. Entomol.,* 4, 801, 1975.

51. **Kellogg, T.,** Egg dispersion patterns and egg avoidance behavior in the butterfly *Pieris sisymbrii* Bdv. (Pieridae), *J. Lepidopt. Soc.,* 39, 268, 1985.

52. **Koslowski, M. W., Lux, S., and Dmoch, J.,** Oviposition behaviour and pod marking in the cabbage seed weevil, *Entomol. Exp. Appl.,* 34, 277, 1983.

53. **Lalonde, R. G. and Roitberg, B. D.,** Host selection behavior of a thistle-feeding fly: choices and consequences, *Oecologia,* 90, 534, 1992.

54. **Leroux, E. J. and Mukerji, M. K.,** Notes on the distribution of immature stages of the apple maggot, *Rhagoletis pomonella* (Walsh) (Diptera: Trypetidae) on apple in Quebec, *Ann. Entomol. Soc. Quebec,* 8, 60, 1963.

55. **Liquido, N.,** *Host Plants of the Mediterranean Fruit Fly,* Entomol. Soc. Am., Lanham, Maryland, 1991.

56. **Lorenz, K. Z.,** *The Foundations of Ethology,* Springer-Verlag, New York, 1981.

57. **Mangel, M.,** Oviposition site selection and clutch size in insects, *J. Math. Biol.,* 25, 1, 1987.

58. **Mangel, M.,** An evolutionary interpretation of the "motivation to oviposit," *J. Evol. Biol.,* 2, 157, 1989.

59. **Mangel, M.,** Evolutionary optimization and neural network models of behavior, *J. Math. Biol.,* 28, 237, 1990.

60. **Mangel, M. and Roitberg, B. D.,** Dynamic information and host acceptance by a tephritid fruit fly, *Ecol. Entomol.,* 14, 181, 1989.

61. **McNeil, J. and Quiring, D.,** Evidence of oviposition-deterring pheromone in the alfalfa blotch leafminer, *Agromyza frontella* (Rond) (Diptera: Agromyzidae), *Environ. Entomol.,* 12, 990, 1982.

62. **Messina, F. J.,** Oviposition deterrent from eggs of *Callosobruchus maculatus*: spacing mechanism or artifact?, *J. Chem. Ecol.,* 13, 218, 1987.

63. **Messina, F. J.,** Host-plant variables influencing spatial distribution of a frugivorous fly, *Rhagoletis indifferens, Entomol. Exp. Appl.,* 50, 287, 1989.

64. **Messina, F. J. and Renwick, J. A. A.,** Mechanism of egg recognition by the cowpea weevil *Callosobruchus maculatus, Entomol. Exp. Appl.,* 37, 241, 1985.

65. **Messina, F. J. and Renwick, J. A. A.,** Ability of ovipositing seed beetles to discriminate between seeds with differing egg loads, *Ecol. Entomol.,* 10, 225, 1985.

66. **Miller, G. E. and Borden, J. H.,** Reproductive behaviour of the Douglas fir cone gall midge, *Contarinia oregonensis* (Diptera: Cecidomyidae), *Can. Entomol.,* 116, 607, 1984.

67. **Mitchell, R.,** The evolution of oviposition tactics in the bean weevil, *Callosobruchus maculatus* (F.), *Ecology,* 56, 696, 1975.

68. **Mitchell, R.,** Behavioral ecology of *Callosobruchus maculatus,* in *Bruchids and Legumes: Economics, Ecology and Coevolution,* Fujii, K., Gatehouse, A. M. R., Johnson, C. D., Mitchell, R., and Yoshida, Y., Eds., Kluwer, The Netherlands, 1990, 317.

69. **Mourikis, P. A.,** Data concerning the development of the immature stages of the Mediterranean fruit fly *Ceratitis capitata* (Weidemann) (Diptera:Trypetidae) on different host fruits and on artificial media under laboratory conditions, *Ann. l'Institut Phytopathol. Benaki,* 7, 59, 1965.

70. **Mudd, A. and Corbet, S. A.,** Mandibular gland secretion of larvae of the stored products pests *Anagasta kuehniella, Ephestia cautella, Plodia interpunctella,* and *Ephestia elutella, Entomol. Exp. Appl.,* 16, 291, 1973.

71. **Oi, D. H. and Mau, F. L.,** Relationship of fruit ripeness to infestation in 'Sharwil' avocados by the Mediterranean fruit fly and the Oriental fruit fly (Diptera: Tephritidae), *J. Econ. Entomol.,* 82, 556, 1989.

72. **Opp, S. B., Ziegner, J., Bui, N., and Prokopy, R. J.,** Factors influencing estimates of sperm competition in *Rhagoletis pomonella* (Walsh) (Diptera: Tephritidae), *Ann. Entomol. Soc. Am.,* 83, 521, 1990.

73. **Oshima, K., Honda, H., and Yamamoto, I.,** Isolation of an oviposition marker from Azuki bean weevil, *Callosobruchus chinensis, Agric. Biol. Chem.,* 37, 2679, 1973.

74. **Papaj, D. R.,** Fruit size and clutch size in *Ceratitis capitata, Entomol. Exp. Appl.,* 54, 195, 1990.

75. **Papaj, D. R. and Prokopy, R. J.,** The effect of prior adult experience on components of habitat preference in the apple maggot fly (*Rhagoletis pomonella*), *Oecologia,* 76, 538, 1988.

76. **Papaj, D. R. and Prokopy, R. J.,** Ecological and evolutionary aspects of learning in phytophagous insects, *Annu. Rev. Entomol.,* 34, 315, 1989.

77. **Papaj, D. R., Roitberg, B. D., and Opp, S. B.,** Serial effects of host infestation on egg allocation by the Mediterranean fruit fly: a rule of thumb and its functional significance, *J. Anim. Ecol.,* 58, 955, 1989.

78. **Papaj, D. R., Hendrichs, J., and Katsoyannos, B. I.,** Use of fruit wounds in oviposition by the Mediterranean fruit fly, *Entomol. Exp. Appl.,* 53, 203, 1989.

79. **Papaj, D. R., Roitberg, B. D., Opp, S. B., Prokopy, R. J., Aluja, M., and Wong, T. T. Y.** Effect of marking pheromone on clutch size in the Mediterranean fruit fly, *Physiol. Entomol.,* 15, 463, 1990.

80. **Papaj, D. R., Averill, A. L., Prokopy, R. J., and Wong, T. T. Y.,** Host-marking pheromone and use of previously established oviposition sites by the Mediterranean fruit fly (Diptera: Tephritidae), *J. Insect Behav.,* 5, 583, 1992.

81. **Pitarra, K. and Katsoyannos, B. I.,** Evidence for a host-marking pheromone in *Chaetorellia australis, Entomol. Exp. Appl.,* 54, 287, 1990.

82. **Price, P. W.,** Biology of and host exploitation by *Pleolophus indistinctus* (Hymenoptera: Ichneumonidae), *Ann. Entomol. Soc. Am.* 63, 1502, 1970.
83. **Price, P., Bouton, C. E., Gross, P., McPheron, B. A., Thompson, J. N., and Weis, E.,** Interactions among trophic levels: influence of plants on interactions between insect herbivores and natural enemies, *Annu. Rev. Ecol. Syst.,* 11, 41, 1980.
84. **Prokopy, R. J.,** Evidence for a pheromone deterring repeated oviposition in apple maggot flies, *Environ. Entomol.,* 1, 326, 1972.
85. **Prokopy, R. J.,** Oviposition-deterring fruit marking pheromone in *Rhagoletis fausta, Environ. Entomol.,* 4, 298, 1975.
86. **Prokopy, R. J.,** Feeding, mating and oviposition activities of *Rhagoletis fausta* flies in nature, *Ann. Entomol. Soc. Am.,* 69, 899, 1976.
87. **Prokopy, R. J.,** Significance of fly marking of oviposition site (in Tephritidae), in *Studies in Biological Control,* Delucchi, V. L., Ed., Cambridge Univ. Press, Cambridge, 1976, 23.
88. **Prokopy, R. J.,** Epideictic pheromones that influence spacing in patterns of phytophagous insects. in *Semiochemicals, Their Role in Pest Control,* Nordlund, D. A., Jones, R. L., and Lewis, W. J., Eds., John Wiley & Sons, New York, 1981, 181.
89. **Prokopy, R. J. and Koyama, J.,** Oviposition site partitioning in *Dacus curcurbitae, Ann. Entomol. Soc. Am.,* 31, 428, 1982.
90. **Prokopy, R. J., Bennett, E. W., and Bush, G. L.,** Mating behavior in *Rhagoletis pomonella* (Diptera: Tephritidae), *Can. Entomol.,* 103, 1405, 1971.
91. **Prokopy, R. J., Reissig, W. H., and Moericke, V.,** Marking pheromones deterring repeated oviposition in *Rhagoletis* flies, *Entomol. Exp. Appl.,* 20, 170, 1976.
92. **Prokopy, R., Greany, P. D., and Chambers, D. L.,** Oviposition-deterring pheromone in *Anastrepha suspensa, Environ. Entomol.,* 6, 463, 1977.
93. **Prokopy, R. J., Ziegler, J. R., and Wong, T. T. Y.,** Deterrence of repeated oviposition by fruit-marking pheromone in *Ceratitis capitata, J. Chem. Ecol.,* 4, 55, 1978.
94. **Prokopy, R. J., Malavasi, A., and Morgante, J. S.,** Oviposition-deterring pheromone in *Anastrepha fraterculus* flies, *J. Chem. Ecol.,* 8, 763, 1982.
95. **Prokopy, R. J., Papaj, D. R., Opp, S. B., and Wong, T. T. Y.,** Intra-tree foraging behavior of *Ceratitis capitata* flies in relation to host fruit density and quality, *Entomol. Exp. Appl.,* 45, 251, 1987.
96. **Raina, A. K.,** Deterrence of repeated oviposition in sorghum shootfly, *Atherigona soccata, J. Chem. Ecol.,* 7, 785, 1981.
97. **Rausher, M. D.,** Egg recognition: its advantages to a butterfly, *Anim. Behav.,* 27, 1034, 1979.
98. **Reissig, W. H. and Smith, D. C.** Bionomics of *Rhagoletis pomonella* in *Crataegus, Ann. Entomol. Soc. Am.,* 71, 155, 1978.
99. **Remund, U., Katsoyannos, B. I., and Boller, E. F.,** Zur Eiverteilung der kirschen fliege, *Rhagoletis cerasi* L. (Dipt., Tephritidae), im Freiland, *Bull. Soc. Entomol. Suisse,* 53, 401, 1980.
100. **Roitberg, B. D. and Prokopy, R. J.,** Experience required for pheromone recognition by the apple maggot fly, *Nature,* 292, 540, 1981.
101. **Roitberg, B. D. and Prokopy, R. J.,** Host deprivation influence on response of *Rhagoletis pomonella* to its oviposition-deterring pheromone, *Physiol. Entomol.,* 8, 69, 1983.
102. **Roitberg, B. D. and Prokopy, R. J.,** Resource assessment by larval and adult codling moths, *J. NY Entomol. Soc.,* 90, 258, 1983.
103. **Roitberg, B. D. and Prokopy, R. J.,** Host discrimination by adult and larval European apple sawflies *Hoplocampa testudinea* (Klug) (Hymenoptera: Tenthredinidae), *Environ. Entomol.,* 13, 1000, 1984.
104. **Roitberg, B. D. and Prokopy, R. J.,** Insects that mark host plants, *BioScience,* 37, 400, 1987.
105. **Roitberg, B. D. and Mangel, M.,** On the evolutionary ecology of marking pheromones, *Evol. Ecol.,* 2, 289, 1988.
106. **Roitberg, B. D., van Lenteren, J. C., van Alphen, J. M., Galis, F., and Prokopy, R. J.,** Foraging behaviour of *Rhagoletis pomonella,* a parasite of hawthorn (*Crataegus viridis*), in nature, *J. Anim. Ecol.,* 51, 307, 1982.
107. **Roitberg, B. D., Mangel, M., Lalonde, R. G., Roitberg, C. A., van Alphen, J. J. M., and Vet, L.,** Seasonal dynamic shifts in patch exploitation by parasitic wasps, *Behav. Ecol.,* 3, 156, 1992.
108. **Rossiter, M. A., Howard, D. J., and Bush, G. L.,** Symbiotic bacteria of *Rhagoletis pomonella,* in *Fruit Flies of Economic Importance,* Cavalloro, R. Ed., Balkema, Rotterdam, The Netherlands, 1983, 77.
109. **Rothschild, M. and Schoonhoven, L. M.,** Assessment of egg load by *Pieris brassicae* (Lepidoptera: Pieridae), *Nature,* 266, 352, 1977.
110. **Schoonhoven, L. M.,** Host-marking pheromones in Lepidoptera, with special reference to two *Pieris* spp., *J. Chem. Ecol.,* 16, 3043, 1990.
111. **Scott, D. R.,** The corn earworm in southwestern Idaho: infestation levels and damage to processing corn and sweet corn seed, *J. Econ. Entomol.,* 70, 709, 1977.

112. **Shapiro, A. M.,** Egg-load assessment and carryover diapause in *Anthocharis* (Pieridae), *J. Lepidopt. Soc.*, 34, 307, 1980.

113. **Shapiro, A. M.,** The pierid red-egg syndrome, *Am. Nat.*, 117, 276, 1981.

114. **Silvestri, F.,** Report of an expedition to Africa in search of the natural enemies of fruit flies (Trypaneidae), with descriptions, observations and biological notes, *Hawaii Board of Agriculture and Forestry Division of Entomology Bull. No. 3.*, 1914.

115. **Singer, M. C. and Mandracchia, J.,** On the failure of two butterfly species to respond to the presence of conspecific eggs prior to oviposition, *Ecol. Entomol.*, 7, 327, 1982.

116. **Smith, D. C. and Prokopy, R. J.,** Mating behavior of *Rhagoletis mendax* (Diptera: Tephritidae) flies in nature, *Ann. Entomol. Soc. Am.*, 74, 388, 1982.

117. **Smith, D. C. and Prokopy, R. J.,** Mating behavior of *Rhagoletis pomonella* (Diptera: Tephritidae) flies in nature, *Can. Entomol.*, 112, 585, 1980.

118. **Stansly, P. and Cate, J.,** Discrimination by ovipositing boll weevils against previously infested *Hampea* flower buds, *Environ. Entomol.*, 13, 1361, 1984.

119. **Stiling, P. D., Brodbeck, B. V., and Strong D. R.,** Intraspecific competition in *Hydrellia valida* (Diptera: Ephydridae), a leaf miner of *Spartina alternifolia*, *Ecology*, 65, 660, 1984.

120. **Stiling, P. D. and Strong, D. R.,** Weak competition among *Spartina* stem borers, by means of murder, *Ecology*, 64, 770, 1983.

121. **Strong, D. R., Lawton, J. H., and Southwood, T. R. E.,** *Insects on Plants*, Harvard University Press, Cambridge, MA, 1984.

122. **Szentesi, A.,** Pheromone-like substances affecting host-related behaviour of larvae and adults in the dry bean weevil, *Acanthoscelides obtectus, Entomol. Exp. Appl.*, 30, 219, 1981.

123. **Takasu, K. and Hirose, Y.,** The parasitoid *Ooencyrtus nezarae* (Hymenoptera: Encyrtidae) prefers hosts parasitized by conspecifics over unparasitized hosts, *Oecologia*, 87, 319, 1991.

124. **Tallamy, D. W.,** "Egg dumping" in lace bugs (*Gargaphia solani*, Hemiptera: Tingidae), *Behav. Ecol. Sociobiol.*, 17, 357, 1985.

125. **Thompson, J. N.,** Selection pressure on phytophagous insects feeding on small host plants, *Oikos*, 40, 438, 1983.

126. **Tsiropoulos, G. J.,** Effects of antibiotics incorporated into defined adult diets on survival and reproduction of the walnut husk fly *Rhagoletis completa* Cress. (Diptera, Trypetidae)., *Z. Ang. Entomol.*, 91, 100, 1981.

127. **Umeya, K.,** Studies on the comparative ecology of bean weevils. I. On the egg distribution and the oviposition behaviors of three species of bean weevils infesting Azuki bean, *Res. Bull. Plant Prot. Jpn.*, 3, 1, 1966.

128. **Urquhart, F. A.,** *The Monarch Butterfly*, University of Toronto Press, Toronto. 1960.

129. **Visser, M. E., van Alphen, J. J. M., and Nell, H. W.,** Adaptive superparasitisim and patch time allocation in solitary parasitoids: the influence of the number of parasitoids depleting a patch, *Behaviour*, 114, 21, 1990.

130. **Visser, M. E., van Alphen, J. J. M., and Hemerik, L.,** Adaptive superparasitism and patch time allocation in solitary parasitoids: an ESS model, *J. Anim. Ecol.*, 61, 93, 1992.

131. **Weismann, R.,** Die Orientierung der Kirschenfliege, *Rhagoletis cerasi* L., bei der Eiablage (Eine sinnesphysiologische Utersuchung), *Landw. Jahrb. Schweiz.*, 51, 1080, 1937.

132. **Wiklund, C. and Ahrberg, C.,** Host plants, nectar source plants, and habitat selection of males and females of *Anthocharis cardamines* (Lepidoptera), *Oikos*, 31, 169, 1978.

133. **Williams, K. S. and Gilbert, L. E.,** Insects as selective agents on plant vegetative morphology: egg mimicry reduces egg laying by butterflies, *Science*, 212, 467, 1981.

134. **Zimmerman, M.,** Oviposition behavior and the existence of an oviposition deterring pheromone in *Hylemya*, *Environ. Entomol.*, 8, 277, 1979.

135. **Zwolfer, H.,** Life systems and strategies of resource exploitation in tephritids, in *Fruit Flies of Economic Importance*, Cavalloro, R., Ed., Balkema, Rotterdam, The Netherlands, 1982, 16.

Chapter 3

FLORAL VOLATILES IN INSECT BIOLOGY

Heidi E. M. Dobson

TABLE OF CONTENTS

0-8493-4125-6/94/$0.00+$.50
© 1994 by CRC Press Inc.

I. INTRODUCTION

The biology of flowers is closely interwoven with the lives of many and diverse insects, providing resources essential to their survival.[156,256] Although the evolutionary history of the class Insecta is older than that of the angiosperms, the two groups have clearly evolved and diversified together.[283] This is especially evident in the interactions between the reproductive organs of angiosperms, namely flowers, and the pollinating insects, which through their visitation to flowers assist the plant in its sexual reproduction. Insects visit flowers primarily to obtain food, most often in the form of nectar and pollen but sometimes as oils or other floral tissues.[8,54,277,278] In addition, flowers may be used as sites for encountering mates, laying eggs, finding shelter, or gathering nest building materials. Associations between flowers and insects range widely, from tightly mutualistic, where both partners benefit strongly and even depend upon each other for survival, to fully parasitic, where only one partner benefits while the other suffers a loss in fitness. Correspondingly, the degree of adaptation displayed by either partner to the association varies widely and is often unequal between the partners.

An insect's perception of its host plant involves the integration of different stimuli, including olfactory, visual, gustatory, and tactile cues. The importance of each varies over space (e.g., distance from the flower) and time (e.g., day versus night), with some stimuli being more dominant than others. This interplay of signal stimuli from the flowers makes it a challenge to isolate and identify the effects that any single stimulus type has on the insect. Floral scents, of main interest here, can be important and in some cases indispensable to the insect's interaction with the flower. They may elicit search behavior, alighting, feeding, and/ or oviposition responses.[168,229] Vision and olfaction are generally the main stimuli that attract insects to flowers; after the insect alights, it may be guided to the food sources within the flower by close-range visual and olfactory cues (i.e., "food guides"), but at this post-alighting stage interactions among different stimulus types can become more complex as tactile stimuli and contact chemicals also enter the picture.[3] The study of how floral stimuli affect insects thus includes both pre- and postalighting effects. Superimposed on these variables are the species effects, whereby the types of responses elicited by floral volatiles vary with both the species of insect and the species of plant. In addition to attracting insects to flowers and guiding them to food or other resources within the flowers, floral volatiles are pivotal in allowing insects to discriminate among plant species and even among individual flowers of a single species. By providing species-specific signals, flower fragrances facilitate an insect's ability to learn particular food sources, thereby increasing its foraging efficiency; at the same time, flower constancy is enhanced, which ensures successful pollen transfer and thus sexual reproduction for the plants.

Insects can show very high levels of sensitivity to airborne floral chemicals. The olfactory threshold of the oriental fruit fly *Dacus dorsalis* (Tephritidae) for methyl eugenol, a component of the fragrances of flowers to which the fly is attracted,[181] reaches 0.01 μg (applied to paper in a 30 cm³ cage).[219] This compound elicits a strong feeding response, and can attract flies from a distance of at least 0.8 km.[281] Male euglossine bees have also been observed to fly at least 1 km over water to filter-paper disks baited with volatiles that occur in the orchid flowers they pollinate (Ackerman, in Reference 311). Furthermore, the orientation of insects to odor sources is greatly influenced by factors such as the wind, which determines the structural characteristics of the odor plume.[224] Olfactory thresholds to floral volatiles have not been determined experimentally for other flower visitors, but they undoubtedly depend on the insect species, chemical properties of the volatiles, and ambient conditions.

The literature is rich in reports and anecdotes of insects being attracted to flowers by their fragrances, but experimental evidence for the role of floral volatiles in shaping flower–insect associations is scant, as is our understanding of the sensory mechanisms involved in flower selection by insects. A century ago, Darwin described insects flying to flowers covered with a muslin net, which he attributed to the effect of flower odors, and he cites earlier, cursory notes by others.[78] The subsequent documentation of flower–insect associations in relation to floral odors has been quite extensive, especially for flowers pollinated by night-flying moths. However, in many of these cases any reference to floral volatiles is generally limited to subjective descriptions of the odors as perceived by the human nose and to inferences of their probable role based on descriptive studies of the insects, flowers, or both. These latter studies consist of compiled name lists and recorded observations of insects visiting flowers, detailed investigations into the structure of flowers leading to interpretations of their probable functional significance to insects, and careful observations of the behavior of insects on flowers. Reports showing links between floral odors and insects are thus largely founded on circumstantial rather than experimental evidence. Nevertheless, it must be recognized that the broad information base provided by this mostly descriptive literature has been central in the formulation of new questions and more experiment-oriented studies.

A major portion of our empirical knowledge of floral volatiles and insects stems from studies that have focused on a few selected groups of plants and insects (e.g., Hymenoptera). These investigations have contributed immensely to our understanding of floral chemistry and its biological effect on flower-visiting insects. They have also alerted us to the frequent complexities and multifaceted nature of the interactions and to the need for carefully planned, controlled experiments. Selected findings are summarized in several general reviews.[22-24,263,282,311] (Detailed reviews are cited in sections below.) In addition, during the last decade a considerable number of studies have addressed other insect groups, including flower-feeding insects associated with crop plants. As new methods used in chemical ecology become more widely accessible, increasing attention is being given to the chemical bases of insect–flower interactions. The most informative studies are those that combine chemical analyses of floral odors with behavioral experiments of the flower-visiting insects. Whereas analyses of floral odors have been performed on relatively many plant species,[165] what is now most lacking are the behavioral tests aimed at identifying which chemicals are active in the insect–flower interactions and at clarifying the behavioral and physiological mechanisms underlying them.[24] With this field currently expanding, there is a need to integrate results from the diverse investigations; we can then look forward to broad syntheses and major advances in our understanding.

The aim of this chapter is to review our present knowledge of the role that floral odors play in insect–flower associations, with special emphasis on empirically obtained evidence; the circumstantial evidence is covered only summarily. The most important goal here is to bring together the experimental evidence, from disparate fields within entomology and botany, that addresses the effects of floral volatiles in orienting and guiding insects to flowers and in eliciting the insects' specific behavioral and physiological responses.

II. FLOWER FRAGRANCE CHEMISTRY

Volatiles emitted by flowers and vegetative plant parts belong to similar major classes of compounds;[125] however, the volatile profiles of flowers are generally distinct from those of a plant's foliage or fruits. Flowers may differ from other plant parts in the identity of major and minor volatiles, as well as in the relative proportions of individual constituent volatiles.[31-33,58,93,124,194,238]

A. COMPOSITION

The chemical composition of flower fragrances varies widely from species to species in terms of both the number and identity of volatiles. Fragrances generally comprise blends of compound classes, including isoprenoids, fatty acid derivatives, benzenoids, and aminoid compounds. Each of these classes is in turn represented by compounds having different functional groups, e.g., hydrocarbons, alcohols, aldehydes, ketones, acids, and esters. In a comprehensive compilation of volatiles identified to date in floral fragrances of 441 taxa, Knudsen et al.[165] list over 700 compounds. Although blends of different compound classes and functional groups are typical in fragrances, many of the volatiles tend to be closely related in their biosynthetic pathways;[24,71] moreover, the blend is usually dominated by one or a few volatiles. For detailed treatment of the chemistry of floral fragrances, the reader is referred to other sources;[23,24,42,125,148,165,217,312] discussion here will be restricted to a few examples selected for illustrative purposes only.

Predominant in many floral fragrances are isoprenoid compounds,[165] especially monoterpenes (e.g., *trans*-β-ocimene, geraniol, linalool) and sesquiterpenes (e.g., α-farnesene, caryophyllene, germacrene), and benzenoid compounds (e.g., benzyl acetate, benzaldehyde, eugenol, 2-phenylethyl alcohol).[33,57,58,87,96,164,197,209,312] Fatty acid-derived volatiles (e.g., fatty acid esters and hydrocarbons) are major components of fragrances in some species, particularly in families of the subclass Magnoliidae.[26,31,42,240,242,289,290] Nitrogen-containing compounds (e.g., indole, amines) are more rarely prominent constituents, as in the Araceae[263,300,301] (Vogel's 1962 work[300] was translated into English in 1990[301] and will be cited as such hereafter), although in many species they tend to occur as minor but essential components of the fragrance.[145,147] Sulfur-containing volatiles (e.g., isothiocyanate, dimethyl trisulfide) are only occasionally reported.[55,62] The genetic basis of floral scent production has not received much attention, but work in progress on the production of specific monoterpenes and benzyl esters in flower odors of *Clarkia* species (Onagraceae) suggests that allelic dominance may be involved in some cases (R. Raguso, personal communication).

The choice of methods used for the collection and analysis of flower volatiles influences the resulting chemical composition of the fragrance.[85,147,311] Also affected is the detection of minor constituents, which may serve as key cues for insects. The selection of analytical procedures is thus of great importance in studies of insect–fragrance interactions. Methods commonly used today and considered to be the most accurate rely upon the collection (i.e., trapping) of airborne volatiles onto an adsorbent material from which they can later be readily desorbed and chemically analyzed by gas chromatography and mass spectrometry.[85,134,147,312]

The quantity of volatiles released from flowers varies widely,[85] with measured amounts ranging from 100 ng/d[28] to 50 μg/h.[233] In fetid-smelling flowers of *Hydnora africana* (Hydnoraceae), the strong odors are described as being easily detected by humans more than 4 m away in still air.[55] Thermogenesis may accompany the emission of floral odors and thereby enhance their volatilization, a phenomenon especially well documented in the Araceae.[213,214,258,276,285,301] Thermogenesis and insect attraction are often restricted to only a short period of floral anthesis; in *Annona* (Annonaceae) the temperature and odor emission rhythms vary with ambient temperature and are less effective on cool nights.[114]

B. TEMPORAL PATTERNS

Floral volatile emissions follow daily cycles, and their temporal correlation with pollinator activity points to an important role of flower aromas in attracting pollinators. Periodicity in fragrance emission is especially striking in species that rely for pollination on night-flying insects. Here the flowers are strongly scented at night, but during the day tend either to be scentless or to emit weaker and qualitatively different aromas.[98] Detailed studies of the emission patterns of individual volatiles within a fragrance indicate that different volatiles follow different rhythms,[5,133,135,198,199,209,229] which raises the question of how these individual changes affect pollinator behavior. Floral fragrance chemistry may also vary over a plant's blooming season, as in alfalfa *Medicago sativa* (Leguminosae).[196] However, changes in floral odors do not always have an effect on the insects, as in this case, where honey bees did not show corresponding changes in their visitation to the alfalfa flowers. Factors controlling the temporal release of fragrance are little understood, but hormones and possibly ethylene have been implicated in some Araceae[215] and environmental conditions in some Leguminosae.[195,262]

Cessation of aroma production in orchids is generally triggered by pollination.[309] In *Platanthera bifolia* (Orchidaceae), changes in volatiles 24 h after pollination are revealed mainly in the total quantity of volatiles emitted; these changes are accompanied by only slight alterations in fragrance composition.[294] As a result of pollination or floral aging, many flowers show localized changes in color, which are thought to serve as specific close-range cues informing pollinators of the reward state of a flower.[111,299,303] In some species these color changes have been described to be associated with changes in odor emission,[176,184] but analytical evidence is lacking.

C. SPATIAL PATTERNS

Different modes of fragrance emission occur in plants. In some species, floral volatiles emanate from highly differentiated glandular regions (i.e., osmophores) within the flowers. This has been extensively studied and well documented by Vogel[301] and appears to be especially prevalent in monocotyledons. More commonly, however, fragrances are believed to be diffused from floral (e.g., petal) tissues, although detailed studies are few.[301,311] While volatiles are typically emitted from the flowers in a near gaseous state, in some species, particularly those with well-defined osmophores, fragrance material may be visible in the flower as a liquid (oil)[301] or more rarely in crystalline form.[313] Liquid fragrances consist entirely of volatile compounds or have fatty oils intermixed,[301] which is similar to pollenkitt, the oily coat on the outside of pollen grains in which pollen odors are contained.[84,88] In certain cases, pollenkitt may dry or rapidly volatilize upon exposure to air,[111] but more frequently it retains its sticky, oily properties. Fatty substances (i.e., glycerides) are commonly present in the pollenkitt[84] and are thought to serve in retarding the evaporation of the odor-providing volatile chemicals, thus lengthening the time over which they are released into the air.

Various parts of the flower may participate in fragrance emission, most particularly the petals and androecium. Furthermore, considerable spatial patterning can occur within a flower both in the kind and amount of volatiles emitted.[87] The potential importance to insects of such olfactory patterns was initially suggested in studies by Lex[184] and von Aufsess.[12] Through training experiments, they demonstrated that honey bees could use olfaction to distinguish among different flower parts, including the visual food guides and pollen.

Odors from anthers and especially from pollen are considered in evolutionary terms to be the oldest olfactory attractants in flowers.[98,252] In some species, pollen odors are a primary contributor of floral volatiles.[26,87,97,164] They are notably strong in many beetle-visited flowers, where they may function in attraction,[252] and they are important stimuli in the flower foraging of other insects, particularly bees.[83,86,104]

Nectar is often described as having odors, but chemical studies of nectar volatiles are lacking. Honey volatiles have been investigated in several species;[30,36,40,118] however, these may not be the same as those present in nectar.[36,284] In spite of the paucity of chemical data, behavioral studies suggest that nectar odors can be used by honey bees and bumble bees to distinguish between rewarding and nonrewarding flowers[204,307] and by flower mites to recognize their host flower species.[132]

III. OVERVIEW OF ODORS IN INSECT–FLOWER INTERACTIONS

A. EVOLUTIONARY CONSIDERATIONS

Some of the earliest members of Insecta are thought to have been phytophagous, feeding mainly on nutritious reproductive organs of Carboniferous and Permian plants,[283] which eventually led to phytophagy on angiosperms during their rise in the Cretaceous.[241] Elaborating on the scheme put forth by Crepet[69] for the evolution of insect pollination, floral volatiles in early angiosperms might have evolved to deter phytophagous insects that fed on flowers (mainly on pollen and ovules).[241] As the benefits derived by the plant from the incidental pollination activity of these insect visitors outweighed the deleterious effects, there would have been selection for a means of increasing insect attraction to flowers. Today, the pollinators of extant archaic angiosperms utilize flowers not only for feeding but also as sites for mating and oviposition;[11,29,112,242,289] the habits of the ancestors of these insects (Coleoptera, Lepidoptera, Diptera, and Thysanoptera) included phytophagy on nonangiosperms. As pointed out by Pellmyr and Thien,[241] insect-vectored pollination of angiosperms may thus have evolved primarily through the interlinking between flowers and the reproductive cycles of phytophagous insects. Thus, floral volatiles would have been used by early pollinators as chemical cues for locating food (e.g., pollen) and sites for both mating and brood rearing.

Morphological evidence from fossil flowers also suggests that the attraction of pollinators was mainly olfactory. Indeed, the earliest angiosperms included not only plants with elaborate magnoliid floral structures but also others with simple, small, and not very showy flowers, where visual cues were probably secondary to olfactory ones.[70,103,237] This is consistent with the presence of prominent floral odors, and more particularly odors of pollen, in primitive plants such as cycads[69,98,240] and members of archaic Magnoliales families.[26,289] Testimony to the central role of pollen in olfactorily attracting insect pollinators is still in evidence today.[83,86] Furthermore, the detrimental impact that pollen-feeding insects can have on plant fitness appears to have led to the selection for deterrent floral volatiles in the pollen, often to the exclusion of other flower parts (see Reference 86, H. Dobson, I. Groth, and G. Bergström, unpublished, and below).

B. EXTANT INTERACTIONS
1. Evidence from Fragrance Chemistry

The active role of floral volatiles in attracting insects seems self evident considering the associations between certain flower odors and insects, the rhythmicity of floral volatile emissions (especially in the case of pollination by night-flying moths), and the general restriction of obvious floral odors to insect-pollinated plants (and to other plants pollinated by mammals that have a well-developed sense of smell).

Comparative studies of fragrance chemistry in relation to the associated flower visitors can shed light on the adaptive significance of floral odors to insects. In particular, fragrances of closely related plant species (i.e., congeneric or confamilial) that rely on different types of insects for pollination may be strikingly different, reflecting the olfactory sensitivities or preferences of the pollinators. Thus, within the Palmae, a diversity of pollination modes can be observed among the different species, which may be pollinated primarily by bees, flies, curculionid and nitidulid beetles, or dynastid beetles.[129,276] Associated with these types of pollinators are characteristic floral odors. Species pollinated by bees and flies are described to have sweet scents; those pollinated by beetles have strong musty, spicy, or fruity odors. Chemical analyses of the fragrances would clarify the odor differences among and within these species categories as well as their possible relation to the particular insects. The genus *Annona* (Annonaceae) includes over 2,000 species, many adapted to pollination by beetles.[114] Fragrance descriptions of selected species range from disagreeably pungent and cyanide- or nutlike, to pleasantly fruity or spicy; beetle species serving as principal pollinators also differ among the flowers. Vogel[301] reports on the production of a pleasant, terpenoid-containing fragrance in *Aristolochia cordiflora* (Aristolochiaceae), whereas a very closely related and morphologically similar species emits an aminoid, carrionlike smell that is more typical of species pollinated by carrion or dung flies. Other notable plant groups that include species having different fragrances in association with different pollination modes include the Orchidaceae and Araceae.[75,301,311]

Plants typically pollinated by a distinctive group of insects tend to display a characteristic assemblage of floral traits (e.g., shape, color, nectar content, time of bloom) regardless of the plant's systematic affinities. These sets of traits, commonly referred to as floral or pollination syndromes, have been described for seven different insect pollinator groups (i.e., beetles, carrion and dung flies, syrphid and bee flies, bees, hawkmoths, small moths, and butterflies).[98] Attempts have been made to assign floral odors to the pollination syndromes, using odor descriptions as perceived by humans.[314] As pointed out by Bergström,[24] however, these are very subjective, creating inherent problems in their interpretation,[287] and are also premature and potentially misleading considering the still early state of our understanding of flower chemistry and insect pollination. Correlation of pollination syndromes with floral odor chemistry has proved to be more appropriate for syndromes with well-defined and distinctive insect groups (i.e., night-flying Lepidoptera or dung and carrion insects) than for others in which wide variation occurs in both fragrance chemistry and insect responses to chemicals.[17] Subdivision of broad syndromes, such as pollination by bees, into well-defined and cohesive syndromes (e.g., euglossine versus other bees) may reveal more consistent patterns.[110] Comparative chemical studies are currently under way to determine whether fragrance syndromes exist in association with certain pollination modes (J. Knudsen and L. Tollsten, unpublished).

The rhythmic release of volatiles in some flowers and the temporally corresponding activity of their pollinators are convincing evidence for the importance of floral scent in attracting insects. This is most clearly exemplified in flowers pollinated by night-flying insects (e.g., moths, dynastid beetles), where floral volatiles are crucial in attracting the insects to the flowers[51,95,114] (discussed further in sections below). In soybean *Glycine max* (Leguminosae), where flowers emit two types of aromas distinguished by the relative proportions of three major volatiles,[262,263] emission of the "sweet-type" aroma is correlated with high visitation by honey bees as well as with climatic conditions and certain floral features that enhance flower attractiveness; however, the effect of aroma alone on the bees remains to be determined.

Further evidence for the key role of scent in shaping flower–insect associations is provided by case studies of species having different scent morphs in which fragrance variation correlates with differences in pollinator behavior or species composition. *Polemonium viscosum* (Polemoniaceae) has a sweet-smelling morph that is preferentially pollinated by bumble bees and a skunky-smelling morph that is pollinated by flies.[109] The skunky-scented morph appears to have been selected in defense against ants, which feed destructively on the flowers. In preference and olfactometry tests, the ants preferred the sweet-scented morph[107] but were repelled by the skunky-smelling glands, located on the floral sepals, when they crawled up the plant stem.[108] Another example, described by Pellmyr,[236] is seen in *Cimicifuga simplex* (Ranunculaceae), where three sympatric morphs are characterized by differences in habitat, pollinators, and floral fragrance chemistry.[122] Nymphalid butterflies were attracted only to one morph, which has a distinctive odor attributed to the presence of methyl anthranilate and isoeugenol. Addition of these two compounds to a different morph that is otherwise unattractive completely reversed its attractiveness to the butterflies. These experimental results clearly point to the floral volatiles as being critical in determining this butterfly–flower association.

2. Evidence from Insect Behavior

Insect responses to odors may be learned or innate. Learned responses occur through insect experience with the volatiles, usually in association with food, and have been well documented in bees;[105,216] innate responses are genetically programmed. The two types of responses are not mutually exclusive and probably operate together in most situations.

Flower fragrances can be divided into two general types, "absolute" and "imitative," as proposed by Faegri and van der Pijl.[98] In absolute fragrances, insect reactions are innate and/or learned, and evolution of the odor involves only the plant and pollinator. By contrast, insect responses to imitative fragrances are predominantly innate, or instinctive, where the odors release a chain of reactions that is the same as that elicited by similar odors unconnected with the flower. The insect is thus attracted by deceit, and evolution of the odor involves three partners rather than two: plant, pollinator, and pre-existing imitated scent. Distinguishing between these two fragrance types can be difficult, since an absolute odor may in fact have evolved initially as an imitative odor, and it is difficult to know which type came first.

Scents of some Lepidoptera-pollinated flowers are described as resembling those released by the insects themselves and, conversely, lepidopteran sex pheromones have been described as flowerlike,[256] which is corroborated by chemical analyses.[15,27] Likewise, secretions of some Hymenoptera are "flowery,"[172] and chemical analyses of hymenopteran pheromones confirm that many do include volatiles typical of flowers.[4,56,92,128,298] In the carpenter bee *Xylocopa varipuncta* (Anthophoridae), the male's flowery pheromone possibly attracts females by mimicking floral food sources.[220] Given the general similarity between flower volatiles and various insect pheromones, perhaps the evolution of these flower and insect odors has been intertwined and complex, with flowers imitating pheromones and pheromones imitating floral scents. Knowledge of which floral volatiles attract a particular insect may then assist in predicting the chemical constituents of the insect's sex attractants and vice versa, as has been pointed out by Rodriguez and Levin;[264] this in fact has been done with some orchids and their pollinators.[42,173]

Special cases of pollinator attraction are those involving deception, whereby flowers mimic food sources, breeding sites, or mates, but provide no rewards.[75,282] The success of these deceptive strategies may rely to varying degrees on olfactory cues to lure the pollinating insects. In food source (i.e., nutritive) deception, insect attraction appears to be based mainly on visual stimuli, with or without olfactory cues. This form of mimicry may also occur within a plant, where female flowers offer no rewards and deceitfully attract pollinators by mimicking various floral traits, including fragrances of the food-rewarding male flowers.[10,157] In breeding-site (i.e., reproductive) and mate (i.e., sexual) deception, olfactory stimuli are dominant and act on the insect's innate behaviors; visual cues are of secondary importance and operate mainly at close range, often together with tactile stimuli. Sexual deceit in orchids is thought to have evolved via food deception.[231]

Many plants pollinated as a result of such "deceit" belong to the Orchidaceae; over a third of the pollination systems in the family employ mimicry to attract pollinators rather than offer edible rewards.[77,231] Furthermore, sexual deception in which the flowers mimic female insects is documented almost exclusively in the Orchidaceae.[75] It reaches its most evolved stages in species pollinated by pseudocopulation, reported on three continents, where the male insect (i.e., bees, wasps, ants, flies, beetles) attempts to copulate with the flower and in the process effects pollination. The flower volatiles show high degrees of specificity for the insects they attract,[42,173] and in *Ophrys* these specific chemical signals have clearly taken part in the sympatric speciation of the genus. In contrast, within the genus *Orchis*, which shows food deception, floral volatiles play a lesser role in pollination and speciation has been mainly allopatric.[76] In flowers that rely on olfactory stimuli to manipulate pollinators and attract them by deceit, selection appears to have led to the evolution of flower species that differ more in fragrance than in morphology,[25] as exemplified in the genus *Cypripedium*.

Trap mechanisms are another strategy displayed by some plants (e.g., Araceae, Aristolochiaceae, Palmae) to ensure pollination. Flies or beetles attracted by chemical and visual cues are trapped for up to a day in flowers or inflorescences.[276,301] A graded increase in fragrance intensity toward the inside of some Aristolochiaceae flowers draws flies into the flower. Floral odors are often emitted mainly during the initial, female flower stage; insects attracted by the volatiles become trapped in the floral chamber, where they generally feed, rest, and/or mate until the onset of the male flower stage, at which time the insects, now dusted with pollen, can escape and fly to another plant in the female stage. The floral volatiles of the initial female stage and those of the later male stage appear to elicit different behaviors in the pollinators. These fragrance changes may involve quantitative differences, such as the cessation of odor release during the transition to the male stage (common in Araceae and Aristolochiaceae),[301] or qualitative differences, such as the emission of different odors during the two sexual stages and sometimes even from different floral structures (Cyclanthaceae).[115] Studies of changes in fragrance and of the specific effects of fragrance volatiles on insect behavior are lacking.

Insects using flowers as oviposition sites may be highly selective in terms of both plant species and floral part. It seems probable that the same cues thought to have evolved to attract pollinating insects are also used by these freeloading flower visitors, which obtain benefits from the flowers often without providing any pollination. The role of floral volatiles in host plant location by flower parasites has received sparse attention, but the high selectivity exhibited by some insects for flowering parts and stages over vegetative ones,[143] or even for flowers of specific ages[245] suggests that flower chemicals are involved. Furthermore, floral stimuli that elicit feeding may differ from those that trigger oviposition, as in insects that show differing flower selectivity in their feeding and in their oviposition behaviors.[232] Likewise, *Meligethes* (Nitidulidae) beetles, which develop in flowers and eat primarily pollen during both adult and larval stages, generally show greater host flower specificity in their mating and oviposition activities than they do in their adult feeding.[73] In *Encephalartos* (Cycadaceae), nitidulid beetles feed on the odoriferous male cones but oviposit in female cones, which to humans are odorless (Rattray, in Reference 98). Involvement of different floral stimuli, including volatiles, in the release of feeding as opposed to oviposition appears to be a recurrent theme in insects using flowers for both activities.

C. EXPERIMENTAL APPROACHES

An insect's perception of its host plant clearly involves the integration of multiple sensory modalities, but olfaction seems most important when visual cues cannot be relied upon (e.g., by night-flying insects) or when floral volatiles elicit strong, innately driven responses. In order to elucidate how volatiles affect insects, their perception by the insects must be manipulated in controlled conditions. Early experimental studies emphasized observations of insect behavior toward flowers, where factors related either to the insect or to the flower were modified and controlled for.[64] Many of these basic approaches are still used today; however, now the added advantages of chemical techniques involving the analysis and manipulation of volatile stimuli have broadened the scope of questions that can be addressed.

Insect responses to floral odors can be studied by isolating the olfactory stimuli (from all nonolfactory cues) or by modifying them. Olfactory stimuli are isolated by visually concealing flowers with paper, cloth, leaves, etc.; or alternatively by removing the olfactory, such as by enclosing flowers in glass containers, thus retaining visual cues. Modification of olfactory stimuli can be variously achieved by removing, exchanging (with other flowers), and adding flower parts to either fresh flowers or artificial flowers (paper, plastic, cloth). Further manipulations include adding volatiles, individually or in mixtures, to fresh or artificial flowers or to other material having no resemblance to flowers (e.g., pieces of cloth, baited traps), and removing volatiles from fresh flowers using solvents.

Insect-related factors, such as age, mating, and learning (experience), that affect the insect's ability to perceive and motivation to respond to stimuli can be manipulated; environmental influences, including climatological factors or diverse stimuli originating from other biotic elements in the insect's surroundings can also be modified. Studies may be performed in the field with free-flying insects; alternatively, studies may be conducted under more controlled conditions, either in the laboratory or outdoors, where tested insects can be preselected on the basis of physiological, developmental, or neurobiological attributes. Controlled conditions include the use of flight tunnels, olfactometers, flight rooms, cages, or other enclosures in which the insects can fly and/or walk to or toward sources of volatiles. Flight behavior can indicate whether the insect is guided primarily by visual cues (i.e., flight direction, independent of wind, follows a straight line) or olfactory cues (i.e., upwind flight is often in a zig-zag pattern). Approaches used less commonly today include modifying the insect's olfactory abilities by coating or amputating the antennae or removing visual input by coating the eyes.

Electrophysiological approaches, using electroantennogram (EAG) and single-cell recording techniques, measure the insect's ability to perceive volatiles. In clarifying insect behavior toward flowers, they have been most valuable in tests involving floral compounds that release strong innate responses (e.g., components of pheromones, for which insects tend to have a very restricted and high sensitivity). Utilization of EAG to elucidate food selection has, however, met with less success. Insects appear to perceive a broad spectrum of chemicals related to food, and specificity to host plant chemicals is more apparent on the behavioral than on the strictly odor receptor level, implying that key integration processes occur in the brain.

IV. EFFECTS OF FLOWER VOLATILES ON INSECTS

Studies providing experimental evidence of interactions between insects and floral volatiles are summarized below, grouped by insect order. Under each order are discussed, in succession, studies of insects that visit flowers 1) to obtain food rewards, 2) to lay eggs, and 3) to fulfill various purposes that, however, all involve attraction to flowers by deceit. These visitation categories are not mutually exclusive.

A. COLEOPTERA

Modern flower-feeding beetles may be divided into those that visit flowers exclusively for adult feeding and those that visit flowers for mating or oviposition and larval development. The first group includes many more families than the second, and its members tend to show more striking adaptations to flowers in the adults.[72]

1. Beetles Feeding on Flowers

Floral odors in beetle-pollinated plants are described as being typically strong, with fruity (especially in the tropics) or aminoid characteristics.[98] Examples of fruity fragrances are provided by various species of Magnoliales, which includes several archaic families. Among these are the two species of *Eupomatia* (Eupomatiaceae), both pollinated exclusively by *Elleschodes* weevils (Curculionidae), which feed, mate, and oviposit in the flowers; the androecium appears to be the principal food of both adults and larvae.[26] In *E. laurina*, weevils begin to arrive at the flowers shortly after anthesis begins in the morning, and olfactory cues, provided by strong odors originating from sticky oils secreted by the androecium, appear to be more important as attractants than is the visual display of petaloid staminodes, since weevils seek flowers even when they are completely hidden within bags.[11]

Cursory studies of *Donatia* (Chrysomelidae) beetles that pollinate *Nuphar lutea* (Nymphaeaceae) suggest that olfactory cues may play a role secondary to that of visual cues. Here, an intense scent is emitted principally from the stigmatic region in first day flowers, but isolation of various floral parts showed the beetles to be attracted mainly to the yellow sepals.[272] However, in *Lysichiton americanum* (Araceae), pollinated diurnally by Staphylinidae, odor from experimental, visually concealed, yellow spathes was sufficient to induce searching and alighting in the beetles, whereas odorless (glass-enclosed) visible spathes elicited no response.[239] The number of alightings increased by a factor of 2.5 when beetles were offered both olfactory and visual stimuli (whole spathes), and search for the yellow-colored spathes was triggered only after the beetles entered the floral fragrance plume. Nitidulid beetles (Nitidulidae) show a greater attraction to floral volatiles from petals than from sexual parts when offered different flower parts of the *Annona* hybrid atemoya (Annonaceae) in both field and laboratory (olfactometer) studies.[251] Reports of *Byturus ochraceus* (Byturidae) beetles feeding almost exclusively on flowers of the most strongly scented specimens of *Actaea spicata* (Ranunculaceae) also suggest that floral volatiles serve as potent elicitors of alighting and feeding,[235] although this evidence is mostly circumstantial and needs experimental confirmation.

In the tropical Araceae[115,318] and Annonaceae[113] pollinated by night-flying dynastid beetles (Scarabaeidae), floral odors appear to be crucial in attracting the beetles, which arrive in a typical zig-zag upwind flight pattern.[113] Strong fragrances produced at dusk by the heated inflorescences of *Philodendron selloum* (Araceae) attract the pollinating dynastids, which often must fly distances of several hundred meters between plants. Experimental modification of visual and olfactory stimuli from the inflorescences showed that volatiles are required to attract beetles to the vicinity of female-stage plants. However, visual cues provided by the plant's bright, light-colored surface are necessary both to guide the beetles to the inflorescence close at hand and to elicit landing responses.[116]

In contrast to the situations described above, where beetles perform some pollination service, many flower-visiting beetles are flower parasites, feeding on the flowers without pollinating. Among these are adult Japanese beetles, *Popillia japonica* (Scarabaeidae), and related species that include flowers in their diet. Studies using baited traps showed them to be strongly attracted to specific floral volatiles, such as eugenol and geraniol,[90,274] but the beetles have not been tested for their responses to whole-flower fragrances.

Relationships between flower-visiting Diabroticite beetles (Chrysomelidae), particularly New World *Diabrotica* species, and their Cucurbitaceae host plants have been well studied.[218] The majority of species feed as larvae on roots of grasses, whereas the adults are polyphagous pollen feeders that often collect on *Cucurbita* blossoms. The beetles show definite preferences when given a choice of flowers from different species, and the frequent preference for male flowers of *Cucurbita maxima* generally corresponds to their higher release rates of floral volatiles such as indole, cinnamaldehyde, cinnamyl alcohol, and β-ionone, and to the presence of cucurbitacin feeding stimulants.[6] Field bioassays to test the attractiveness of *C. maxima* floral volatiles have met with mixed success when using extracts of flowers, perhaps due to the choice of extraction solvents and methods, but baited traps containing whole flowers were clearly attractive.[210] Results from studies using traps baited with a spectrum of volatiles indicate that the beetle species differ in their responses to the volatiles.[177,182] One volatile blend developed to bait the beetles, TIC, mimics *Cucurbita* floral odors and comprises 1,2,4-trimethoxybenzene, indole, and *trans*-cinnamaldehyde. It has been suggested that the attraction of Diabroticite beetles to cucurbit flowers and to individual floral volatiles may have arisen as a means to locate ephemeral sources of pollen.

2. Beetles Using Flowers for Feeding and Oviposition

Palms are predominantly entomophilous, pollinated most commonly by beetles, bees, and flies.[129,276] Evidence suggesting that floral odor is a key factor in their pollination includes observations of insects visiting flowers only after odor release is initiated. In addition, the two beetle groups most closely associated with palms, Nitidulidae and Curculionidae, feed mainly in the male flowers (on pollen) and their visits to female flowers, which offer no food rewards, appear to result from their mistaking them for male flowers on the basis of similarity in fragrances. Among the insects that pollinate the oil palm, *Elaeis quineensis* (Palmae) is the weevil *Elaeidobius kamerunicus* (Curculionidae), which is host-specific, completing its entire life cycle on the palm.[157] The weevils feed and breed exclusively in male flowers, to which they are attracted by strong odors emitted during anthesis. Olfactometry tests indicate that the beetles use estragole, the major component of the floral odors, to locate oil palms in bloom.[140]

Many weevils (Curculionidae) feed on pollen, for which the adults of some species show considerable host plant specificity. In other species, plant selectivity may be low, in spite of differing nutritional values of the pollen.[144] In the red sunflower seed weevil *Smicronyx fulvus*, the adults feed on sunflower pollen (necessary for oocyte maturation) and oviposit in developing florets at the onset of anthesis.[61] Using field traps baited with volatiles, Roseland et al.[266] tested the attraction of adults to synthetic mixtures of sunflower floral volatiles. Only a single mixture comprising five volatiles showed high attractivity; baiting with fewer than these essential compounds or altering their relative abundance in the mixture resulted in substantial reductions in beetle attraction. Volatiles specifically from pollen have been implicated as possible attractants to pollen-feeding beetles. In bioassays aimed at evaluating the response of *Meligethes aeneus* (Nitidulidae) to chemical stimuli from different parts of its host plants, *Brassica* spp. (Cruciferae), the beetles showed a preference for stamens/pollen chemicals, and this appeared to be predominantly based on olfaction.[62] The weevil *Bruchus pisorum* (Bruchidae) feeds on nectar and pollen of *Pisum* (Leguminosae), which are essential in stimulating oogenesis and extending life span.[63] During pilot studies, beetles offered fresh *Pisum* pollen were strongly attracted to it and stimulated to feed, but when offered the same pollen in lyophylized form, they showed no response, suggesting that volatile cues essential for attractivity were lost during the lyophylization process (S. L. Clement, personal communication).

3. Pseudocopulatory Associations

Casual reports that various scarabaeid beetles visiting flowers of *Ophrys* (Orchidaceae) are attracted and even sexually excited by the floral odor were confirmed in field bioassay tests on one beetle, *Phyllopertha horticola*.[41] Blends of floral volatiles were applied to pieces of black velvet covered with green netting, and beetle responses to odors were studied under field conditions. Only males were attracted to the scented "dummies," and on several occasions they exhibited behaviors that were interpreted as early stages of copulation, thus implying that their attraction to the *Ophrys* volatiles has some sexual basis. In *Ophrys litigiosa*, visitation by male *Agriotes* (Elateridae) may be due to chemical similarities between volatiles emitted from the flowers (citronellyl and farnesyl esters) and sexual attractants released by the female beetles (geranyl esters), to which the males show high EAG responses.[44]

B. DIPTERA

Flowers visited by flies fall into two major groups; 1) those offering rewards (i.e., general fly pollination syndrome) and which are variously described as having imperceptible to heavy-sweet scents,[98,256] and 2) those pollinated by carrion and dung flies (i.e., sapromyophily pollination syndrome), where flies are deceptively attracted by fetid floral odors that trigger innate behaviors for feeding or oviposition.[98]

1. Flies Feeding on Flowers

Field experiments aimed at determining the effects of floral volatiles on typical flower-visiting flies have not yielded conclusive results, but intimate that the flies tend to rely mostly on visual stimuli, especially color, to locate flowers. To test the effect of removing olfactory cues, Knoll[163] placed glass cylinders over inflorescences of the blue-flowered *Muscari* (Liliaceae) visited by *Bombylius* (Bombyliidae). The bombyliids appeared unaffected; they continued to fly directly to the flowers and pushed against the glass. However, when offered uncovered, scentless, shrivelled flowers, their visits were very hasty compared to those on scented, open flowers. In studies of *Eristalis* (Syrphidae), Kugler[170] applied artificial scent to flowers being visited by the flies. In response to this new scent, the syrphids showed an initial, temporary decrease in their number of alightings, implying that the olfactory cues are integrated into their search image of their host flowers.

Midges of the Cecidomyiidae and Ceratopogonidae are thought to be the principal pollinators of cacao, *Theobroma* species (Sterculiaceae). In confirmation, they were the most abundant insects captured in field traps baited with steam-distilled floral oils.[317] The volatiles appeared to contain chemicals generally attractive to Diptera, which comprised most of the insects trapped. Although there are clear differences in the chemical composition of floral volatiles among *Theobroma* species and their cultivars, the species composition of insects attracted to the traps tended to be very similar, suggesting that certain behaviorally active chemicals may have been lost during the steam-distillation.[315,316] When the major fragrance volatiles were individually tested, some were as effective in attracting flies as were the whole-flower fragrances.[315]

Many mosquito (Culicidae) species, both males and females, visit flowers to feed on nectar, and in some cases they effect pollination.[98] In both field and laboratory studies, mosquitoes were shown to prefer some flowers over others,[7] and the pale color of flowers visited during crepuscular hours appears to play a minor role in attraction compared to olfactory floral stimuli.[7,52] By offering *Aedes aegypti* cloth-covered flower or honey samples, Thornsteinson and Brust[292] demonstrated that mosquitoes aggregate in response to floral volatiles, and later studies with volatile extracts of honey established that the volatiles attracted the mosquitoes and stimulated them to feed.[306] Similar behavior-releasing activity from fresh flowers has been confirmed in wind tunnel experiments using volatiles of fresh flowers[141] and in field studies using cloth-covered flowers.[7,52] But extracts of flowers elicited inconsistent responses, perhaps due to methodological artifacts.[7,141,297] As with other insects, mosquito responses to floral volatiles increase with the lengthening period of food deprivation.[141,297,306]

Floral fragrances emitted during the female and male floral stages of *Peltandra virginica* (Araceae) differ in the ratio of two major volatiles, and experimental manipulations of inflorescences in the field suggest that these differences trigger specific behavioral responses in the plant's mutualistic pollinator, *Elachiptera formosa* (Chloropidae) (J. Patt, unpublished). Adult flies feed and mate on male-stage inflorescences, but the female flies oviposit only on younger inflorescences still in the female stage. This enables the larvae to feed on pollen released during the ensuing male stage. The fly's life cycle is thus closely tied to the plant's two flowering stages, and volatile-activated movement between the stages is essential for both the fly's survival and the plant's pollination.

2. Sapromyophilous Associations

Sapromyophily is the pollination syndrome in which pollinator flies are attracted to flowers that emit foul odors, which mimic the flies' normal food and oviposition sites. It occurs in a variety of plant families, and the floral traits associated with the syndrome are very similar to those in flowers pollinated by carrion and dung beetles. The insects can locate food sources from a long distance, but it is uncertain whether this is due to an acute olfactory sense or due to their habit of constantly moving around and thus increasing the chances of encountering food sources.[98]

Calliphoridae flies frequently visit nectariferous flowers to feed, but they are also deceitfully attracted to sapromyophilous flowers for oviposition. Newly emerged individuals of *Lucilia*, *Calliphora*, and *Sarcophaga* show little attraction to colors alone. However, when odorous material is placed under colored cloth, strong interactions between visual and olfactory stimuli become apparent: the colors to which the flies are attracted depend on the nature of associated odors.[171] In the presence of floral odors, which signal food (nectar) sources, flies showed a clear preference for yellow and to a lesser extent orange, but when presented with odor of excrement, which is typical of their egg-laying sites, they preferred the color purple-brown. *Calliphora vicina* perceives odors of decaying meat or of flowers through separate antennal receptors, among which different types can be recognized, each responding to a particular class of compounds.[146] Flowers of *Rafflesia* species (Rafflesiaceae), famous for their enormous size, are pollinated by carrion flies in the genera *Lucilia* and *Chrysomya* (Calliphoridae). These are lured to the flowers by the elaborate visual presentation of concentrically arranged white spots against a brick-red background, as well as by olfactory cues.

Field experiments in which visual and olfactory stimuli were independently masked suggest that olfactory cues may be the more important in enabling flies to locate a flower. Visitation by flies decreased to approximately 30% of the control upon exclusion of visual cues, but decreased to less than 10% when olfactory cues were excluded.[18]

Some cases of sapromyophily involve trap flower mechanisms, commonly reported in Aristolochiaceae and Araceae.[301] In the voodoo lily, *Sauromatum guttatum* (Araceae), the plant simultaneously releases both sweet and vile-smelling odors from different parts of the inflorescence, which also heats up; it has been suggested that these odors serve distinct functions in the plant's pollination.[215] The foul odors, produced by the sterile appendix of the inflorescence that protrudes above the floral chamber, attract fly pollinators (mostly carrion and dung flies). The sweet odors, consisting of terpenoid compounds emitted from small organs inside the floral chamber, act on the flies' feeding and mating instincts and keep the flies inside the chamber until the flowers enter their male phase, at which time the pollen-dusted insects fly to another plant.

C. HYMENOPTERA

1. Bees

The literature on bees is vast and will be only summarized here, without covering in depth the numerous studies that have led to our present understanding of interactions between bees and floral fragrances. Although a good portion of the literature addresses the honey bee *Apis mellifera* (Apidae), considerable attention has been given to other bees, both social and solitary. Due to their special nature, two well-documented types of pollination in the orchids are discussed under separate headings below. These pollination types involve, on the one hand, fragrance-collecting male euglossine bees and, on the other, sexually attracted male insects, especially bees and wasps, which pollinate flowers by pseudocopulation.

Flower volatiles are generally considered to be effective in orienting bees at short range (e.g., within 1 m) and in influencing whether a bee will alight on a flower; visual cues, however, operate primarily at long range (e.g., several meters). This has been confirmed for honey bees in training experiments conducted mainly with feeder tables,[39,104,105] for honey bees and bumble bees using both fresh and model flowers,[203,169] and for bumble bees in field experiments.[86] Odors thus provoke much stronger discrimination between flowers than do visual stimuli. Correspondingly, honey bees and bumble bees learn odors more rapidly and with greater retention than they do colors or other visual cues.[104,105,169,187,188,203,216] The relative reliance on olfactory and visual stimuli can change with the bee's foraging experience as it forms a more integrated search image of the flower. Thus, while colors are considered to be important reinforcers of odor-based search images, they may become prominent orientation cues following experience, as is documented in honey bees,[167] bumble bees,[74] and solitary bees.[83] However, in *Chelostoma florisomne* (Megachilidae), individuals that have not previously foraged on flowers respond to odors only when they are presented together with the appropriate color stimuli (H. Dobson, unpublished).

Honey bees can be trained readily to particular scents when these are associated with food.[104,187,246,302] They learn flower-like odors more rapidly than other odors, but can be conditioned even to volatiles that are otherwise repellent.[216] Just as some volatiles are more attractive than others, so do volatiles vary in the concentrations at which they elicit maximal visitations at the feeders.[131] The ability of bees other than honey bees to be conditioned to food-associated odors is less well known, but Kugler[168] successfully trained bumble bees to specific scents. Pham et al.[246] showed that while bumble bees can be trained to individual volatiles as rapidly as honey bees, they respond less rapidly to changes in food-associated odors and thus probably to changing resources in the field. In honey bees, the learning of odors has been shown to have a genetic component, with different races, different hives, or even different individuals within a hive showing different predispositions to learn certain chemical cues and different rates of learning.[47,79,166] Neurobiological and behavioral aspects of odor learning have been the focus of considerable research.[35,48,208,216,279]

The key role of floral scents, carried on honey bees returning to the hive, in the recruitment of new foragers was established by von Frisch;[104] but there is controversy about the relative importance of olfactory cues in forager recruitment.[117,305] The importance of floral odors is demonstrated by the fact that a bee returning to the hive not only provides information about the direction and distance of a flower patch in her dance, but also communicates the flower odor by its presence on her body surface and in her pollen or nectar loads. This helps newly recruited bees locate the flowers. Recruitment rate can be increased by intensifying the odor at a food source.[105,117,142]

Behavioral studies coupled with chemical analyses of floral volatiles indicate that honey bees rely on only a portion of the volatiles to recognize flower species, as in alfalfa,[130,304] or to discriminate among different flower races, as in sunflower.[247,248] Volatiles released from vegetative plant parts may also influence flower selection and discrimination of bees.[19]

Vareschi,[296] measuring the electrical responses of single olfactory cells of honey bees to different volatiles, found that the cells fall into different reaction groups. Each cell group reacts to only a certain spectrum of substances and shows little or no overlap with the spectra of other groups. This has received corroboration in behavioral experiments, where honey bees trained to individual monoterpenes (i.e., ocimene, myrcene, and limonene) that, according to Vareschi, are perceived by the same reaction cell group did not distinguish among them in behavioral choice tests.[303] However, attempts to find correlations between the behavioral responses of bees to floral volatiles and the relative amplitudes of EAG responses to the same volatiles have met with low success; EAG responses are not always good predictors of behavioral response. In a study of honey bees and individual volatiles in alfalfa fragrances, EAG responses to the behaviorally attractive compound were comparatively high and to the repellent compound low, but no congruence between the two investigative approaches was evident for the other three volatiles tested.[131] A similar lack of correspondence between behavioral and EAG results has been found in studies of solitary bee responses to floral odors (L. Ågren and H. Dobson, unpublished). Moreover, attempts by these authors to condition the bees' olfactory responses (by exposing them to volatiles during larval and pupal stages) likewise yielded inconsistent behavioral and EAG patterns. However, other data show significant changes in antennal sensitivity in relation to learning, where EAG responses of honey bees to floral odors increased significantly after the bees underwent conditioning to the odors.[79] We hope that future studies will provide clarification of these contradictory findings.

Volatiles from pollen are clearly perceived by bees, and their role in flower visitation deserves further study. Honey bees can olfactorily distinguish among pollen of different plant species[12,104,180,200,271] and among volatile-containing extracts of pollen,[137,138,179,200] as can solitary bees.[83] Furthermore, both von Frisch[104] and von Aufsess[12] were able to train honey bees separately to the odors of pollen and of whole flowers from the same species, indicating that these are chemically distinct to bees. Behavioral choice studies with two solitary bees, *Colletes fulgidus* (Colletidae) and *Chelostoma florisomne* (Megachilidae) corroborate this (H. Dobson, 1987, unpublished).

Additional evidence indicates that pollen odor can be of key importance in flower discrimination by bees. Besides its role in providing olfactory information for recruiting new workers in honey bees,[104] it can influence flower selection in solitary bees and bumble bees. In the flower specialist *C. florisomne*, foraging-naive bees can recognize their host plant *Ranunculus* (Ranunculaceae) among other flower species more effectively when they are tested with odors of pollen than with odors of whole flowers, suggesting that it is the pollen that carries key chemicals used in flower recognition; these are detected only at very close range, after alighting (H. Dobson, unpublished). Complementary chemical studies of *Ranunculus* revealed that the volatile profile of pollen is distinctively different from that of the whole flowers (H. Dobson, G. Bergström, and I. Groth, unpublished). The floral volatiles consist of a blend of 30 compounds dominated by the monoterpene *trans*-β-ocimene, whereas the pollen profile contains only 2-phenyl ethanol, α-farnesene, and protoanemonin; the latter occurs mainly in pollen. Another flower-specialist, solitary bee, *C. fulgidus*, can olfactorily recognize its pollen host plant among other species, based on either flower or pollen odors.[83] Comparative studies make it clear that parameters characterizing one bee–flower association do not necessarily apply to others; each association must be examined within its own biological and ecological context. In field studies of bumble bees that forage for pollen on *Rosa rugosa* (Rosaceae), Dobson (Reference 86, unpublished) chemically modified the olfactory signals from pollen by variously removing anthers or pollen and adding volatile compounds singly or in mixtures to the flowers. Resulting changes in landing responses by the bees indicated not only that pollen odors can modify flower visitation but also that they can assist bees in distinguishing between rewarding and nonrewarding flowers. Bees may thus be able to evaluate the food reward content of a flower based on its volatile emissions. Reported observations of bumble bees discriminating among flowers that have differing amounts of pollen[202,291,321] or of bees showing close-range preferences for flowers in the male floral stage over the female stage[155,227] are intriguing. Perhaps further investigations will show that these cases likewise involve olfactory evaluation of pollen odors in the flowers.

2. Male Euglossine Bees

The remarkable association between male Euglossini (Apidae) bees and the orchids they pollinate has received extensive attention, although some aspects relating to the bees' biology still remain unclear.[41,131] The male bees do not visit orchid flowers to collect food, since the orchids are nectarless, but rather to gather floral volatiles. During a typical visit, the bees scrape the odorous petal surface with their front leg tarsi and transfer the volatile oils to specialized organs in their enlarged hind leg tibiae, where the volatiles are stored. Over 600 species of orchids, mainly of the subtribes Stanhopeinae and Catasetinae, are associated with male euglossines, and both the orchids and the bees occur exclusively in the Neotropics. The bees also collect floral volatiles from several plant species in other families.[9,312]

The relationships between the bees and the orchids are often highly specific, with only one or a few bee species being attracted to one orchid species within a particular habitat. Such specificity can be important in orchid speciation by acting as an isolating mechanism among sympatric species (e.g., see References 120, 136, 311). Up to 60 terpenoid and benzenoid compounds have been identified in the orchid fragrances, with individual species having unique mixtures of 2 to 18 volatiles. The specific attraction of bees to the fragrance volatiles has been established through field tests using baits, where papers saturated with fragrance compounds are placed in various habitats. Individual bee species are usually attracted to several compounds when these are offered singly, but combinations of two or more compounds modifies the attraction. The number of attracted bee species decreases as the complexity of mixtures increases (e.g., see References 89, 311, 312). Geographical differences in fragrance preference within bee species are also common, but the basis of this is unclear.[1]

Functions served by the floral volatiles in the biology of male euglossines, although much debated, have not been established.[91,158] It is generally thought that the volatiles are used in the bees' reproductive activities, either as pheromones, such as in territorial displays[270] and attraction of conspecific males in leks, or as precursors to sex pheromones that are stored in the mandibular glands.[312]

3. Wasps

The close relationship between figs, *Ficus* species (Moraceae), and their wasp (Agaonidae) pollinators is a classic example of mutualism between flowers and insects, because both partners benefit from the association and in fact depend upon each other for their livelihood. Yet, in spite of the attention this relationship has been given in the literature, experimental studies on the role that volatiles play in this interaction are scant. Upon leaving their natal fig, the small wasps may be required to fly long distances in order to successfully locate another plant of the appropriate fig species for oviposition, and there is general agreement that wasps use olfactory cues. With few exceptions, one species of wasp pollinates one species of fig, and data from sticky traps placed at trees indicate that wasps make few mistakes in their flights.[53] Furthermore, release of volatile attractants from the receptive inflorescences (synconia) is strongly implied by the coordination between the arrival of wasps, which is often concentrated over a short period, and the presence of unpollinated receptive synconia. Comparisons between the quantity of volatiles emitted from inflorescences in *Ficus ingens* and the number of arriving wasps showed a close correspondence.[16] Volatile emissions increased with synconia age and then sharply dropped, and parallel wasp capture rates followed a similar trend, thus providing support for the notion that wasps are guided by synconia volatiles.

Females of the alfalfa and clover seed chalcid wasps (Eurytomidae) are thought to rely primarily on olfaction to locate and select host plants. Both respond to extracts of their host flowers,[152] and laboratory bioassays established the attraction and oviposition-stimulating activity of individual floral volatiles.[151,153] Response of newly emerged females to flower fragrances was also tested in the alfalfa seed chalcid, *Bruchophagus roddi*, which oviposits exclusively in ovules and developing seeds of alfalfa, *Medicago sativa* (Leguminosae) but may use flowers of other plants for nectar foraging.[149] When offered a choice of two to four flower fragrances (flowers hidden in open vials), wasps significantly preferred alfalfa in terms of landing responses (preceded by hovering, olfactory orientation); however, volatiles from the alfalfa buds and pods were more attractive. Moreover, evaluation of chemical stimuli while in flight (as evidenced by in-flight orientation behavior) requires the presence of natural light.[150] EAG responses of the wasps to individual alfalfa volatiles did not show a consistent correlation with either the volatiles' percent representations in the plant's odor or the wasp's previously documented behavioral responses.[185]

4. Pollination by Pseudocopulation

Remarkable among the interactions between insects and flowers are those in which flowers are pollinated by male insects attracted to flower odors that mimic female sex pheromones and that release in the male an instinctive chain of behaviors leading to copulation. In its attempted copulation on the flower, the male insect effects pollination by picking up or depositing pollen in an exact manner. Pollination by pseudocopulation is exclusive to the Orchidaceae. It has reached various degrees of elaboration and specificity and has involved mainly Hymenoptera, although there have been some reports of Coleoptera and Diptera being sexually attracted to the flowers.[41,75,234]

Most thoroughly studied are the relationships of bees and sphecid wasps with flowers of the genus *Ophrys*, in which pseudocopulatory pollination is highly developed (see References 42, 173, 231, 234). Chemical and visual stimuli are intricately involved; flowers attract the hymenopteran males not only by the volatiles they emit but also by their shape, color, and pilosity. This process has been largely elucidated by Kullenberg and associates.[173] The insect–flower associations are often very specific, involving bees of nine genera (four families) and wasps of two genera (two families).[234] In a very comprehensive, landmark work on *Ophrys* pollination, Kullenberg[173] explained the Hymenoptera–flower associations of many *Ophrys* species based on extensive field observations, chemical baiting experiments of male pollinators to determine the chemicals to which they are attracted, and experimental studies on the role of visual stimuli. These early studies established that while visual and tactile cues are important, floral fragrances play a central role in male attraction. Following upon this, chemical aspects were investigated in greater detail for many of the associations, taking advantage of newly developed experimental and analytical techniques. Field bioassay tests were performed by applying volatile extracts of flowers and of the insects' secretory glands onto "dummies" (pieces of black velvet covered by green mesh).[46,174,286] These were carried out in parallel with chemical analyses of bee and flower odors (e.g., see References 42, 43, 45, 46, 175). These investigations confirmed and clarified the chemical basis of the attraction and subsequent copulatory response of males to the flowers: the floral fragrances of *Ophrys* include compounds also present in the secretions (i.e., pheromones) of the bee pollinators. Although this chemical mimesis involved only a portion of the volatiles present in an odor, the strong behavioral activity released by the odors implied that only a few key chemicals are necessary for a specific insect–flower association. These chemicals include fatty acid derivatives (e.g., 1- and 2-alcohols) and terpenes (e.g., geraniol, geranial, linalool, farnesol), which are present in certain *Ophrys* and in several *Andrena* bee species. These studies were complemented by investigations of the effect of other floral stimuli, including tactile cues.[3]

EAG investigations of the responses by male pollinators to *Ophrys* volatiles[2,255] corroborated the relationship patterns described by Kullenberg.[173] Priesner[255] tested extracts of labella from 18 *Ophrys* forms, as well as single odor components, on about 50 bee species, both males and females. High levels of electrophysiological activity were registered only in males and, furthermore, only in those that had been observed to be pollinators of the particular *Ophrys* species in question.

Pollination exclusively by male Hymenoptera has been reported in other orchids where sexual attraction appears to be involved (see Reference 234). Bino et al.[34] showed by concealing *Orchis galilaea* flowers with gauze that the floral volatiles suffice to attract the pollinator male bees of *Halictus marginatus* (Halictidae). Although no copulatory responses have been reported in this species, the exclusive visitation to flowers by the male bees suggests that the musty odors emitted by the flowers resemble pheromones of the females and that this pollination mechanism represents a less advanced stage on the evolutionary path toward pseudocopulatory pollination found in *Ophrys*. Another case, which however is not exclusively restricted to males, concerns the orchid *Cymbidium pumilum* in Japan.[269] Workers, drones, and swarms of the oriental honey bee, *Apis cerana japonica*, are strongly attracted to the flowers, and this attraction has an olfactory basis, since flowers concealed with black cloth and extracts of floral fragrances are both highly attractive. The attraction of drones mainly during their mating flights suggests that the active fragrance components resemble a species-specific pheromone, possibly one associated with the queen.

D. LEPIDOPTERA
1. Day-flying Butterflies and Moths

Floral fragrances associated with butterflies and other diurnally active Lepidoptera are described as weak and agreeable to humans. Some butterfly species show flower constancy,[38,260] but evidence of their use of olfactory cues in flower location and selection is rather meager compared to the numerous studies on moths. The butterflies' primarily diurnal activity suggests that they generally rely more heavily on visual stimuli in their orientation than do night-active moths, although floral volatiles may still be decisive in host plant discrimination.

Flowers of certain species, including Compositae and Orchidaceae, serve as major sources of pyrrolizidine alkaloids sought by some butterflies (subfamilies Ithomiinae and Danainae) and moths (Arctiidae and Ctenuchidae). The alkaloids are used by both males and females for defense and by males for courtship (as pheromone precursors). Attraction of these Lepidoptera to the alkaloid-containing flowers, which they also pollinate, appears to be primarily olfactory, based on their behavior toward various plant and alkaloid baits tested in the field.[81,249,250] Furthermore, male butterflies associated with *Epidendrum paniculatum* (Orchidaceae) not only showed similar behaviors when they approached flowers and pyrrolizidine alkaloid baits, but they also landed on flowers that were visually concealed in bags.[81]

Floral volatiles can act as critical stimuli in determining whether or not butterflies visit flowers, as was demonstrated in the differently scented flower morphs of *Cimicifuga simplex* (Ranunculaceae). By adding two principal volatiles from a scent morph that is actively visited by Nymphalid butterflies to an unattractive scent morph, Pellmyr[236] made the latter highly attractive to butterflies, which alighted on the flowers at a frequency equal to that demonstrated toward the naturally attractive morph. *Danaus gilippus berenice* (Danaidae) butterflies will extend their proboscides in response to honey volatiles, pointing to the involvement of floral volatiles in feeding elicitation;[225] covering different parts of their antennae established that only the short, thin-walled peg sensilla were necessary for realeasing the response.

All species of the diurnally flying burnet moths, *Zygaena* (Zygaenidae), exhibit host plant specificity to flowers of Dipsacaceae. Using a combination of electrophysiological and behavioral approaches, Naumann et al. [226] elegantly demonstrated the key role of floral volatiles in this association. Extracts of floral volatiles from *Knautia arvensis*, a member of this family, evoked in moths strong EAG activity, which was traced to the monoterpenoid compound verbenone. Follow-up field tests using artificial flowers confirmed the moths' attraction to the extract and to verbenone, but only when the odors were offered in combination with specific color cues. The volatiles and visual cues together elicited attraction, landing, and feeding, but either stimulus alone failed to elicit responses.

2. Night-flying Moths

Floral fragrances in moth-pollinated plants are typically strong, heavy-sweet to humans, and emitted mainly at night.[98] However, some species also keep their flowers open during the day, suggesting that other, diurnally active insects may participate in the pollination.[243] This has been well documented in several species of Onagraceae, where both moths and bees may serve as pollinators;[121,191,192,193,201] examples such as these would be well worth investigating from the chemical perspective. Existence of a general moth-flower fragrance syndrome is suggested by the lack of flower specificity in the nectar feeding of moths and by the general attraction of naive moths to floral fragrances.[51] This is exploited in the pollination of *Plumeria rubra* (Apocynaceae), which relies on its strong, moth-flower type fragrance to attract naive hawkmoths to visit its rewardless flowers.[123] Night-blooming, white flowers visited by moths often contain acyclic terpenes (e.g., linalool, nerolidol, farnesol, and their corresponding hydrocarbons) accompanied by benzenoid (e.g., *cis*-3-hexenyl benzoate) and nitrogen-containing (especially indole) compounds.[148] Studies on the role of olfaction in host plant location by moths (see Reference 257) include several that touch on the specific role of floral volatiles.

Moths use flowers as sites for feeding, oviposition, or both. In the latter situation, they can benefit the plant by serving as pollinators during the adult stage and yet have a detrimental impact through the seed predation effected during the larval stage (see Reference 244). An example of narrow mutualism between flowers and insects that is as clear as the fig–wasp relationship is that between *Yucca* flowers (Agavaceae) and their lepidopteran pollinators and seed predators, primarily *Tegeticula* species (Incurvariidae).[13] However, also equally surprising is the scarce documentation of the attraction mechanisms involved,[254] restricted to casual observations that attribute an essential role to floral odors.[259]

Many studies have led to general conclusions on the role of floral fragrances in attracting moths, but few have involved investigations of the interactions between floral volatiles and moth behavior. In his summary of olfactory orientation by night-flying moths to flowers, Brantjes[51] divided the moths into small moths (e.g., Geometridae) and large moths (e.g., Noctuidae, Sphingidae). Evidence, often obtained by visually concealing flowers, indicates that all can olfactorily orient to flower sources, and some investigations that have addressed small moths point to the involvement of odor in eliciting subsequent landing and feeding responses.[52,231a] Most behavioral studies, discussed below, have focused on large moths.

Studies at the beginning of the 19th century established that large moths (e.g., hawkmoths) are guided to flowers at long range mostly by sight but rely on chemical stimuli for close-range flower selection and host plant recognition (see references in Clements and Long[64]). From subsequent work, summarized by Brantjes,[49,51] it is now clear that odor releases a food-seeking behavior in the moths. This is displayed in the moth's characteristic searching flight pattern and in its positive response (approach) to objects, whether white or colored. Flower odor also releases feeding behavior.

In a caged situation, nonflying noctuids and sphingids show a prompt change in behavior, displayed by warming up and flight, upon presentation of flower odors.[49,50,273] Flying moths also change behavior; their flight switches from aimless to searching, but then the searching behavior ceases upon removal of odors.[49] To locate flowers, moths primarily use volatile cues in combination with visual ones, although they can orient to flowers by odor alone.[50,273] After experience with the flowers, visual cues may largely replace olfactory ones, although odor still remains of primary importance in eliciting landing responses, and alighting frequency varies in proportion to scent intensity.[50,51,229] The noctuid *Autographa gamma* relies increasingly, and even principally, on visual cues with experience, but conditioning remains stronger to olfactory than to visual stimuli.[273] When offered artificial flowers, noctuid moths pollinating *Platanthera chlorantha* (Orchidaceae) often approached but rarely alighted on the unscented flowers, whereas scented ones elicited approaches, followed by landing and eager attempts to feed.[229] In Sphingidae, the relative importance of smell and vision in attracting moths varies among species,[49] but after experience, visual orientation always prevails; nevertheless, odor is still required to initiate food-seeking behavior, and preference for certain odors will override any visually learned preference.[49] Noctuidae respond to a wider variety of flower odors than do Sphingidae.[51] While discrimination between flowers at close range is olfactorily based, final guidance into the flower opening is visual in Sphingidae (which generally hover while feeding) but olfactory in Noctuidae (which generally land to feed).[49] Effects of flower alteration on noctuid feeding suggest that these moths use spatial odor patterns on the flower surface, or odor guides, to successfully insert their proboscis into the floral tube.[51]

Electroantennogram tests on noctuids, using fresh leaves and flowers of *Abelia grandiflora* (Caprifoliaceae), yielded responses to leaves that were only slightly higher than to the control, but the high responses to flowers clearly demonstrated the ability of moths to perceive floral volatiles.[119] Zhu et al.[320] tested the electroantennogram response of black cutworm *Agrotis ipsilon* (Noctuidae) to whole-flower volatiles of 25 species and obtained a wide range of response levels, although some plant species observed to be attractive to the moths in the field also evoked comparatively high EAG responses. Females of the tortricid *Lobesia botrana* are attracted to tansy flowers (*Tanacetum* — Compositae), and EAG screening of the volatile components in the floral fragrance showed that only nine volatiles release responses consistently.[106] Discrimination between sympatric *Platanthera* (Orchidaceae) species by pollinating sphingids and noctuids, based on differences in fragrance chemistry, appears to contribute strongly to the orchids' low rate of interbreeding.[230]

Considerable attention has been directed to the role of volatiles in attracting moths to cotton, *Gossypium* (Malvaceae). Using olfactometer chambers, Salama et al.[268] found volatile fractions of dried flower extracts to be attractive to both sexes of the cotton moth, *Spodoptera littoralis* (Noctuidae), but this attraction was equal to that of leaves. Pink bollworm moths, *Pectinophora gossypiella* (Gelechiidae), oviposit on the leaves and bolls of cotton; during the night they visit the closed flowers to feed on externally located bracteal nectaries but rarely use the flowers as oviposition sites. Nevertheless, when offered different plant parts in a wind tunnel bioassay, more moths approached and landed on flowers than on vegetative parts, and this attraction increased after mating.[308] Follow-up elimination of the flowers' attractiveness by washing them in pentane and subsequent restoration of attractiveness by adding the pentane extract to washed flowers localized the attractive olfactory cues to the entire flower. In wind tunnel bioassays of the tobacco budworm *Heliothis virescens*,[293] mated females landed more frequently in responses to volatiles from different flower parts than did unmated males or females.

Early attempts to determine which floral volatiles are active in moth attraction resulted in the identification of amyl salicylate and amyl benzoate as strong attractants to moths visiting flowers of tobacco, *Nicotiana* (Solanaceae), and jimson weed, *Datura* (Solanaceae).[221] Subsequent studies showed that phenylacetaldehyde is attractive to night-flying Lepidoptera,[68] and Cantelo and Jacobson[59] identified it as the single most active attractant to diurnal- and nocturnal-flying moth species visiting the bladder flower plant, *Araulia sericofera* (Asclepiadaceae). Both this compound and 2-phenylethanol, when tested singly, show activities equal to those they have when present in floral blends, as measured by the responses of cabbage looper moths *Trichoplusia ni* (Noctuidae) to floral volatiles of *Abelia grandiflora*.[126] More recently, Heath et al.[127] confirmed the high attraction of *T. ni* to phenylacetaldehyde and added to the list a third compound, benzyl acetate, based on flight tunnel studies using floral volatiles of jessamine, *Cestrum nocturnum* (Solanaceae). Curiously, phenylacetaldehyde has also been shown to be a contact inhibitor of oviposition in several moth species, testifying to the different effects a single volatile can have on different insect species.[101,102]

Lepidoptera that use flowers as oviposition sites seem to rely in their orientation on typical, flower volatiles, as suggested by the preference of ovipositing corn earworms, *Heliothis zea*, for the flowering stages of four different crops, offered at different phenological stages.[143] In the noctuid *Hadena bicruris*, which both feeds and oviposits on flowers of *Melandrium album* (Caryophyllaceae), feeding moths were attracted to various flowers, but for oviposition, naive individuals showed an innate preference for their host plant, implying that different chemicals are involved in feeding versus oviposition.[50] This plant is dioecious, and while moths feed at both male and female flowers, female moths oviposit only in pistillate flowers, for which they show a preference when offered a choice. The means by which the two flower forms are distinguished is unclear, but behavioral studies with altered floral parts suggest that petals are the source of the cues.[30]

The first evidence that volatiles from pollen can affect not only the behavior but also the physiology of insects was provided in recent studies on the sunflower moth *Homoeosoma electellum* (Pyralidae). Females oviposit preferentially in newly opened heads of sunflower, *Helianthus* (Compositae), which ensures that neonate larvae have access to freely exposed pollen, an essential dietary component. This behavior is facilitated by the presence of an oviposition stimulant in the pollen, and its perception by antennal sensilla implies that the chemicals are volatiles.[80] Furthermore, when virgin females are in the presence of pollen volatiles, they initiate calling behavior earlier, spend more time calling, and show a higher rate of egg maturation.[211] Release of calling behavior by host plant volatiles has also been reported in other Lepidoptera,[212] but the mechanisms by which the responses are elicited are not understood. The active chemicals involved in sunflower moth oviposition, calling, and ovarian development are all contained in the ethanolic extract of the pollen but have not been identified. The ability of females to initiate or delay the onset of reproduction in response to pollen availability is clearly of adaptive value in permitting females to adjust to the temporal and spatial unpredictability of a food resource that is essential to larvae.

E. OTHER GROUPS

1. Thysanoptera

Thrips are common visitors to flowers, where they feed on pollen, oviposit in and around flowers, and in some cases act as pollinators.[14,159,242,288] Some thrips show narrow flower selectivity.[162] To evaluate the effect of flower scents on host flower selection, Kirk[160] applied four floral volatiles as bait in field traps. Thrip species differed in their degrees of attraction to the volatiles, and their responses were species specific. However, anisaldehyde traps received a significantly higher number of flower thrips than did myrcene, eugenol, or geraniol, confirming previous evidence that aromatic aldehydes are especially attractive to thrips (see Reference 160). This is further corroborated by reports that large numbers of *Thrips major* and *Thrips fuscipennis* spend the night feeding and mating in flowers of *Filipendula ulmaria* (Rosaceae),[161] the fragrances of which have been shown to contain anisaldehyde.[189]

Pollen is the principal food of many flower-visiting thrips, and they feed on the grains by ingesting only the contents, a procedure that can be completed in only a few seconds.[159] Working with a flower-specific species, Kirk[161] offered the thrips pollen from different plants to determine if they are able to distinguish their host plant(s) on the basis of pollen chemistry. Thrips spent a greater proportion of time feeding when given pollen from their host plant as compared to pollen from other species. Moreover, they appeared to recognize their host plant pollen without probing (they paused with their heads above, but not touching, the pollen grains before initiating feeding), which suggests that they use pollen odors to discriminate between plant species.

2. Parasitoids

Adults of many parasitoids feed on flowers and some, especially those with short mouthparts that restrict their access to nectar, are known to exhibit flower preferences.[178] In olfactometer bioassays females of *Eucelatoria* (Tachinidae) flies, which are parasitoids on *Heliothis*, respond positively to volatile extracts of both flowers and leaves of some plant species but only to flowers of others.[206,228] In olfactometer tests using fresh flowers, a positive response to flower volatiles of several plant species was also observed in *Campoletis sonorensis* (Ichneumonidae), another parasitoid of *Heliothis*, and contact responses measured by antennation and ovipositor thrusts were also greater for flowers than for green plant parts.[94]

The olfactory preferences for floral odors[275] exhibited by parasitoids may assist them in locating the host plants of the host insects and therefore in locating the host insects themselves. This may be similar to the preferences shown by parasitoids for the leaf volatiles of their host insect's host plant, which increase host location (e.g., see References 82, 183). Shahjahan[275] found that nymphs of the tarnished plant bug *Lygus lineolaris* (Lygaeidae) suffer a higher parasitization by *Peristenus pseudopallipes* (Braconidae) when they feed on *Erigeron* (Compositae) than on other host plants. Bioassays using a Y-olfactometer revealed that the female parasitoids show a higher attraction to odors of *Erigeron* flowers than to flowers of other host plant species.

3. Noninsect Arthropods

It is appropriate here to mention some studies on hummingbird flower mites, whose associations with flowers are not unlike those of some insects. The mites of the genera *Rhinoseius* and *Proctolaepas* (Ascidae) inhabit flowers of many species of hummingbird-pollinated plants, where they feed on nectar and pollen, mate, and lay eggs.[66,67] Each species depends on only one or a few plant species to provide it with a reliable, year-round supply of flowers. When the flowers of an inflorescence cease to bloom, the mites must disperse to new inflorescences, which they do as passengers in the nostrils of hummingbirds. They then run off the birds at flowers of their appropriate plant species, showing a strict host plant fidelity. The rarity of mistakes in plant selection suggests that flower recognition is made at a distance, using olfactory stimuli. To determine if mites use nectar cues to recognize their host plants, different nectars and artificial sugar solutions were offered to them in choice experiments using T-shaped chambers.[132] Mites exhibited a greater attraction to nectars of their own host plants compared to matched sugar solutions, nectar of sympatric nonhost plants, and nectar of hummingbird flowers generally devoid of mites. Although it was not always possible in these tests to firmly distinguish between the mites' use of olfactory and gustatory stimuli, the results imply that mites can identify their appropriate host flowers by olfaction using nectar volatiles. These preferences have also been shown to be subject to shifts as a result of conditioning by adult mites to new host plants.[66,67]

F. VOLATILES DETERRENT TO INSECTS

The role of floral volatiles in enhancing pollination or insect attraction to flowers has overshadowed the role these chemicals may play in defense against destructive, flower-feeding insects and microbial pathogens. Evidence that floral chemicals have antiherbivore activity, summarized in Reference 223, has been increasing in recent years. However, most studies have focused on compounds of low volatility that presumably act on contact chemo-receptors, rather than on chemicals perceived by olfaction, although in many cases it is not clear which sensory modalities are actually involved.

Larvae of the sunflower moth *Homoeosoma electellum* feed primarily on pollen and other floral tissue of *Helianthus,* but show decreased feeding in flowers of certain wild species that have high densities of glandular trichomes containing sesquiterpene lactones.[267] Feeding trials suggest that the larvae are deterred by these chemicals, which, when mixed in a synthetic diet, are toxic to the larvae and cause decreased growth and increased mortality.[265,267] The glandular trichomes are appropriately located on the apex of the anthers, where they deter feeding on pollen. Gossypol-containing glands on the anthers of cotton *Gossypium hirsutum* appear to function in a similar fashion on larvae of *Heliothis virescens*; however, while the larvae avoid feeding on the glands, they are still able to penetrate into the anthers and feed on the pollen.[20]

The many volatiles that comprise a floral fragrance can have differing effects on insect behavior when tested singly; but these effects are not apparent when testing the whole fragrance. In behavioral studies of honey bee responses to alfalfa floral volatiles, only one of five tested components (linalool) was found to be attractive to the bees, while the others were considered to be either neutral in their effect or repellent.[130] The two repellent compounds were methyl salicylate and 3 octanone; parallel EAG studies yielded lower antennal responses only in the case of the ketone.[131] Several aliphatic ketones appear to be generally deterrent or even toxic to many insects: they are repellent and toxic to the Colorado potato beetle *Leptinotarsa decemlineata* (Chrysomelidae)[154] and to *Heliothis zea,*[99,100] and are constituents of alarm pheromones in some social insects, including honey bees and stingless bees.[37,295] A series of odd numbered 2-alkanones have been identified in the mandibular glands of *Bombus lapidarius* (Apidae), where they reportedly serve in defense and as pheromonal deterrents.[60] Attempts to deter honey bees from flowers marked with 2-heptanone, a repellent component of their mandibular gland secretions, met with mixed success.[261] While 2-heptanone effec-tively repelled bees when presented at a feeder table, when sprayed on a field, the duration of the repellence was short term, and the spray even became attractive. It appears that behavioral responses to repellent pheromones may show considerable plasticity as a result of learning during a particular encounter event; attractiveness or deterrence of a chemical may also depend upon its concentration.

Aliphatic ketones are not common as constituents of floral fragrances, with the exception of 6-methyl-5-hepten-2-one, which has been reported in amounts of less than 1% in several species belonging to different families (e.g., see References 31–33, 42, 58, 164, 240). Given this compound and other methylated heptanones are common components of alarm phero-mones in various ant species,[37] it is possible that one of its functions in flowers is to repel ants, since some have pollenicidal secretions.[139] When field traps were baited with ketones from the floral fragrances of red clover *Trifolium pratense* (Leguminosae), 2-hexanone and 2-heptanone were weak attractants for *Lygus* bugs (Lygaeidae) but failed to attract *Meligethes* beetles (Nitidulidae) or seed chalcids, *Bruchophagous gibbus* (Chalcidae).[58]

Pollen, however, may be a relatively common repository of volatile aliphatic ketones, as was suggested by chemical analyses of pollen odors (H. Dobson, I. Groth, and G. Bergström, unpublished). Moreover, in some wind-pollinated species, higher amounts have been found in male as opposed to female plants[240] (H. Dobson, I. Groth, and G. Bergström, unpublished). Our pollen studies indicate that heptadecanone is the overwhelmingly dominant volatile in pollen of *Filipendula vulgaris* (Rosaceae); it is also detected in major quantities in the floral fragrance. This species is known to attract many fewer insects than does its congener *F. ulmaria*, for which no ketones (besides methyl heptanone) were reported to be present in the floral fragrances.[189]

This circumstantial evidence that pollen contains chemicals with deterrent activity is corroborated by experimental studies with *Rosa rugosa* (Rosaceae). Close to 20% of the pollen volatiles in this species comprise ketones, as compared to only trace representations in the petals.[87] In behavioral bioassays on field-foraging bumble bees, application of a mixture of two pollen volatiles, 2-tridecanone and tetradecyl acetate, to antherless flowers had a repellent effect on bee visitation, compared to control flowers with no added volatiles.[86] A second ketone, 2-undecanone, is also present in *R. rugosa* pollen volatiles, and in equal representation. Perhaps not surprisingly, 2-tridecanone is a component of the defense-serving mandibular secretions of *Bombus lapidarius*[60] and, especially when combined with 2-undecanone, is toxic to larvae of several Lepidoptera[99,186] and ovicidal in the fruit fly *Dacus dorsalis* (Tephritidae).[205] Bumble bees are, however, attracted to the pollen of *R. rugosa*, which suggests that the deterrent properties of the ketones are masked by other, highly attractive compounds. This is consistent with the observation that the two dominant volatiles in pollen, eugenol and methyl eugenol, are very attractive to the bees, and when applied to antherless flowers, they elicit an increased visitation rate.[86]

In closing, it should be noted that many volatiles typically found in floral fragrances are also found in the essential oils of vegetative plant parts and have been shown to have antimicrobial activity.[207,222,319] Similarly, ketones, discussed above, can exhibit antifungal properties.[65] In another twist, jasmonate, a fragrance constituent of several moth-pollinated flowers, can act as a growth regulator in plants and can also modify the expression of specific plant genes.[280] This reminds us that floral volatiles apparently involved in insect attraction or deterrence may in fact serve multiple functions for both the plant and the insect.

V. CONCLUSION

As the end of most reviews of this nature, one is left with the sense that many loose ends are left hanging, that there are many more questions than one began with, and that in fact we know but little about the interplay between insects and flower odors. The topic of insect–flower interactions, examined from the perspectives of insect olfaction and flower volatiles, is a very broad one, touching on most insect orders and almost all entomophilous angiosperms. Although discussion of flowers and their flower-visiting insects is often assigned to the field of pollination biology, it is time to dissolve all such disciplinary boundaries and bring together all endeavors that aim to shed light on the evolutionary significance of floral volatiles to both the plant and the insect.

From the material presented here, it is apparent that only few examples of insect–flower associations have been examined in terms of olfactory interactions; and most, with the exception of bees and possibly moths, have been studied only cursorily, leaving diverse questions ripe for investigation. Many insect groups have not been touched upon. The landmark attempts to understand certain associations, such as those between bees and orchids, have provided important insights into the complexities of interactions that are to be expected in studies of insects and flower odors, and furthermore remind us of the need for multidisciplinary efforts. With the foundations laid by these pioneering investigators, we can now turn an inquiring eye toward other insect–flower associations, embracing the integrative approaches of chemical ecology, which bridge the fields of chemistry and behavior/physiology. The discipline is very open, from the molecular to the community level, and it is hoped that the recent diversification of inquiry will continue unhindered in both depth and breadth.

ACKNOWLEDGMENTS

I dedicate this paper to Professor Bertil Kullenberg on the occasion of his 80th birthday. His broad perspectives and multidisciplinary research approaches, as well as our numerous stimulating discussions, have been a strong source of inspiration.

Gettysburg College provided support and library facilities during the early stages of manuscript preparation and Lenore Barkan kindly offered logistical assistance. I thank Lennart Ågren, Gunnar Bergström, Elizabeth Bernays, Inga Groth, and Robbin Thorp for their helpful comments on the manuscript.

REFERENCES

1. **Ackerman, J. D.,** Geographic and seasonal variation in fragrance choices and preferences of male euglossine bees, *Biotropica*, 21, 340, 1989.
2. **Ågren, L. and Borg-Karlson, A.-K.,** Responses of *Argogorytes* (Hymenoptera: Sphecidae) males to odor signals from *Ophrys insectifera* (Orchidaceae); preliminary EAG and chemical investigation, *Nova Acta Reg. Soc. Sci. Ups. ser V.C*, 3, 111, 1984.
3. **Ågren, L., Kullenberg, B., and Sensenbaugh, T.,** Congruences in pilosity between three species of *Ophrys* (Orchidaceae) and their hymenopteran pollinators, *Nova Acta Reg. Soc. Sci. Ups. ser V:C*, 3, 15, 1984.
4. **Altenburger, R. and Matile, P.,** Further observations on rhythmic emission of fragrance in flowers, *Planta*, 180, 194, 1990.
5. **Andersen, J. F., Buchmann, S. L., Weisleder, D., Plattner, R. D., and Minckley, R. L.,** Identification of thoracic gland constituents from male *Xylocopa* spp. Latreille (Hymenoptera: Anthophoridae) from Arizona, *J. Chem. Ecol.*, 14, 1153, 1988.
6. **Andersen, J. F. and Metcalf, R. L.,** Factors influencing distribution of *Diabrotica* spp. in blossoms of cultivated *Cucurbita* spp., *J. Chem. Ecol.*, 13, 681, 1987.
7. **Andersson, I. H.,** *Nectar Feeding Behaviour and the Significance of Sugar Meals in Mosquitoes (Diptera: Culicidae)*, Ph.D. dissertation, University of Uppsala, 1991, chap. 4.
8. **Armbruster, W. S.,** The role of resin in angiosperm pollination: ecological and chemical considerations, *Am. J. Bot.*, 71, 1149, 1984.
9. **Armbruster, W. S., Keller, S., Matsuki, M., and Clausen, T. P.,** Pollination of *Dalechampia magnoliifolia* (Euphorbiaceae) by male euglossine bees, *Am. J. Bot.*, 76, 1279, 1989.
10. **Armstrong, J. E. and Drummond, B. A.,** Floral biology of *Myristica fragrans* Houtt. (Myristicaceae), the nutmeg of commerce, *Biotropica*, 18, 32, 1986.
11. **Armstrong, J. E. and Irvine, A. K.,** Functions of staminodia in the beetle-pollinated flowers of *Eupomatia laurina*, *Biotropica*, 22, 429, 1990.
12. **von Aufsess, A.,** Geruchliche Nahorientierung der Biene bei entomophilen und ornithophilen Blüten, *Z. Vergl. Physiol.*, 43, 469, 1960.
13. **Baker, H. G.,** Yuccas and Yucca moths — a historical commentary, *Ann. Mo. Bot. Gard.*, 73, 556, 1986.
14. **Baker, J. D. and Cruden, R. W.,** Thrips-mediated self-pollination of two facultatively xenogamous wetland species, *Am. J. Bot.*, 78, 959, 1991.
15. **Baker, T. C., Nishida, R., and Roelofs, W. L.,** Close-range attraction of female oriental fruit moths to herbal scent of male hairpencils, *Science*, 214, 1359, 1981.
16. **Barker, N. P.,** Evidence of a volatile attractant in *Ficus ingens* (Moraceae), *Bothalia*, 15, 607, 1985.
17. **Barkman, T. J., Gage, D. A., and Beaman, J. H.,** Implications of floral fragrance compounds in the systematics of *Paphiopedilum* (Orchidaceae), (Abstract), *Am. J. Bot.*, 79 (Suppl.), 135, 1992.
18. **Beaman, R. S., Decker, P. J., and Beaman, J. H.,** Pollination of *Rafflesia* (Rafflesiaceae), *Am. J. Bot.*, 75, 1148, 1988.

19. **Beker, R., Dafni, A., Eisikowitch, D., and Ravid, U.,** Volatiles of two chemotypes of *Majorana syriaca* L. (Labiatae) as olfactory cues for the honeybee, *Oecologia*, 79, 446, 1989.

20. **Belcher, D. W., Schneider, J. C., Hedin, P. A., and French, J. C.,** Impact of glands in cotton anthers on feeding behavior of *Heliothis virescens* (F.) (Lepidoptera: Noctuidae) larvae, *Environ. Entomol.*, 12, 1478, 1983.

21. **Bergström, G.,** Role of volatile chemicals in *Ophrys*-pollinator interactions, in *Biochemical Aspects of Plant and Animal Coevolution*, Harborne, J. B., Ed., Academic Press, New York, 1978, 207.

22. **Bergström, G.,** On the role of volatile chemical signals in the evolution and speciation of plants and insects: why do flowers smell and why do they smell differently?, in *Insects–Plants*, Labeyrie, V., Fabres, G., and Lachaise, D., Eds., Dr. W. Junk, Dordrecht, 1987, 321.

23. **Bergström, G.,** Role of volatile chemicals in the evolution and coadaptation of flowering plants and insects, in *Evolution and Coadaptation in Biotic Communities*, Kawano, S., Connell, J. H., and Hidaka, T., Eds., University of Tokyo Press, 1987, 151.

24. **Bergström, G.,** Chemical ecology of terpenoid and other fragrances of angiosperm flowers, in *Ecological Chemistry and Biochemistry of Plant Terpenoids*, Harborne, J. B. and Tomas-Barberan, F. A., Eds., Proc. Phytochem. Soc. Europe, Clarendon Press, Oxford, 1991, 277.

25. **Bergström, G., Birgersson, G., Groth, I., and Nilsson, L. A.,** Floral fragrance disparity between three taxa of Lady's slipper *Cypripredium calceolus* (Orchidaceae), *Phytochemistry*, 31, 2315, 1992.

26. **Bergström, G., Groth, I., Pellmyr, O., Endress, P. K., Thien, L. B., Hubener, A., and Francke, W.,** Chemical basis of a highly specific mutualism: chiral esters attract pollinating beetles in Eupomatiaceae, *Phytochemistry*, 30, 3221, 1991.

27. **Bergström, G. and Lundgren, L.,** Androconial secretion of three species of butterflies of the genus *Pieris* (Lep., Pieridae), *Zoon*, Suppl. 1, 67, 1973.

28. **Bergström, J. and Bergström, G.,** Floral scents of *Bartsia alpina* (Scrophulariaceae): chemical composition and variation between individual plants, *Nord. J. Bot.*, 9, 363, 1989.

29. **Bernhardt, P. and Thien, L. B.,** Self-isolation and insect pollination in the primitive angiosperms: new evaluations of older hypotheses, *Plant Syst. Evol.*, 156, 159, 1987.

30. **Bicchi, C., Belliardo, F., and Frattini, C.,** Identification of the volatile components of some piedmontese honeys, *J. Apic. Res.*, 22, 130, 1983.

31. **Binder, R. G., Benson, M. E., and Flath, R. A.,** Volatile components of safflower, *J. Agric. Food Chem.*, 38, 1245, 1990.

32. **Binder, R. G., Turner, C. E., and Flath, R. A.,** Volatile components of purple starthistle, *J. Agric. Food Chem.*, 38, 1053, 1990.

33. **Binder, R. G., Turner, C. E., and Flath, R. A.,** Comparison of yellow starthistle volatiles from different plant parts, *J. Agric. Food Chem.*, 38, 764, 1990.

34. **Bino, R. J., Dafni, A., and Meeuse, A. D. J.,** The pollination ecology of *Orchis galilaea* (Bornm. et Schulze) Schltr. (Orchidaceae), *New Phytol.*, 90, 315, 1982.

35. **Bitterman, M. E.,** Vertebrate-invertebrate comparisons, in *Intelligence and Evolutionary Biology*, Jenson, H. J. and Jenson, I., Eds., NATO ASI ser., Vol. G17, Springer-Verlag, Berlin, 1988, 251.

36. **Blank, I., Fischer, K. H., and Grosch, W.,** Intensive neutral odourants of linden honey, *Z. Lebensm. Unters. Forsch.*, 189, 426, 1989.

37. **Blum, M. S.,** Alarm pheromones, *Annu. Rev. Entomol.*, 14, 57, 1969.

38. **Boggs, C. L.,** Ecology of nectar and pollen feeding in Lepidoptera, in *Nutritional Ecology of Mites, Spiders, and Related Invertebrates*, Slansky, F. and Rodriguez, J. G., Eds., John Wiley & Sons, New York, 1987, 369.

39. **Bolwig, N.,** The role of scent as a nectar guide for honeybees on flowers and an observation on the effect of colour on recruits, *Br. J. Anim. Behav.*, 1, 81, 1954.

40. **Bonaga, G. and Giumanini, A. G.,** The volatile fraction of chestnut honey, *J. Apic. Res.*, 25, 113, 1986.

41. **Borg-Karlson, A.-K.,** Attraction of *Phyllopertha horticola* (Coleoptera, Scarabaeidae) males to fragrance components of *Ophrys* flowers (Orchidaceae, section Fuciflorae), *Entomol. Tidskr.*, 109, 105, 1989.

42. **Borg-Karlson, A.-K.,** Chemical and ethological studies of pollination in the genus *Ophrys* (Orchidaceae), *Phytochemistry*, 29, 1359, 1990.

43. **Borg-Karlson, A.-K. and Tengö, J.,** Odor mimetism? Key substances in *Ophrys lutea–Andrena* pollination relationship (Orchidaceae: Andrenidae), *J. Chem. Ecol.*, 12, 1927, 1986.

44. **Borg-Karlson, A.-K., Ågren, L., Dobson, H., and Bergström, G.,** Identification and electroantennographic activity of sex-specific geranyl esters in an abdominal gland of female *Agriotes obscurus* (L.) and *A. lineatus* (L.) (Coleoptera, Elateridae), *Experientia*, 44, 531, 1988.

45. **Borg-Karlson, A.-K., Bergström, G., and Groth, I.,** Chemical basis for the relationship between *Ophrys* orchids and their pollinators, *Chem. Scripta*, 25, 283, 1985.

46. **Borg-Karlson, A.-K., Bergström, G., and Kullenberg, B.,** Chemical basis for the relationship between *Ophrys* orchids and their pollinators, *Chem. Scripta*, 27, 303, 1987.

47. **Brandes, C., Frisch, B., and Menzel, R.,** Time course of memory formation differs in honeybee lines selected for good and bad learning, *Anim. Behav.*, 36, 981, 1988.

48. **Brandes, C. and Menzel, R.,** Common mechanisms in proboscis extension conditioning and visual learning revealed by genetic selection in honeybees (*Apis mellifera capensis*), *J. Comp. Physiol. A*, 166, 545, 1990.
49. **Brantjes, N. B. M.,** Sphingophilous flowers, function of their scent, in *Pollination and Dispersal*, Brantjes, N. B. M. and Linskens, H. F., Eds., Publ. Botany, Univ. Nijmegen, 1973, 27.
50. **Brantjes, N. B. M.,** Riddles around the pollination of *Melandrium album* (Mill.) Garcke (Caryophyllaceae) during the oviposition by *Hadena bicruris* Hufn. (Noctuidae, Lepidoptera). *Proc. Kon. Nederl. Akad. Wet. Ser. C*, 79, 1 and 127, 1976.
51. **Brantjes, N. B. M.,** Sensory responses to flowers in night-flying moths, in *The Pollination of Flowers by Insects*, Richards, A. J., Ed., Academic Press, London, Linn. Soc. Symp. Ser. No. 6, 13, 1978.
52. **Brantjes, N. B. M. and Leemans, J. A. A. M.,** *Silene otites* (Caryophyllaceae) pollinated by nocturnal Lepidoptera and mosquitoes, *Acta Bot. Neerl.*, 25, 281, 1976.
53. **Bronstein, J. L.,** Maintenance of species-specificity in a neotropical fig–pollinator wasp mutualism, *Oikos*, 48, 39, 1987.
54. **Buchmann, S. L.,** The ecology of oil flowers and their bees, *Annu. Rev. Ecol. Syst.*, 18, 343, 1987.
55. **Burger, B. V., Munro, Z. M., and Visser, J. H.,** Determination of plant volatiles. 1. Analysis of the insect-attracting allomone of the parasitic plant *Hydnora africana* using Grob-Habich activated charcoal traps, *J. High Res. Chrom. Chrom. Comm.*, 11, 496, 1988.
56. **Butler, C. G. and Calam, D. H.,** Pheromones of the honey bee — the secretion of the Nassanoff gland of the worker, *J. Insect Physiol.*, 15, 237, 1969.
57. **Buttery, R. G., Kamm, J. A., and Ling, L. C.,** Volatile components of alfalfa flowers and pods, *J. Agric. Food Chem.*, 30, 739, 1982.
58. **Buttery, R. G., Kamm, J. A., and Ling, L. C.,** Volatile components of red clover leaves, flowers, and seed pods: possible insect attractants, *J. Agric. Food Chem.*, 32, 254, 1984.
59. **Cantelo, W. W. and Jacobson, M.,** Phenylacetaldehyde attracts moths to Bladder Flower and to blacklight traps, *Environ. Entomol.*, 8, 444, 1979.
60. **Cederberg, B.,** Chemical basis for defense in bumble bees, in *Proc. VIII Int. Cong. I. U. S. S. I.*, Wageningen, Netherlands, 1977, 77.
61. **Charlet, L. D., Kopp, D. D., and Oseto, C. Y.,** Sunflowers: their history and associated insect community in the northern great plains, *Bull. Entomol. Soc. Am.*, 33, 69, 1987.
62. **Charpentier, R.,** Host plant selection by the pollen beetle *Meligethes aeneus*, *Entomol. Exp. Appl.*, 38, 277, 1985.
63. **Clement, S. L.,** On the function of pea flower feeding by *Bruchus pisorum*, *Entomol. Exp. Appl.*, 63, 115, 1992.
64. **Clements, F. E. and Long, F. L.,** *Experimental Pollination: An Outline of the Ecology of Flowers and Insects*, Carnegie Institute Washington, Washington, D.C., 1923.
65. **Cole, L. K., Blum, M. S., and Roncadori, R. W.,** Antifungal properties of the insect alarm pheromones, citral, 2-heptanone, and 4-methyl-3-heptanone, *Mycologia*, 67, 701, 1975.
66. **Colwell, R. K.,** Community biology and sexual selection: lessons from hummingbird flower mites, in *Community Ecology*, Diamond, J. and Case, T. J., Eds., Harper & Row, New York, 1985, 406.
67. **Colwell, R. K.,** Population structure and sexual selection for host fidelity in the speciation of hummingbird flower mites, in *Evolutionary Processes and Theory*, Karlin, S. and Nevo, E., Eds., Academic Press, New York, 1986, 475.
68. **Creighton, C. S., McFadden, T. L., and Cuthbert, E. R.,** Supplementary data on phenylacetaldehyde: an attractant for Lepidoptera, *J. Econ. Entomol.*, 66, 114, 1973.
69. **Crepet, W. L.,** The role of insect pollination in the evolution of angiosperms, in *Pollination Biology*, Real, L., Ed., Academic Press, Orlando, FL, 1983, 29.
70. **Crepet, W. L. and Friis, E. M.,** The evolution of insect pollination in angiosperms, in *The Origin of Angiosperms and their Biological Consequences*, Friis, E. M., Chaloner, W. G., and Crane, P. R., Eds., Cambridge University Press, Cambridge, 1987, 181.
71. **Croteau, R. and Karp, F.,** Origin of natural odorants, in *Perfumes: Art, Science and Technology*, Müller, P. M. and Lamparsky, D., Eds., Elsevier Applied Science, London, 1991, 101.
72. **Crowson, R. A.,** *The Biology of the Coleoptera*, Academic Press, London, 1981, 7 and 45.
73. **Crowson, R. A.,** Meligethinae as possible pollinators (Coleoptera: Nitidulidae), *Entomol. Gener.*, 14, 61, 1988.
74. **Cumber, R. A.,** Some aspects of the biology and ecology of humble-bees bearing upon the yields of red-clover seed in New Zealand, *N. Z. J. Sci. Technol.*, 34, 227, 1953.
75. **Dafni, A.,** Mimicry and deception in pollination, *Annu. Rev. Ecol. Syst.*, 15, 259, 1984.
76. **Dafni, A.,** Pollination in *Orchis* and related genera: evolution from reward to deception, in *Orchid Biology: Reviews and Perspectives 4*, Arditti, J., Ed., Cornell University Press, Ithaca, NY, 1987, 79.
77. **Dafni, A. and Bernhardt, P.,** Pollination of terrestrial orchids of southern Australia and the Mediterranean region, in *Evolutionary Biology*, Vol. 24, Hecht, M. K., Wallace, B., and Macintyre, R. J., Eds., Plenum Press, NY, 1990, 193.

78. **Darwin, C.,** *The Effects of Cross and Self Fertilisation in the Vegetable Kingdom,* Appleton and Co., New York, 1898, chap. 10, 11.
79. **DeJong, R. and Pham-Delegue, M.-H.,** Electroantennogram responses related to olfactory conditioning in the honey bee *(Apis mellifera ligustica), J. Insect Physiol.,* 37, 319, 1991.
80. **Delisle, J., McNeil, J. N., Underhill, E. W., and Barton, D.,** *Helianthus annuus* pollen, an oviposition stimulant for the sunflower moth, *Homoeosoma electellum, Entomol. Exp. Appl.,* 50, 53, 1989.
81. **DeVries, P. J. and Stiles, F. G.,** Attraction of pyrrolizidine alkaloid seeking Lepidoptera to *Epidendrum paniculatum* orchids, *Biotropica,* 22, 290, 1990.
82. **Ding, D., Swedenborg, P. D., and Jones, R. L.,** Plant odor preferences and learning in *Macrocentrus grandii* (Hymenoptera: Braconidae), a larval parasitoid of the European corn borer, *Ostrinia nubilalis* (Lepidoptera: Pyralidae), *J. Kan. Entomol. Soc.,* 62, 164, 1989.
83. **Dobson, H. E. M.,** Role of flower and pollen aromas in host-plant recognition by solitary bees, *Oecologia,* 72, 618, 1987.
84. **Dobson, H. E. M.,** Survey of pollen and pollenkitt lipids — chemical cues for flower visitors?, *Am. J. Bot.,* 75, 170, 1988.
85. **Dobson, H. E. M.,** Analysis of flower and pollen volatiles, in *Essential Oils and Waxes,* Modern Methods of Plant Analysis, New Ser. Vol. 12, Linskens, H. F. and Jackson, J. F., Eds., Springer-Verlag, Berlin, 1991, 231.
86. **Dobson, H. E. M.,** Pollen and flower fragrances in pollination, *Acta Hort.,* 288, 313, 1991.
87. **Dobson, H. E. M., Bergström, G., and Groth, I.,** Differences in fragrance chemistry between flower parts of *Rosa rugosa* Thunb. (Rosaceae), *Isr. J. Bot.,* 39, 143, 1990.
88. **Dobson, H. E. M., Bergström, J., Bergström, G., and Groth, I.,** Pollen and flower volatiles in two *Rosa* species, *Phytochemistry,* 26, 3171, 1987.
89. **Dodson, C. H., Dressler, R. L., Hills, H. G., Adams, R. M., and Williams, N. H.,** Biologically active compounds in orchid fragrances, *Science,* 164, 1243, 1969.
90. **Donaldson, J. M. I., McGovern, T. P., and Ladd, T. L.,** Trapping techniques and attractants for Cetoniinae and Rutelinae (Coleoptera: Scarabaeidae), *J. Econ. Entomol.,* 79, 374, 1986.
91. **Dressler, R. L.,** Biology of the orchid bees (Euglossini), *Annu. Rev. Ecol. Syst.,* 13, 373, 1982.
92. **Duffield, R. M., Wheeler, J. W., and Eickwort, G. C.,** Sociochemicals of bees, in *Chemical Ecology of Insects,* Bell, W. J. and Cardé, R. T., Eds., Chapman and Hall, London, 1984, 387.
93. **Elakovich, S. D. and Oguntimein, B. O.,** The essential oil of *Lippia adoensis* leaves and flowers, *J. Nat. Prod.,* 50, 503, 1987.
94. **Elzen, G. W., Williams, H. J., and Vinson, S. B.,** Response by the parasitoid *Campoletis sonorensis* (Hymenoptera: Ichneumonidae) to chemicals (synomones) in plants: implications for host habitat location, *Environ. Entomol.,* 12, 1873, 1983.
95. **Erhardt, A.,** Pollination of *Dianthus superbus* L., *Flora,* 185, 99, 1991.
96. **Etiévant, P. X., Azar, M., Pham-Delegue, M.-H., and Masson, C. J.,** Isolation and identification of volatile constituents of sunflowers *(Helianthus annuus* L.), *J. Agric. Food Chem.,* 32, 503, 1984.
97. **Faden, R. B.,** Floral attraction and floral hairs in the Commelinaceae, *Ann. Mo. Bot. Gard.,* 79, 46, 1992.
98. **Faegri, K. and van der Pijl, L.,** *The Principles of Pollination Ecology,* 3rd ed., Pergamon Press, Oxford, 1979.
99. **Farrar, R. R. and Kennedy, G. G.,** 2-Undecanone, a constituent of the glandular trichomes of *Lycopersicon hirsutum* f. *glabratum:* effects on *Heliothis zea* and *Manduca sexta* growth and survival, *Entomol. Exp. Appl.,* 43, 17, 1987.
100. **Farrar, R. R., Kennedy, G. G., and Rose, R. M.,** The protective role of dietary unsaturated fatty acids against 2-undecanone-induced pupal mortality and deformity in *Helicoverpa zea, Entomol. Exp. Appl.,* 62, 191, 1992.
101. **Flint, H. M., Noble, J. M., and Shaw, D.,** Phenylacetaldehyde: tests for control of the pink bollworm and observations on other Lepidoptera infesting cotton, *J. Ga. Entomol. Soc.,* 13, 284, 1978.
102. **Flint, H. M., Smith, R. L., Pomonis, J. G., Forey, D. E., and Horn, B. R.,** Phenylacetaldehyde: oviposition inhibitor for the pink bollworm, *J. Econ. Entomol.,* 70, 547, 1977.
103. **Friis, E. M. and Endress, P. K.,** Origin and evolution of angiosperm flowers, *Adv. Bot. Res.,* 17, 99, 1960.
104. **von Frisch, K.,** Über die "Sprache" der Bienen, *Zool. Jahrb. Abt. Allg. Zool. Physiol.,* 40, 1, 1923.
105. **von Frisch, K.,** *Bees: Their Vision, Chemical Senses, and Language,* Cornell University Press, Ithaca, NY, 1971.
106. **Gabel, B., Thiery, D., Suchy, V., Marion-Poll, F., Hradsky, P., and Farkas, P.,** Floral volatiles of *Tanacetum vulgare* L. attractive to *Lobesia botrana* Den. et Schiff. females, *J. Chem. Ecol.,* 18, 693, 1992.
107. **Galen, C.,** The effects of nectar thieving ants on seedset in floral scent morphs of *Polemonium viscosum, Oikos,* 41, 245, 1983.
108. **Galen, C.,** The smell of success, *Nat. Hist.,* 94, 28, 1985.
109. **Galen, C. and Kevan, P. G.,** Scent and color, floral polymorphisms and pollination biology in *Polemonium viscosum* Nutt., *Am. Midl. Nat.,* 104, 281, 1980.

110. **Gerlach, G. and Schill, R.,** Composition of orchid scents attracting euglossine bees, *Bot. Acta,* 104, 379, 1991.

111. **Gori, D. F.,** Post-pollination phenomena and adaptive floral changes, in *Handbook of Pollination Biology,* Jones, E. C. and Little, R. J., Eds., Sci. Acad. Ed., New York, 1983, 31.

112. **Gottsberger, G.,** Some aspects of beetle pollination in the evolution of flowering plants, *Plant Syst. Evol.,* Suppl. 1, 211, 1977.

113. **Gottsberger, G.,** Comments on flower evolution and beetle pollination in the genera *Annona* and *Rollinia* (Annonaceae), *Plant Syst. Evol.,* 167, 189, 1989.

114. **Gottsberger, G.,** Beetle pollination and flowering rhythm of *Annona* spp. (Annonaceae) in Brazil, *Plant Syst. Evol.,* 167, 165, 1989.

115. **Gottsberger, G.,** Flowers and beetles in the South American tropics, *Bot. Acta,* 103, 360, 1990.

116. **Gottsberger, G. and Silberbauer-Gottsberger, I.,** Olfactory and visual attraction of *Erioscelis emarginata* (Cyclocephalini, Dynastinae) to the inflorescence of *Philodendron selloum* (Araceae), *Biotropica,* 23, 23, 1991.

117. **Gould, J. L.,** The dance-language controversy, *Q. Rev. Biol.,* 51, 211, 1976.

118. **Graddon, A. D., Morrison, J. D., and Smith, J. F.,** Volatile constituents of some unifloral Australian honeys, *J. Agric. Food Chem.,* 27, 832, 1979.

119. **Grant, G. G.,** Feeding activity of adult cabbage loopers on flowers with strong olfactory stimuli, *J. Econ. Entomol.,* 64, 315, 1971.

120. **Gregg, K. B.,** Variation in floral fragrances and morphology: incipient speciation in *Cycnodes, Bot. Gaz.,* 144, 566, 1983.

121. **Gregory, D. P.,** Hawkmoth pollination in the genus *Oenothera, Aliso* 5, 357, 1963.

122. **Groth, I., Bergström, G., and Pellmyr, O.,** Floral fragrances in *Cimicifuga:* chemical polymorphism and incipient speciation in *Cimicifuga simplex, Biochem. Syst. Ecol.,* 15, 441, 1987.

123. **Haber, W. A.,** Pollination by deceit in a mass-flowering tropical tree *Plumeria rubra* L. (Apocynaceae), *Biotropica,* 16, 269, 1984.

124. **Hamilton-Kemp, T. R., Loughrin, J. H., and Andersen, R. A.,** Identification of some volatile compounds from strawberry flowers, *Phytochemistry,* 29, 2847, 1990.

125. **Harborne, J. B. and Turner, B. L.,** *Plant Chemosystematics,* Academic Press, London, 1984.

126. **Haynes, K. F., Zhao, J. Z., and Latif, A.,** Identification of floral compounds from *Abelia grandiflora* that stimulate upwind flight in Cabbage looper moths, *J. Chem. Ecol.,* 17, 637, 1991.

127. **Heath, R. R., Landolt, P. J., Dueben, B., and Lenczewski, B.,** Identification of floral compounds of night-blooming Jessamine attractive to Cabbage looper moths, *Environ. Entomol.,* 21, 854, 1992.

128. **Hefetz, A.,** The role of Dufour's gland secretions in bees, *Physiol. Entomol.,* 12, 243, 1987.

129. **Henderson, A.,** A review of pollination studies in the Palmae, *Bot. Rev.,* 52, 221, 1986.

130. **Henning, J. A., Peng, Y.-S., Montague, M. A., and Teuber, L. R.,** Honey bee (Hymenoptera: Apidae) behavioral response to primary alfalfa (Rosales: Fabaceae) floral volatiles, *J. Econ. Entomol.,* 85, 233, 1992.

131. **Henning, J. A. and Teuber, L. R.,** Combined gas chromatography-electroantennogram characterization of alfalfa floral volatiles recognized by honey bees (Hymenoptera: Apidae), *J. Econ. Entomol.,* 85, 226, 1992.

132. **Heyneman, A. J., Colwell, R. K., Naeem, S., Dobkin, D. S., and Hallet, B.,** Host plant discrimination experiments with hummingbird flower mites, in *Plant–Animal Interactions: Evolutionary Ecology in Tropical and Temperate Regions,* Price, P. W., Lewinsohn, T. M., Fernandes, G. W., and Benson, W. W., Eds., John Wiley & Sons, 1991, 455.

133. **Hills, H. G.,** Fragrance cycling in *Stanhopea pulla* (Orchidaceae, Stanhopeinae) and identification of *trans*-limonene oxide as a major fragrance component, *Lindleyana,* 4, 61, 1989.

134. **Hills, H. G. and Schutzman, B.,** Considerations for sampling floral fragrances, *Phytochem. Bull.,* 22, 2, 1990.

135. **Hills, H. G. and Williams, N. H.,** Fragrance cycle of *Clowesia rosea, Orquidea,* 12, 19, 1990.

136. **Hills, H. G., Williams, N. H., and Dodson, C. H.,** Floral fragrances and isolating mechanisms in the genus *Catasetum* (Orchidaceae), *Biotropica,* 4, 61, 1972.

137. **Hohmann, H.,** Über die Wirkung von Pollenextrakten und Duftstoffen auf das Sammel- und Werbeverhalten Hoselnder Bienen (*Apis mellifera* L.), *Apidologie,* 1, 157, 1970.

138. **Hügel, M.-F.,** Étude de quelques constituents du pollen, *Ann. Abeille,* 5, 97, 1962.

139. **Hull, D. A. and Beattie, B. J.,** Adverse effects on pollen exposed to *Atta texana* and other North American ants: implications for ant pollination, *Oecologia,* 75, 153, 1988.

140. **Hussein, M. Y., Lajis, N. H., and Ali, J. H.,** Biological and chemical factors associated with the successful introduction of *Elaeisobius kamerunicus* Faust, the oil palm pollinators in Malaysia, *Acta Hort.,* 288, 81, 1991.

141. **Jepson, P. C. and Healy, T. P.,** The location of floral nectar sources by mosquitoes: an advanced bioassay for volatile plant odours and initial studies with *Aedes aegypti* (L.) (Diptera: Culicidae), *Bull. Entomol. Res.,* 78, 641, 1988.

142. **Johnson, D. L. and Wenner, A. M.,** Recruitment efficiency in honeybees: studies on the role of olfaction, *J. Apic. Res.,* 9, 13, 1970.

143. **Johnson, M. W., Stinner, R. E., and Rabb, R. L.,** Ovipositional response of *Heliothis zea* (Boddie) to its major hosts in North Carolina, *Environ. Entomol.*, 4, 291, 1975.

144. **Jones, R. W., Cate, J. R., Hernandez, E. M., and Sosa, E. S.,** Pollen feeding and survival of the boll weevil (Coleoptera: Curculionidae) on selected plant species in northeastern Mexico, *Environ. Entomol.*, 22, 99, 1993.

145. **Joulain, D.,** The composition of the headspace from fragrant flowers: further results, *Flav. Fragr. J.*, 2, 149, 1987.

146. **Kaib, M.,** Die Fleisch-und Blumenduftrezeptoren auf der Antenne der Schmeissfliege *Calliphora vicina*, *J. Comp. Physiol.*, 95, 105, 1974.

147. **Kaiser, R.,** Trapping, investigation and reconstitution of flower scents, in *Perfumes: Art, Science and Technology*, Müller, P. M. and Lamparsky, D., Eds., Elsevier Applied Science, London, 1991, 213.

148. **Kaiser, R.,** *The Scents of Orchids: Olfactory and Chemical Investigations*, Elsevier, Amsterdam, 1993.

149. **Kamm, J. A.,** In-flight assessment of host and nonhost odors by alfalfa seed chalcid (Hymenoptera: Eurytomidae), *Environ. Entomol.*, 18, 56, 1989.

150. **Kamm, J. A.,** Control of olfactory-induced behavior in alfalfa seed chalcid (Hymenoptera: Eurytomidae) by celestial light, *J. Chem. Ecol.*, 16, 291, 1990.

151. **Kamm, J. A. and Buttery, R. G.,** Response of the alfalfa seed chalcid, *Bruchophagus roddi*, to alfalfa volatiles, *Entomol. Exp. Appl.*, 33, 129, 1983.

152. **Kamm, J. A. and Buttery, R. G.,** Response of the alfalfa and clover seed chalcids (Hymenoptera: Eurytomidae) to host plant components, *Environ. Entomol.*, 15, 1244, 1986.

153. **Kamm, J. A. and Buttery, R. G.,** Ovipositional behavior of the alfalfa seed chalcid (Hymenoptera: Eurytomidae) in response to volatile components of alfalfa, *Environ. Entomol.*, 15, 388, 1986.

154. **Kennedy, G. G. and Sorenson, C. F.,** Role of glandular trichomes in the resistance of *Lycopersicon hirsutum* f. *glabratum* to Colorado potato beetle (Coleoptera: Chrysomelidae), *J. Econ. Entomol.*, 78, 547, 1985.

155. **Kevan, P. G.,** How honey bees forage for pollen at skunk cabbage, *Symplocarpus foetidus* (Araceae), *Apidologie*, 20, 485, 1989.

156. **Kevan, P. G. and Baker, H. G.,** Insects as flower visitors and pollinators, *Annu. Rev. Entomol.*, 28, 407, 1983.

157. **Kevan P. G., Hussein, M. Y., Hussey, N., and Wahid, M. B.,** Modelling the use of *Elaeidobius kamerunicus* for pollination of oil palm, *Planter*, 62, 89, 1986.

158. **Kimsey, L. S.,** The behavior of male orchid bees (Apidae, Hymenoptera, Insecta) and the question of leks, *Anim. Behav.*, 28, 996, 1980.

159. **Kirk, W. D. J.,** Pollen-feeding in thrips (Insecta: Thysanoptera), *J. Zool.*, 204, 107, 1984.

160. **Kirk, W. D. J.,** Effect of some floral scents on host finding by thrips (Insecta: Thysanoptera), *J. Chem. Ecol.*, 11, 35, 1985.

161. **Kirk, W. D. J.,** Aggregation and mating of thrips in flowers of *Calystegia sepium*, *Ecol. Entomol.*, 10, 433, 1985.

162. **Kirk, W. D. J.,** Pollen-feeding and the host specificity and fecundity of flower thrips (Thysanoptera), *Ecol. Entomol.*, 10, 281, 1985.

163. **Knoll, F.,** *Bombylius fuliginosus* und die Farbe der Blumen, *Abh. zool.-bot. Ges. Wien*, 12, 17, 1921.

164. **Knudsen, J. T. and Tollsten, L.,** Floral scent and intrafloral scent differentiation in *Moneses* and *Pyrola* (Pyrolaceae), *Plant Syst. Evol.*, 177, 81, 1991.

165. **Knudsen, J. T., Tollsten, L., and Bergström, L. G.,** Floral scents — a checklist of volatile compounds isolated by head–space techniques, *Phytochemistry*, 33, 253, 1993.

166. **Koltermann, R.,** Rassen- bzw. artspezifische Duftbewertung bei der Honigbiene und ökologische Adaptation, *J. Comp. Physiol.*, 85, 327, 1973.

167. **Kriston, I.,** Die Bewertung von Duft- und Farbsignalen als Orientierungshilfen an der Futterquelle durch *Apis mellifera* L., *J. Comp. Physiol.*, 84, 77, 1973.

168. **Kugler, H.,** Blütenökologische Untersuchungen mit Hummeln. IV. Der Duft als chemischer Nahfaktor bei duftenden und "duftlosen" Blüten, *Planta*, 16, 543, 1932.

169. **Kugler, H.,** Hummeln als Blütenbesucher, *Ergb. Biol.*, 19, 143, 1943.

170. **Kugler, H.,** Der Blütenbesuch der Schlammfliege (*Eristalomyia tenax*), *Z. vergl. Physiol.*, 32, 328, 1950.

171. **Kugler, H.,** Über die optische Wirkung von Fliegenblumen auf Fliegen, *Deutsch. Bot. Gesell. Ber.*, 69, 387, 1956.

172. **Kullenberg, B.,** Some observations on scents among bees and wasps (Hymenoptera), *Entomol. Ts. Arg.*, 74, 1, 1953.

173. **Kullenberg, B.,** Studies in *Ophrys* pollination, *Zool. Bidr. Upps.*, 34, 1, 1961.

174. **Kullenberg, B.,** Field experiments with chemical sexual attractants on aculeate Hymenoptera males, II. *Zoon*, Suppl. 1, 31, 1973.

175. **Kullenberg, B. and Bergström, G.,** Hymenoptera aculeata males as pollinators of *Ophrys* orchids, *Zool. Scripta*, 5, 13, 1976.

176. **Lamont, B. B. and Collins, B. G.,** Flower colour change in *Banksia ilicifolia*: a signal for pollinators, *Aust. J. Ecol.*, 13, 129, 1988.
177. **Lance, D. R., Scholtz, W., Stewart, J. W., and Fergen, J. K.,** Non-pheromonal attractants for Mexican corn rootworm beetles, *Diabrotica virgifera zeae* (Coleoptera: Chrysomelidae), *J. Kan. Entomol. Soc.*, 65, 10, 1992.
178. **Leius, K.,** Attractiveness of different foods and flowers to the adults of some hymenopterous parasites, *Can. Entomol.*, 92, 369, 1960.
179. **Lepage, M. and Boch, R.,** Pollen lipids attractive to honeybees, *Lipids*, 3, 530, 1968.
180. **Levin, M. D. and Bohart, G. E.,** Selection of pollens by honey bees, *Am. Bee J.*, 95, 392, 1955.
181. **Lewis, J. A., Moore, C. J., Fletcher, M. T., Drew, R. A., and Kitching, W.,** Volatile compounds from the flowers of *Spathiphyllum cannaefolium*, *Phytochemistry*, 27, 2755, 1988.
182. **Lewis, P. A., Lampman, R. L., and Metcalf, R. L.,** Kairomonal attractants for *Acalymma vittatum* (Coleoptera: Chrysomelidae), *Environ. Entomol.*, 19, 8, 1990.
183. **Lewis, W. J. and Tumlinson, J. H.,** Host detection by chemically mediated associative learning in a parasitic wasp, *Nature*, 331, 257, 1988.
184. **Lex, T.,** Duftmale an Blüten, *Z. Vergl. Physiol.*, 36, 212, 1954.
185. **Light, D. M., Kamm, J. A., and Buttery, R. G.,** Electroantennogram response of alfalfa seed chalcid, *Bruchophagus roddi* (Hymenoptera: Eurytomidae) to host- and nonhost-plant volatiles, *J. Chem. Ecol.*, 18, 333, 1992.
186. **Lin, S. Y. H., Trumble, J. T., and Kumamoto, J.,** Activity of volatile compounds in glandular trichomes of *Lycopersicon* species against two insect herbivores, *J. Chem. Ecol.*, 13, 837, 1987.
187. **Lindauer, M.,** Lernen und Gedächtnis der Honigbiene, *Bienenwelt*, 14, 2, 1971.
188. **Lindauer, M.,** Fundamentals of information storage in honeybees, in *Fundamentals of Memory Formation: Neuronal Plasticity and Brain Function*, Rahmann, Ed., Gustav Fischer Verlag, Stuttgart, 1989, 299.
189. **Lindeman, A., Jounela-Eriksson, P., and Lounasmaa, M.,** The aroma composition of the flower of meadowsweet (*Filipendula ulmaria* (L.) Maxim.), *Lebensm. Wiss. Technol.*, 15, 286, 1982.
190. **Linsley, E. G., MacSwain, J. W., and Raven, P. H.,** Comparative behavior of bees and Onagraceae I *Oenothera* bees of the Colorado desert, *Univ. Calif. Publ. Entomol.*, 33, 1, 1963.
191. **Linsley, E. G., MacSwain, J. W., and Raven, P. H.,** Comparative behavior of bees and Onagraceae II *Oenothera* bees of the Great Basin, *Univ. Calif. Publ. Entomol.*, 33, 25, 1963.
192. **Linsley, E. G., MacSwain, J. W., and Raven, P. H.,** Comparative behavior of bees and Onagraceae III *Oenothera* bees of the Mojave desert, California, *Univ. Calif. Publ. Entomol.*, 33, 59, 1964.
193. **Linsley, E. G., MacSwain, J. W., Raven, P. H., and Thorp, R. W.,** Comparative behavior of bees and Onagraceae V *Camissonia* and *Oenothera* bees of cismontane California and Baja, California, *Univ. Calif. Publ. Entomol.*, 71, 1, 1973.
194. **Lokar, L. C., Moneghini, M., and Mellerio, G.,** Taxonomical studies on *Seseli elatum* L. and allied species. 2. Essential oil variation among three populations, *Webbia*, 40, 279, 1986.
195. **Loper, G. M.,** Factors affecting the quantity of alfalfa flower aroma and a subsequent influence on honey bee selection, *Agron. Abstr.*, 15, 1972.
196. **Loper, G. M. and Berdel, R. L.,** Seasonal emanation of ocimene from alfalfa flowers with three irrigation treatments, *Crop Sci.*, 18, 447, 1978.
197. **Loughrin, J. H., Hamilton-Kemp, T. R., Andersen, R. A., and Hildebrand, D. F.,** Headspace compounds from flowers of *Nicotiana tabacum* and related species, *J. Agric. Food Chem.*, 38, 455, 1990.
198. **Loughrin, J. H., Hamilton-Kemp, T. R., Andersen, R. A., and Hildebrand, D. F.,** Volatiles from flowers of *Nicotiana sylvestris*, *N. otophora* and *Malusxdomestica*: headspace components and day/night changes in their relative concentrations, *Phytochemistry*, 29, 2473, 1990.
199. **Loughrin, J. H., Hamilton-Kemp, T. R., Andersen, R. A., and Hildebrand, D. F.,** Circadian rhythm of volatile emission from flowers of *Nicotiana sylvestris* and *N. suaveolens*, *Physiol. Plant.*, 83, 492, 1991.
200. **Louveaux, J.,** Recherches sur la récolte du pollen par les abeilles (*Apis mellifica* L.), *Ann. Abeille*, 2, 13, 1959.
201. **MacSwain, J. W., Raven, P. H., and Thorp, R. W.,** Comparative behavior of bees and Onagraceae IV *Clarkia* bees of the western United States, *Univ. Calif. Publ. Entomol.*, 70, 1, 1973.
202. **Mamood, A. N. and Schmidt, J. O.,** Pollination and seed set in pearl millet by caged honey bees (Hymenoptera: Apidae), *Bee Sci.*, 1, 151, 1991.
203. **Manning, A.,** Some evolutionary aspects of the flower constancy of bees, *Proc. R. Phys. Soc. Edinb.*, 25, 67, 1957.
204. **Marden, J. H.,** Remote perception of floral nectar by bumblebees, *Oecologia*, 64, 232, 1984.
205. **Marr, K. L. and Tang, C. S.,** Volatile insecticidal compounds and chemical variability of Hawaiian *Zanthoxylum* (Rutaceae) species, *Biochem. Syst. Ecol.*, 20, 209, 1992.
206. **Martin, W. R., Nordlund, D. A., and Nettles, W. C.,** Response of parasitoid *Eucelatoria bryani* to selected plant material in an olfactometer, *J. Chem. Ecol.*, 16, 499, 1990.

207. **Maruzella, J. C.,** The germicidal properties of perfume oils and perfumery chemicals, *Am. Perf. Cosmet.,* 77, 167, 1962.
208. **Masson, C. and Arnold, G.,** Ontogeny, maturation and plasticity of the olfactory system in the workerbee, *J. Insect Physiol.,* 30, 7, 1984.
209. **Matile, P. and Altenburger, R.,** Rhythms of fragrance emission in flowers, *Planta,* 174, 242, 1988.
210. **McAuslane, H. J., Ellis, C. R., and Teal, P. E. A.,** Chemical attractants for monitoring for adult northern and western corn rootworms (Coleoptera: Chrysomelidae) in Ontario, *Proc. Entomol. Soc. Ontario,* 117, 49, 1986.
211. **McNeil, J. N. and Delisle, J.,** Host plant pollen influences calling behavior and ovarian development of the sunflower moth, *Homoeosoma electellum, Oecologia,* 80, 201, 1989.
212. **McNeil, J. N. and Delisle, J.,** Are host plants important in pheromone-mediated mating systems of Lepidoptera?, *Experientia,* 45, 236, 1989.
213. **Meeuse, B. J. D.,** The physiology of some sapromyophilous flowers, in *The Pollination of Flowers by Insects,* Richards, A. J., Ed., Linn. Soc. Acad. Press, London, 1978, 97.
214. **Meeuse, B. J. D. and Raskin, I.,** Sexual reproduction in the arum lily family, with emphasis on thermogenicity, *Sex Plant Reprod.,* 1, 3, 1988.
215. **Meeuse, B. J. D., Schneider, E. L., Hess, C. M., Kirkwood, K., and Patt, J. M.,** Activation and possible role of the "food-bodies" of *Sauromatum* (Araceae), *Acta Bot. Neerl.,* 33, 483, 1984.
216. **Menzel, R.,** Learning in honey bees in an ecological and behavioral context, in *Experimental Behavioral Ecology and Sociobiology,* Hölldobler, B. and Lindauer, M., Eds., Gustav Fischer Verlag, Stuttgart, 1985, 55.
217. **Metcalf, R. L.,** Plant volatiles as insect attractants, *CRC Crit. Rev. Plant Sci.,* 45, 251, 1987.
218. **Metcalf, R. L. and Lampman, R. L.,** The chemical ecology of Diabroticites and Cucurbitaceae, *Experientia,* 45, 240, 1989.
219. **Metcalf, R. L., Mitchell, W. C., Fukuto, T. R., and Metcalf, E. R.,** Attraction of the oriental fruit fly, *Dacus dorsalis,* to methyl eugenol and related olfactory stimulants, *Proc. Natl. Acad. Sci. U.S.A.,* 72, 2501, 1975.
220. **Minckley, R. L., Buchmann, S. L., and Wcislo, W. T.,** Bioassay evidence for a sex attractant pheromone in the large carpenter bees, *Xylocopa varipuncta* (Anthophoridae: Hymenoptera), *J. Zool. Lond.,* 224, 285, 1991.
221. **Morgan, A. C. and Lyon, S. C.,** Notes on amyl salicylate as an attractant to the tobacco hornworm moth, *J. Econ. Entomol.,* 21, 189, 1928.
222. **Morris, J. A., Khettry, A., and Seitz, E. W.,** Antimicrobial activity of aroma chemicals and essential oils, *J. Am. Oil Chem. Soc.,* 56, 595, 1979.
223. **Mullin, C. A., Alfatafta, A. A., Harman, J. L., Serino, A. A., and Everett, S. L.,** Corn rootworm feeding on sunflower and other Compositae: influence of floral terpenoid and phenolic factors, in *Naturally Occurring Pest Bioregulators,* Hedin, P. A., Ed., Am. Chem. Soc. Symp. Ser. 449, 1991, 278.
224. **Murlis, J., Elkinton, J. S., and Cardé, R. T.,** Odor plumes and how insects use them, *Annu. Rev. Entomol.,* 37, 505, 1992.
225. **Myers, J. H. and Walter, M.,** Olfaction in the Florida Queen Butterfly: honey odour receptors, *J. Insect Physiol.,* 16, 573, 1970.
226. **Naumann, C. M., Ockenfels, P., Schmitz, J., Schmidt, F., and Francke, W.,** Reactions of *Zygaena* moths to volatile compounds of *Knautia arvensis* (Lepidoptera: Zygaenidae), *Entomol. Gener.,* 15, 255, 1991.
227. **Neff, J. L. and Simpson, B. B.,** The roles of phenology and reward structure in the pollination biology of wild sunflower (*Helianthus annuus* L., Asteraceae), *Isr. J. Bot.,* 39, 197, 1990.
228. **Nettles, W. C.,** *Eucelatoria* sp. females: factors influencing response to cotton and okra plants, *Environ. Entomol.,* 8, 619, 1979.
229. **Nilsson, L. A.,** Pollination ecology and adaptation in *Platanthera chlorantha* (Orchidaceae), *Bot. Notiser,* 131, 35, 1978.
230. **Nilsson, L. A.,** Processes of isolation and introgressive interplay between *Platanthera bifolia* (L.)Rich. and *P. chlorantha* (Custer)Reichb. (Orchidaceae), *Bot. J. Linn. Soc.,* 87, 325, 1983.
231. **Nilsson, L. A.,** Orchid pollination biology, *TREE,* 8, 255, 1992.
231a. **Nilsson, L. A., Rabakonandrianina, E., Pettersson, B., and Ranaivo, J.,** "Ixoroid" secondary pollen presentation and pollination by small moths in the Malagasy treelet *Ixora platythyrsa* (Rubiaceae), *Plant Syst. Evol.,* 170, 161, 1990.
232. **Patrock, R. J. and Schuster, D. J.,** Feeding, oviposition and development of the pepper weevil (*Anthonomus eugenii* Cano) on selected species of Solanaceae, *Trop. Pest Manage.,* 38, 65, 1992.
233. **Patt, J. M., Rhoades, D. F., and Corkill, J. A.,** Analysis of the floral fragrance of *Platanthera stricta, Phytochemistry,* 27, 91, 1988.
234. **Paulus, H. F. and Gack, C.,** Pollinators as prepollinating isolation factors: evolution and speciation in *Ophrys* (Orchidaceae), *Isr. J. Bot.,* 39, 43, 1990.
235. **Pellmyr, O.,** The pollination ecology of *Actaea spicata* (Ranunculaceae), *Nord. J. Bot.,* 4, 443, 1984.
236. **Pellmyr, O.,** Three pollination morphs in *Cimicifuga simplex*; incipient speciation due to inferiority in competition, *Oecologia,* 68, 304, 1986.

237. **Pellmyr, O.,** Evolution of insect pollination and angiosperm diversification, *TREE*, 7, 46, 1992.
238. **Pellmyr, O., Bergström, G., and Groth, I.,** Floral fragrances in *Actaea*, using differential chromatograms to discern between floral and vegetative volatiles, *Phytochemistry*, 26, 1603, 1987.
239. **Pellmyr, O. and Patt, J. M.,** Function of olfactory and visual stimuli in pollination of *Lysichiton americanum* (Araceae) by a staphylinid beetle, *Madroño*, 33, 47, 1986.
240. **Pellmyr, O., Tang, W., Groth, I., Bergström, G., and Thien, L. B.,** Cycad cone and angiosperm floral volatiles: inferences for the evolution of insect pollination, *Biochem. Syst. Ecol.*, 19, 623, 1991.
241. **Pellmyr, O. and Thien, L. B.,** Insect reproduction and floral fragrances: keys to the evolution of the angiosperms, *Taxon*, 35, 76, 1986.
242. **Pellmyr, O., Thien, L. B., Bergström, G., and Groth, I.,** Pollination of New Caledonian Winteraceae: opportunistic shifts or parallel radiation with their pollinators?, *Plant Syst. Evol.*, 173, 143, 1990.
243. **Perkins, G., Estes, J. R., and Thorp, R. W.,** Pollination of *Cnidoscolus texanus* (Euphorbiaceae) in south-central Oklahoma, *Southwestern Nat.*, 20, 391, 1975.
244. **Pettersson, M. W.,** Flower herbivory and seed predation in *Silene vulgaris* (Caryophyllaceae): effects of pollination and phenology, *Holarctic Ecol.*, 14, 45, 1991.
245. **Pettersson, M. W.,** Taking a chance on moths: oviposition by *Delia flavifrons* (Diptera: Anthomyiidae) on the flowers of bladder campion, *Silene vulgaris* (Caryophyllaceae), *Ecol. Entomol.*, 17, 57, 1992.
246. **Pham, M.-H., Fonta, C., and Masson, C.,** L'apprentissage olfactif chez l'abeille et le bourdon: une étude comparée par conditionnement associatif, *C. R. Acad. Sc. Paris Ser. III*, 296, 501, 1983.
247. **Pham-Delegue, M.-H., Masson, C., Etiévant, P., and Azar, M.,** Selective olfactory choices of the honeybee among sunflower aromas: a study by combined olfactory conditioning and chemical analysis, *J. Chem. Ecol.*, 12, 781, 1986.
248. **Pham-Delegue, M.-H., Etiévant, P., Guichard, E., and Masson, C.,** Sunflower volatiles involved in honeybee discrimination among genotypes and flowering stages, *J. Chem. Ecol.*, 15, 329, 1989.
249. **Pliske, T. E.,** Attraction of Lepidoptera to plants containing pyrrolizidine alkaloids, *Environ. Entomol.*, 4, 455, 1975.
250. **Pliske, T. E.,** Pollination of pyrrolizidine alkaloid-containing plants by male Lepidoptera, *Environ. Entomol.*, 4, 474, 1975.
251. **Podoler, H., Galon, I., and Gazit, S.,** The role of nitidulid beetles in natural pollination of annona in Israel, *Acta Oecol. Oecol. Appl.*, 5, 369, 1984.
252. **Porsch, O.,** Geschlechtgebundener Blütenduft, *Österr. Bot. Z.*, 101, 359, 1954.
253. **Porsch, O.,** Windpollen und Blumeninsekt, *Österr. Bot. Z.*, 103, 1, 1956.
254. **Powell, J. A.,** Interrelationships of Yuccas and Yucca moths, *TREE*, 7, 10, 1992.
255. **Priesner, E.,** Reaktionen von Riechrezeptoren männlicher Solitärbienen (Hymenoptera, Apoidea) auf Inhaltsstoffe von *Ophrys*-Blüten, *Zoon*, Suppl. 1, 43, 1973.
256. **Proctor, M. C. F. and Yeo, P. F.,** *The Pollination of Flowers*, Collins, London, 1972.
257. **Ramaswamy, S. B.,** Host finding by moths: sensory modalities and behaviours, *J. Insect Physiol.*, 34, 235, 1988.
258. **Raskin, I., Ehmann, A., Melander, W. R., and Meeuse, B. J. D.,** Salicylic acid: a natural inducer of heat production in *Arum* lilies, *Science*, 237, 1601, 1987.
259. **Rau, P.,** The Yucca plant, *Yucca filamentosa*, and Yucca moth, *Tegeticula (Pronuba) yuccasella* Riley: an ecologico-behavior study, *Ann. Mo. Bot. Gard.*, 32, 373, 1945.
260. **Reddi, C. S. and Bai, G. M.,** Butterflies and pollination biology, *Proc. Indian Acad. Sci. (Anim. Sci.)*, 93, 391, 1984.
261. **Rieth, J. P., Wilson, W. T., and Levin, M. D.,** Repelling honeybees from insecticide-treated flowers with 2-heptanone, *J. Apic. Res.*, 25, 78, 1986.
262. **Robacker, D. C., Flottum, P. K., Sammataro, D., and Erickson, E. H.,** Why soybeans attract honey bees, *Am. Bee J.*, 122, 481, 1982.
263. **Robacker, D. C., Meeuse, B. J. D., and Erickson, E. H.,** Floral aroma, *BioScience*, 38, 390, 1988.
264. **Rodriguez, E. and Levin, D. A.,** Biochemical parallelisms of repellents and attractants in higher plants and arthropods, in *Biochemical Interaction between Plants and Insects*, Wallace, J. W. and Mansell, R. L., Eds., Rec. Adv. Phytochem., 10, 214, 1976.
265. **Rogers, C. E., Gershenzon, J., Ohno, N., Mabry, T. J., Stipanovic, R. D., and Kreitner, G. L.,** Terpenes of wild sunflowers (*Helianthus*): an effective mechanism against seed predation by larvae of the sunflower moth, *Homoeosoma electellum* (Lepidoptera: Pyralidae), *Environ. Entomol.*, 16, 586, 1987.
266. **Roseland, C. R., Bates, M. B., Carlson, R. B., and Oseto, C. Y.,** Discrimination of sunflower volatiles by the red sunflower seed weevil, *Entomol. Exp. Appl.*, 62, 99, 1992.
267. **Rossiter, M., Gershenzon, J., and Mabry, T. J.,** Behavioral and growth responses of specialist herbivore, *Homoeosoma electellum*, to major terpenoid of its host, *Helianthus* spp., *J. Chem. Ecol.*, 12, 1505, 1986.
268. **Salama, H. S., Rizk, A. F., and Sharaby, A.,** Chemical stimuli in flowers and leaves of cotton that affect behaviour in the cotton moth *Spodoptera littoralis* (Lepidoptera: Noctuidae), *Entomol. Gener.*, 10, 27, 1984.

269. **Sasaki, M., Ono, M., Asada, S., and Yoshida, T.,** Oriental orchid (*Cymbidium pumilum*) attracts drones of the Japanese honeybee (*Apis cerana japonica*) as pollinators, *Experientia*, 47, 1229, 1991.

270. **Schemske, D. W. and Lande, R.,** Fragrance collection and territorial display by male orchid bees, *Anim. Behav.*, 32, 935, 1984.

271. **Schmidt, J. O.,** Pollen foraging preferences of honey bees, *Southwestern Entomol.*, 7, 255, 1982.

272. **Schneider, E. L. and Moore, L. A.,** Morphological studies of the Nymphaeaceae. VII. The floral biology of *Nuphar lutea* subsp. *macrophylla*, *Brittonia*, 29, 88, 1977.

273. **Schremmer, F.,** Sinnesphysiologie und Blumenbesuch des Falters von *Plusia gamma* L., *Zool. Jahrb. Syst.*, 74, 375, 1941.

274. **Schwartz, P. H. and Hamilton, D. W.,** Attractants for the Japanese beetle, *J. Econ. Entomol.*, 62, 516, 1969.

275. **Shahjahan, M.,** *Erigeron* flowers as a food and attractive odor source for *Peristenus pseudopallipes*, a braconid parasitoid of the tarnished plant bug, *Environ. Entomol.*, 3, 69, 1974.

276. **Silberbauer-Gottsberger, I.,** Pollination and evolution in Palms, *Phyton*, 30, 213, 1990.

277. **Simpson, B. B. and Neff, J. L.,** Floral rewards: alternatives to pollen and nectar, *Ann. Mo. Bot. Gard.*, 68, 301, 1981.

278. **Simpson, B. B. and Neff, J. L.,** Evolution and diversity of floral rewards, in *Handbook of Pollination Biology*, Jones, C. E. and Little, R. J., Eds., Sci. Acad. Ed., New York, 1983, 142.

279. **Smith, B. H.,** The olfactory memory of the honeybee *Apis mellifera*, *J. Exp. Biol.*, 367, 1991.

280. **Staswick, P. E.,** Jasmonate, genes, and fragrant signals, *Plant Physiol.*, 99, 804, 1992.

281. **Steiner, L. F.,** Methyl eugenol as an attractant for oriental fruit fly, *J. Econ. Entomol.*, 45, 241, 1952.

282. **Stowe, M. K.,** Chemical mimicry, in *Chemical Mediation of Coevolution*, Spencer, K. C., Ed., Academic Press, San Diego, 1988, 513.

283. **Strong, D. R., Lawton, H., and Southwood, R.,** *Insects on Plants: Community Patterns and Mechanisms*, Harvard University Press, Cambridge, 1984.

284. **Tan, S.-T., Holland, P. T., Wilkins, A. L., and Molan, P. C.,** Extractives from New Zealand honeys. 1. White clover, Manuka, and Kanuka unifloral honeys, *J. Agric. Food Chem.*, 36, 453, 1988.

285. **Tang, W.,** Heat production in cycad cones, *Bot. Gaz.*, 148, 165, 1987.

286. **Tengö, J.,** Odour-released behaviour in *Andrena* male bees (Apoidea, Hymenoptera), *Zoon*, 7, 15, 1979.

287. **Thiboud, M.,** Empirical classification of odours, in *Perfumes: Art, Science and Technology*, Müller, P. M. and Lamparsky, D., Eds., Elsevier Applied Science, London, 1991, 253.

288. **Thien, L. B.,** Patterns of pollination in the primitive angiosperms, *Biotropica*, 12, 1, 1980.

289. **Thien, L. B., Bernhardt, P., Gibbs, G. W., Pellmyr, O., Bergström, G., Groth, I., and McPherson, G.,** The pollination of *Zygogynum* (Winteraceae) by a moth, *Sabatinca* (Micropterigidae): an ancient association?, *Science*, 227, 540, 1985.

290. **Thien, L. B., Heimermann, W. H., and Holman, R. T.,** Floral odors and quantitative taxonomy of *Magnolia* and *Liriodendron*, *Taxon*, 24, 557, 1975.

291. **Thomson, J. D.,** Effects of variation in inflorescence size and floral rewards on the visitation rates of traplining pollinators of *Aralia hispida*, *Evol. Ecol.*, 2, 65, 1988.

292. **Thorsteinson, A. J. and Brust, R. A.,** The influence of flower scents on aggregations of caged adult *Aedes aegypti*, *Mosquito News*, 22, 349, 1962.

293. **Tingle, F. C. and Mitchell, E. R.,** Attraction of *Heliothis virescens* (F.) (Lepidoptera: Noctuidae) to volatiles from extracts of cotton flowers, *J. Chem. Ecol.*, 18, 907, 1992.

294. **Tollsten, L. and Bergström, J.,** Variation and post-pollination changes in floral odours released by *Platanthera bifolia* (Orchidaceae), *Nord. J. Bot.*, 9, 359, 1989.

295. **Vallet, A., Cassier, P., and Lensky, Y.,** Ontogeny of the fine structure of the mandibular glands of the honeybee (*Apis mellifera*) workers and the pheromonal activity of 2-heptanone, *J. Insect Physiol.*, 37, 789, 1991.

296. **Vareschi, E.,** Duftunterscheidung bei der Honigbiene — Einzelzell-Ableitungen und Verhaltensreaktionen, *Z. Vergl. Physiol.*, 75, 143, 1971.

297. **Vargo, A. M. and Foster, W. A.,** Responsiveness of female *Aedes aegypti* (Diptera: Culicidae) to flower extracts, *J. Med. Entomol.*, 19, 710, 1982.

298. **Vinson, S. B., Williams, H. J., Frankie, G. W., and Coville, R. E.,** Comparative morphology and chemical constituents of male mandibular glands of several *Centris* species (Hymenoptera: Anthophoridae) in Costa Rica, *Comp. Biochem. Physiol.*, 77A, 685, 1984.

299. **Vogel, S.,** Farbwechsel und Zeichnungsmuster bei Blüten, *Österr. Botan. Z.*, 97, 44, 1950.

300. **Vogel, S.,** Duftdrüsen im Dienste der Bestäubung, *Abh. Math.-Naturwiss. Klasse Akad. Wiss. Lit. Mainz*, 10, 599, 1962.

301. **Vogel, S.,** *The Role of Scent Glands in Pollination: on the Structure and Function of Osmophores* (translated from above), Smithsonian Institute Library, Washington, D.C., 1990.

302. **Waller, G. D.,** The effect of citral and geraniol conditioning on the searching activity of honeybee recruits, *J. Apic. Res.*, 12, 53, 1973.

303. **Waller, G. D., Loper, G. M., and Berdel, R. L.,** Olfactory discrimination by honeybees of terpenes identified from volatiles of alfalfa flowers, *J. Apic. Res.*, 13, 191, 1974.

304. **Weiss, M. R.,** Floral colour changes as cues for pollinators, *Nature*, 354, 227, 1991.

305. **Wenner, A. M., Wells, P. H., and Johnson, D. L.,** Honey bee recruitment to food sources: olfaction or language?, *Science*, 164, 84, 1969.

306. **Wensler, R. J. D.,** The effect of odors on the behavior of adult *Aedes aegypti* and some factors limiting responsiveness, *Can. J. Zool.*, 50, 415, 1972.

307. **Wetherwax, P. B.,** Why do honeybees reject certain flowers?, *Oecologia*, 69, 567, 1986.

308. **Wiesenborn, W. D. and Baker, T. C.,** Upwind flight to cotton flowers by *Pectinophora gossypiella* (Lepidoptera: Gelechiidae), *Environ. Entomol.*, 19, 490, 1990.

309. **Williams, N. H.,** The biology of orchids and euglossine bees, in *Orchid Biology, Reviews and Perspectives, II.*, Arditti, J., Ed., Cornell University Press, Ithaca, 1982, 119.

310. **Williams, N. H.,** Floral fragrances as cues in animal behavior, in *Handbook of Experimental Pollination Biology*, Jones, E. C. and Little, R. J., Eds., Sci. Acad. Ed., New York, 1983, 50.

311. **Williams, N. H. and Dodson, C. H.,** Selective attraction of male euglossine bees to orchid floral fragrances and its importance in long distance pollen flow, *Evolution*, 26, 84, 1972.

312. **Williams, N. H. and Whitten, W. M.,** Orchid floral fragrances and male euglossine bees: methods and advances in the last sesquidecade, *Biol. Bull.*, 164, 355, 1983.

313. **Williams, N. H., Whitten, W. M., and Pedrosa, L. F.,** Crystalline production of fragrance in *Gongora quinquenervis*, *Am. Orchid Soc. Bull.*, 54, 598, 1985.

314. **Wyatt, R.,** Pollinator–plant interactions and the evolution of breeding systems, in *Pollination Biology*, Real, L., Ed., Academic Press, Orlando, FL, 1983, 51.

315. **Young, A. M.,** Pollination biology of *Theobroma* and *Herrania* (Sterculiaceae). IV. Major volatile constituents of steam-distilled floral oils as field attractants to cacao-associated midges (Diptera: Cecidomyiidae and Ceratopogonidae) in Costa Rica, *Turrialba*, 39, 454, 1989.

316. **Young, A. M.,** Comparative attractiveness of floral fragrance oils of "RIM" and "Catongo" cultivars of cacao (*Theobroma cacao* L.) to Diptera in a Costa Rican cacao plantation, *Turrialba*, 39, 137, 1989.

317. **Young, A. M., Erickson, B. J., and Erickson, E. H.,** Pollination biology of *Theobroma* and *Herrania* (Sterculiaceae). III. Steam-distilled floral oils of *Theobroma* species as attractants to flying insects in a Costa Rican cacao plantation, *Insect Sci. Appl.*, 10, 93, 1989.

318. **Young, H.,** Beetle pollination of *Dieffenbachia longispatha* (Araceae), *Am. J. Bot.*, 73, 931, 1986.

319. **Zaika, L. L.,** Spices and herbs: their antimicrobial activity and its determination, *J. Food Saf.*, 9, 97, 1988.

320. **Zhu, Y., Keaster, A. J., and Gerhardt, K. O.,** Field observations on attractiveness of selected blooming plants to noctuid moths and electroantennogram responses of black cutworm (Lepidoptera: Noctuidae) moths to flower volatiles, *Environ. Entomol.*, 22, 167, 1993.

321. **Zimmerman, M.,** Optimal foraging: random movement by pollen collecting bumblebees, *Oecologia*, 53, 394, 1982.

Chapter 4

ENDOPHYTIC FUNGI AS MEDIATORS OF PLANT–INSECT INTERACTIONS

Alison J. Popay and Daryl D. Rowan

TABLE OF CONTENTS

I. INTRODUCTION

Endophytic organisms are widespread in nature, occurring in both marine and terrestrial environments, where they exist in a range of associations, from parasitic to mutualistic, with their hosts.[26] The "true endophytes" are those that grow entirely within the plant, relying on their host not only for nutrition but also for transmission, either vegetatively or within the seed. The fungal endophytes considered in this review grow in mutualistic associations almost entirely within the tissues of living plants. This definition excludes the mycorrhiza and smuts, which can live both internally and externally, but includes fungi that may produce external mycelia or sexual spores that are disseminated to produce new infections.

Taxonomically, the fungal endophytes of plants belong almost exclusively to the Ascomycetes and are harbored by two major plant groups, the woody perennials and the grasses and sedges. Among the woody perennials, conifers, in particular, are frequent hosts to a number of different endophytes,[26,28] but many other tree species also contain endophytes. Carroll sets out criteria for identifying mutualistic associations between plants and endophytes and proposes a list of ten species of trees infected with a variety of endophytes that are likely candidates for endophytic mutualism.[26] The second group of plants to host endophytic fungi are the grasses and sedges. These endophytes all belong to the tribe Balansiae within the Clavicipitacea. Genera within this tribe include the *Balansia, Balansiopsis, Atkinsonella,* and *Myriogenospora* which infect a wide range of tropical grasses, many of which are weeds or grasses of minor agricultural importance.[118] Other genera infect a number of economically important agricultural and turf grasses. *Epichloe,* which usually produces characteristic choke symptoms around the inflorescence of its host, infects species in the genera *Agrostis, Dactylis, Festuca, Holcus, Hordeum,* and *Lolium.* The true endophytes include the genera *Acremonium, Sphacelia, Gliocladium*-like, and *Phialophora*-like, whose host ranges include the grasses *Festuca, Lolium, Bromus, Poa,* and *Stipa.*[118] A list of reported host genera for clavicipitaceous endophytes of grasses is given in Clay.[39]

A link between endophytes infecting grasses and toxicity to mammalian herbivores has long been recognized. In the United States, the agricultural importance of such links emerged in the 1970s, when an anamorphic state of the endophyte *Epichloe typhina* (now called *Acremonium coenophialum* Morgan-Jones and Gams) infecting tall fescue (*Festuca arundinacea*) was identified as the cause of toxic symptoms in cattle.[6] Similarly in New Zealand, the occurrence of ryegrass staggers syndrome in grazing animals was found to be associated with the widespread infection of commercial cultivars of perennial ryegrass (*Lolium perenne*) with the endophyte *Acremonium lolii* Latch, Christensen, and Samuels.[53] The recognition in 1982 that infection with *A. lolii* also protected perennial ryegrass from attack by Argentine stem weevil, *Listronotus bonariensis* (Coleoptera: Curculionidae), added another dimension to the research.[106] Such discoveries have provided the stimulus for much of the recent surge of interest in grass endophytes.

A number of recent reviews consider the relationships between fungal grass endophytes, their hosts, and herbivory.[27,34,44,80,117,128] This review concentrates entirely on the current state of our knowledge of the insect herbivore–plant–endophyte relationships, highlighting the areas of which we know little and examining the interactions in the wider context of existing theories on plant–insect interactions.

II. SECONDARY METABOLITES PRODUCED IN ENDOPHYTE–PLANT ASSOCIATIONS

The response of insects to endophyte infection of plants is mediated by secondary metabolites, generally alkaloids of fungal origin, present in endophyte-infected tissues. Before discussing insect interactions with endophytes, it is appropriate to describe the known metabolites from grass endophyte systems for which biological activity against insects has been demonstrated. No information is currently available concerning the nature of secondary compounds involved in the interactions of tree endophytes with insects.

To date, four main classes of secondary metabolites, the indole diterpenes, peramine, and the loline and ergot alkaloids, have been identified as insect-active constituents of endophyte-infected grasses. These metabolites have been identified using bioassay-directed fractionation of plant extracts or by examination of endophyte-infected tissues or cultures for metabolites known from other sources. It is now some ten years since the discovery that endophytes in grasses could mediate insect resistance. Despite this, there is still little published information on the occurrence and concentrations of these metabolites in various grass–endophyte associations or detailed studies of their effects on insects.

A. INDOLE DITERPENE MYCOTOXINS

Indole diterpenes comprise a group of mycotoxins from diverse fungal species known principally for their neurotoxic and tremorgenic activities in mammals.[21] Members of this group share the common structural and biosynthetic features of an indole ring system bonded onto a cyclized 20-carbon diterpene unit. Further elaboration and adjustment of oxidation state then gives rise to the various indole diterpene mycotoxins.

1. Paxilline

Paxilline was originally isolated from *Penicillium paxilli* Bain[40] and is the simplest member of this class to exhibit tremorgenic activity. This compound is also produced by several members of the genus *Emericella*[92] and by a wide range of *Acremonium* endophytes.[94] Concentrations of approximately 3 µg/g have been reported from endophyte-infected ryegrass seed[130] and from 2 to 13 µg/g dry matter in ryegrass herbage.[63] Paxilline is the obvious biosynthetic precursor to the more complex lolitrem mycotoxins.[123] The related metabolites, paspaline, paspalicine, paspalinine, and paspalitrem A, are produced in the sclerotia of *Claviceps paspali*[41] and are implicated in the livestock disorder of paspallum staggers, but to date these paxilline derivatives have not been identified in endophyte-infected grasses.

2. Lolitrems and Lolitriol

The lolitrems are more complex indole diterpenes, formally derived from paxilline by the addition of further isoprene equivalents. The lolitrems are potent tremorgens causing sustained trembling when administered to sheep or mice.[57] Lolitrem B is the major lolitrem found in endophyte-infected seed and herbage of ryegrass.[59,60] Additional lolitrems, A and C to E, have been identified in endophyte-infected ryegrass seed and herbage.[60,87] Structurally and biosynthetically, the lolitrems are clearly related to the known fungal tremorgens, paxilline, aflatrem, penitrem A, and janthitrem E.[21] Lolitrem neurotoxins occur widely in endophyte-infected perennial ryegrass, in Italian ryegrass, *L. multiflorum*,[103] and occasionally in endophyte-infected *Festuca* sp.[30,120] However, lolitrems are not produced by all endophytic fungi in their natural hosts and do not occur in tall fescue infected with *A. coenophialum*[120] or in grasses infected with *Phialophora* or *Gliocladium*-like endophytes.[50]

Concentrations of lolitrem B in endophyte-infected perennial ryegrass from 0.2 to 10 µg/g dry weight have been reported.[61] Within the plant itself, lolitrems are localized in the basal leaf sheaths, where the concentrations of fungal mycelium are highest.[61,75] High concentrations of lolitrem B are also found in the older leaves and may exceed 25 µg/g in the outer leaf sheath.[61]

Lolitriol, a presumed biosynthetic precursor to the lolitrems, has recently been identified in endophyte-infected ryegrass, where it occurs in concentrations comparable to those of lolitrem B.[88] Lolitriol is toxic but not tremorgenic to mice; however, its activity against insects has not been reported.

B. ERGOT ALKALOIDS

The ergot alkaloids are generally divided into the ergolines, the lysergides, and the ergopeptine classes.[19] All have been reported from endophyte-infected grasses and sedges,[7,85] as have other unique structural types.[97] Attention has been focused on the better characterized ergopeptine alkaloids, principally ergovaline, for which mass spectroscopic and high performance liquid chromatography analytical methods are available.[110,111,132,133] Ergovaline concentrations in the 0.1 to 1.0 μg/g range have been reported for endophyte-infected tall fescue and ryegrass,[110,113] but higher concentrations under glasshouse conditions have also been reported.[113,120]

C. PERAMINE

Peramine was identified by bioassay-guided fractionation as the major feeding deterrent to Argentine stem weevil endophyte-infected perennial ryegrass.[112,114] The molecule consists of a pyrrolopyrazine heterocyclic ring system to which is attached a short guanidinyl-terminated side chain. Peramine occurs widely in endophyte-infected grasses of the genera *Festuca, Lolium, Poa, Bromus, Elymus, Sitanion,* and *Agrostis* and also occurs in ryegrasses and fescues infected with *E. typhina.* As with the lolitrems and ergovaline, not all *Acremonium* species or biotypes of *E. typhina* produce peramine in their hosts.[50,120] Biosynthetically, peramine appears to be derived from the diketopiperazine derivative of proline and arginine, but as yet, no related metabolites have been isolated from endophyte-infected grasses to support this. Peramine concentrations in endophyte-infected ryegrass plants typically range between 10 and 30 μg/g dry weight.[126] In contrast to the lolitrems and the endophyte itself, peramine is not localized within the ryegrass plant.[75]

D. LOLINE ALKALOIDS

The loline alkaloids occur in various endophyte-infected grasses of the genera *Lolium, Hordeum,* and *Festuca* but are not found in endophyte-infected perennial ryegrass (but see Huizing et al.[70]). *N*-acetyl and *N*-formyl lolines are the major loline isomers reported in American fescues, although *N*-acetyl norloline predominates in some European material (D.D. Rowan, unpublished results). Unlike the indole diterpenes, ergot alkaloids and peramine, lolines are produced in large quantities in infected plants; up to 8000 μg/g has been reported in *A. coenophialum*-infected tall fescue.[44] Lolines are the only endophyte-associated alkaloids which have not been reported to occur in fungal culture; they may possibly be synthesized by the grass in response to endophyte infection.

Most of the intermediates in the biosynthesis of the lolitrems, the ergopeptine alkaloids, peramine, and the loline alkaloids, remain unknown or are not readily available for testing with insects. In addition, there have been few serious attempts[134] to quantify the contribution of particular chemicals or chemical fractions to the effect on the insect. A number of other biologically active metabolites have been isolated from endophyte-infected tissues or cultures,[45,79,135] but in no case were effects on insects reported.

III. INSECT RESPONSES TO ENDOPHYTES

The effect of endophytes on invertebrate herbivory was first noted by Cubit,[43] who showed that presence of a fungal endophyte, *Turgidiosculum* sp., greatly reduced grazing of the marine green alga, *Enteromorpha vexata,* by marine mollusks. Subsequently in 1981, two reports indicated that endophytic fungi in trees also affected insects.[46,129] In the last ten years, a flurry of research has produced a list of more than 30 species of insects that are adversely affected by endophytes in grasses and sedges (Table 1). Several different orders of insects are involved with various modes of feeding, including stem borers and phloem, foliage, and root feeders.

TABLE 1
Responses of Insects Adversely Affected by *Acremonium* Endophyte in Tall Fescue and/or Perennial Ryegrass

Species	Host/Endophyte	Stage	Response	Refs.
Lepidoptera				
Spodoptera frugiperda	Tall fescue/*A. coenophialum*	Larvae	T and D	68
	Ryegrass/*A. lolii*	Larvae	T and D	67
	Paspalum dilatatum/			
	Myriogenospora atramentosa	Larvae	T and D	35
	Cyperus virens/Balansia cyperus	Larvae	ND	36
	C. pseudovegetus/B. cyperus	Larvae	ND	36
S. eridania	Ryegrass/*A. lolii*	Larvae	T	3
Mythimna convecta	Ryegrass/*A. lolii*	Larvae	T	128
Crambus roman	Ryegrass/*A. lolii*	Larvae	D	56
	Tall fescue/*A. coenophialum*	Larvae	ND	
Parapediasa teterella	Ryegrass/*A. lolii*	Larvae	ND	74
Agrotis segetum	*Dactylis glomerata/E. typhina*	Larvae	ND	46
A. infusa	Ryegrass/*A. lolii*	Larvae	ND	128
Graphania mutans	Ryegrass/*A. lolii*	Larvae	T and D	48
Coleoptera				
Listronotus bonariensis	Ryegrass/*A. lolii*	Adults	D	12
		Larvae	D and T	13
	Tall fescue/*A. coenophialum*	Adults	ND	49
	Bromus anomalus/A. starii	Adults	ND	50
Heteronychus arator	Ryegrass/*A. lolii*	Adults	D	18
Chaetocnema pulicaria	Tall fescue/*A. coenophialum*	Mixed	ND	18
Sphenophorus parvulus	Ryegrass/*A. lolii*			77
S. inaequalis	and tall fescue/	Adults	T	2, 71
S. minimus	*A. coenophialum*			
S. venatus				71
Popillia japonica	Tall fescue/*A. coenophialum*	Larvae	ND	93
Costelytra zealandica	Tall fescue/*A. coenophialum*	Larvae	ND	100
Tribolium castaneum	Tall fescue/*A. coenophialum*	Adults	ND	31
Hemiptera				
Balanococcus poae	Ryegrass/*A. lolii*	Larvae	ND	96
Blissus leucopterus	Tall fescue/ *A. coenophialum*	Mixed	T and D	115
hirtus	Ryegrass/*A. lolii*	Larvae	T and D	86
		Adults	T and D	
Oncopeltus fasciatus	Tall fescue/*A. coenophialum*	Nymphs	T	73
Draeculacephala antica				
Agallia constricta				
Endria inmica	Tall fescue/*A. coenophialum*	Mixed	ND	77, 91
Exitianus exitiosus		Age		91
Graminella nigrofons				91
Prosapia bicincta				91
Rhopalosiphum padi	Tall fescue/*A. coenophialum*	Mixed	T and D	83,51, 7
R. maidis	Ryegrass/*A. lolii*	Mixed	T and D	73
Schizaphis graminum	Tall fescue/*A. coenophialum*	Mixed	T and D	73
	Ryegrass/*A. lolii*			
Diuraphis noxia	Ryegrass/*A. lolii*	Nymphs	T	38
	Tall fescue/*A. coenophialum*	and adults	T and D	76

TABLE 1 (Continued)
Responses of Insects Adversely Affected by *Acremonium* Endophyte
in Tall Fescue and/or Perennial Ryegrass

Species	Host/Endophyte	Stage	Response	Refs.
	Diptera			
Drosophila melanogaster	Tall fescue/*A. coenophialum*	Adults	T	42
	Orthoptera			
Acheta domesticus	Ryegrass/*A. lolii*	Nymphs	T	1

Note: T = toxic, D = deterrent, ND = not determined.

A. THREE EXAMPLES SHOW DIVERSITY OF RESPONSE

A stem miner, Argentine stem weevil (*Listronotus bonariensis*), a foliage feeder, the fall armyworm (*Spodoptera frugiperda*), and several species of phloem-feeding aphids have been studied more fully than most of the other insects known to be sensitive to endophytes. Their responses to the different alkaloids produced by the endophyte–grass association give an indication of the complexity of the interactions, something we are far from fully understanding.

1. Argentine Stem Weevil (*Listronotus bonariensis*)

This weevil is a major pasture pest in New Zealand, where perennial ryegrass is a basic component of pastures. The resistance of endophyte-infected ryegrass to this weevil was first reported by Prestidge et al.[106] Shortly afterward, Mortimer and di Menna described a key grazing trial where, during regrowth of ryegrass plots at the end of a grazing period, there was a three- to four-fold greater dry matter production from the high endophyte plots compared with that from endophyte-free plots.[90] The poor regrowth and low yield in the low endophyte plots was found to be due to predation by the Argentine stem weevil. The prospect of controlling this pest by exploiting the natural plant resistance conferred by the endophyte has provided the impetus for the extensive research conducted into the interactions between endophytes and the Argentine stem weevil.

The Argentine stem weevil maintains a close association with its host plant throughout its life cycle. Adults feed on the foliage of pasture grasses and cereal crops and oviposit in the stem, or occasionally on leaves, by biting a hole and inserting eggs into the epidermal tissue. Once hatched, larvae begin mining the stem of their host, transferring to new tillers as the need arises.

In the field, the population dynamics of Argentine stem weevil are closely linked with the availability of *A. lolii*-free ryegrass tillers.[16] The effect is mediated primarily through the mobile adult stage, which exhibits a marked preference for feeding and ovipositing on endophyte-free ryegrass.[13,64] This is probably due at least in part to the presence of peramine,[112] which has been found to deter adult weevils from feeding on diet at a concentration of 0.1 µg/g.[113] The ergopeptine alkaloids may also contribute to adult deterrency, with ergovaline and ergotamine reducing feeding in choice tests at 0.1 and 1.0 µg/g, respectively.[48,99] Paxilline (unpublished results) has also been identified as an adult feeding deterrent at 1 µg/g concentrations. However, oviposition by gravid females is only reduced by high concentrations of peramine,[99] and it therefore seems likely that the reduction in egg laying recorded on endophyte-infected grass is a result of reduced adult feeding. Certainly, exposure to endophytes over a period of time has profound physiological effects in the adults, causing a regression in, or lack of, reproductive development coinciding with increased flight muscle development.[15] The effects are analagous to those which occur when weevils are maintained in crowded conditions and are thus presumably attributable to a lack of suitable food rather than directly to chemical effects. Antibiosis in the adult appears not to be a factor.

Endophyte-infected ryegrass is also resistant to larvae,[14,98] with effects again mediated through the deterrency of peramine at concentrations above 2 μg/g.[48,113] Peramine is not toxic to larvae at concentrations up to 25 μg/g, whereas both lolitrem B at a concentration of 5 μg/g[49,104,105] and paxilline at 10 μg/g[58] cause reductions in growth and development. Lolitrem B was shown to act via the insect midgut rather than by adsorption through the integument of the burrowing larvae.[105]

When infected with either *A. coenophialum* or *E. typhina*, in which peramine and the loline and ergopeptine alkaloids (but not the lolitrems) are present, tall fescue is also resistant to Argentine stem weevil.[50] Annual ryegrass seedlings infected with endophytes show resistance to Argentine stem weevil, even if the endophyte is nonviable,[122] but this resistance disappears as the seedlings grow and peramine levels drop below 1 μg/g dry weight.[50] Another *Acremonium* endophyte, *A. starii*, which produces high concentrations of peramine in its host grass, *Bromus anomalus*, has also been shown to reduce feeding and oviposition of the weevil.[18]

2. Fall Armyworm (*Spodoptera frugiperda*)

Larvae of the fall armyworm, which are generalist feeders on species of Poaceae, are sensitive to a range of endophytes in various hosts under laboratory conditions[35,36,67,68] (Table 1). Typically, growth, development, and survival of larvae feeding on the endophyte-infected plants are adversely affected. Early stage larvae show a clear preference for uninfected plants, but by the fourth instar, no preferences are apparent.[68] The chemical bases of the responses are not well understood. Both antibiotic and antifeedant effects have been demonstrated for various ergot alkaloids but only at high concentrations.[37] These authors attributed at least some of the insecticidal activity of endophyte-infected grasses to the presence of ergot alkaloids, despite the fact that the concentrations of between 77 to 100 μg/g required to affect armyworm larvae were up to two orders higher than those occurring naturally in tall fescue infected with *Acremonium coenophialum* or in smutgrass infected with *Balansia epichloe*. Paxilline (and paspaline) reduces the growth of neonate fall armyworm larvae at 25 μg/g, and three other tremorgenic mycotoxins of fungal origin, dihydroxyaflavinine, roseotoxin B, and penitrem A, which are structurally analogous to those produced in endophyte-infected plants, are also toxic to fall armyworm at microgram concentrations.[47] *N*-acetyl loline is toxic to larvae[108] but cannot be the compound responsible for the antibiosis observed on endophyte-infected ryegrass, where it does not occur. It is also unlikely that these pyrrolizidine alkaloids are responsible for deterrency to endophyte-infected tall fescue.[68,108] There is no information on the effect of peramine or lolitrem B on fall armyworm, and no reports of the feeding preferences of this insect in the field have been published.

3. Aphids

If readers are hoping for predictability in insect reactions to endophytes, an examination of the aphid response will both dispel such hopes and show that it is not possible to make generalizations regarding plant–insect interactions, even within closely related species. Aphid responses to endophytes are species-specific and may even be biotype-specific. The bird-cherry oat aphid, *Rhopalosiphum padi*, avoids tall fescue infected with *A. coenophialum* but is not affected by perennial ryegrass infected with *A. lolii* or chewings fescue infected with *E. typhina*.[81] However, biotypes of *R. padi* in Japan do appear to be sensitive to *A. lolii*-infected ryegrass (M. Christensen, personal communication), as is its close relative, *R. maidis*, although this species shows no apparent preference for endophyte-free tall fescue.[23,73] Two other aphid species, *Schizaphis graminum*[22,120] and *Diuraphis noxia*,[38,76,121] are affected by *Acremonium* endophytes in both tall fescue and ryegrass. *Diuraphis noxia*, however, is not affected by *E. typhina* in tall fescue.[76]

Both toxic and deterrent factors are apparent in aphid responses to endophyte infection of grasses. For instance, *R. padi* and *S. graminum* not only avoid endophyte-infected tall fescue but are unable to survive if confined to these plants.[73] The loline alkaloids, *N*-acetyl loline and *N*-formyl loline, have been determined as the cause of mortality of *R. padi*,[51,120] while both the lolines and peramine are mediators of the effects of endophyte infection on *S. graminum*.[22,120] *Diuraphis noxia* exhibits both toxicity and deterrency on *Acremonium*-infected tall fescue,[121] but only antibiosis factors are apparent in its response to *Acremonium*-infected perennial ryegrass.[39,76] The chemical basis of the response of this aphid species is not known. Three species of aphid, *Metopolophium dirhodum*, *Sitobion fragariae*, and *Macrosiphon avenae*, are not affected by endophytes.[73,83]

B. ADVERSE EFFECTS OF GRASS ENDOPHYTES ON OTHER INSECTS

Simple population and feeding damage studies have provided much of the evidence for adverse effects of fungal endophytes on other graminivorous insects. In the field, fewer larvae, adults, and eggs of sod webworms (*Crambus* spp.)[56] and bluegrass billbugs (*Sphenophorus parvulus*)[2] were found on *A. lolii*-infected perennial ryegrass than were found on uninfected ryegrass. The abundance of several leafhopper species (Table 1), the sharpshooter, *Draeculacephala antica*, a froghopper, *Prosapia bicuncta*, and the corn flea beetle, *Chaetocnema pulicaria*, all decreased with increasing endophyte levels in tall fescue pasture.[77,91] Populations of the sap-sucking hairy chinch bug (*Blissus leucopterous hirtus*) were lower on endophyte-infected fine fescues.[115] Other laboratory-based studies have shown that survival and mass gain of *Agrotis segetum* are reduced by *E. typhina* in *Dactylis glomerata*[116] and that survival of black beetle (*Heteronychus arator*) adults is adversely affected by endophytes in perennial ryegrass.[8] The source of the resistance to black beetle appears to be the ergot alkaloids, which deter adult feeding at relatively low concentrations, while peramine has no effect.[9] Feeding by the cutworm, *Graphania mutans*, is reduced on endophyte-infected ryegrass but is not affected by peramine, although this alkaloid disrupts the larval development of this species.[48]

Seed feeders are also sensitive to the presence of endophytes. Survival and population growth of the flour beetle (*Tribolium castaneum*) were reduced by endophytes in ground perennial ryegrass seed.[31] Given that the secondary metabolites can be found in relatively high concentrations in the seed of endophyte-infected grasses, it would not be surprising to find that other insect predators of field or stored seed are sensitive to endophytes.

There is now some evidence to suggest that although the endophyte mycelium does not extend below ground, its presence in the plant will affect root feeding insects. Six of nine fescue–endophyte combinations reduced survival of third instar grass grub (*Costelytra zealandica*) larvae.[100] The loline alkaloids were implicated as the cause of this effect and have been shown to be deterrent and toxic to the closely related Japanese beetle, *Popillia japonica*.[95,119] However, field and laboratory studies of this root-feeding scarab have provided inconclusive evidence of an effect. Larvae of *P. japonica* were found to be more prevalent in field plots of endophyte-free perennial ryegrass than in endophyte-infected plots, but the differences were not significant, and a reduction in growth and survival of third instar grubs demonstrated in one pot trial did not occur in a second similar trial.[93] Likewise, Potter et al. obtained inconsistent results when investigating the effects of endophytes on the feeding ecology of the Japanese beetle and southern masked chafer grub.[101] However, further evidence that the effects of endophytes do extend to the root system has been provided by experiments using nongraminivorous insects. Survival and fecundity of the collembolan, *Folsomia candida*, are affected by tall fescue root tissues,[20] and root material of tall fescue infected with *A. coenophialum* has also been shown to be toxic to *Drosophila melanogaster*.[42]

1. Toxicity and Deterrency

Both toxic and deterrent factors are implicated in the responses of most of the insects for which we have sufficient information, with some notable exceptions. The southern army-worm, *Spodoptera eridania,* and the house cricket, *Acheta domesticus,* were apparently highly susceptible to endophyte toxins in perennial ryegrass but were not deterred from feeding by the presence of endophytes.[1,3] For both these species, sudden death coincided with their feeding on the basal leaf sheath material where endophyte mycelium is concentrated.[1,3] In the case of house crickets, gross morphological changes to the gut of the dead insects were observed and were attributed to the feeding on endophyte-infected grass.[1] No morphological differences were apparent with *S. eridania.* These are the only reports of such rapid and complete mortality of insects feeding either on endophyte-infected ryegrasses or their metabo-lites, although endophytes have been suggested as the cause of acute mortality in house crickets fed ground tall fescue forage, without any apparent deterrency.[4] Surprisingly, *S. eridania* appears not to be susceptible to endophyte-infected tall fescue,[73] which suggests that further study of this species, with accompanying biochemical analyses to determine its sensitivity to alkaloids, would be useful.

Other insects also appear to succumb to toxic effects of endophytes, apparently without being deterred. The Russian wheat aphid, *Diuraphis noxia,* showed no apparent preference for endophyte-free perennial ryegrass in laboratory choice tests, but populations rapidly declined if it was confined to endophyte-infected ryegrass, indicating that antibiosis factors were present.[39] The milkweed bug, *Oncopeltus fasciatus,* cannot survive on endophyte-infected tall fescue seed extracts, or some alkaloids from them, but would apparently ingest them.[73,134] It is worth noting that, with the exception of the highly polyphagous southern armyworm, the insect species which succumb to toxic factors apparently without being deterred from feeding do not utilize the grass as a natural host. For those species for which grass is a natural host, e.g., Argentine stem weevil, the fall armyworm, *R. padi, S. graminum* (referenced above), and the hairy chinch bug,[86] antibiosis is accompanied by deterrency to endophyte infection. Thus, acute mortality is rarely observed for these species, as a general reduction in feeding occurs first. This lends support to the theory that deterrency is a result of coevolution of plant and herbivore. However, deterrency may not be a factor in the response of the graminivorous billbugs (*Sphenophorus* spp.), which are affected by endophytes. Johnson-Cicalese and Funk observed little difference in adult feeding between endophyte-infected and uninfected ryegrass and tall fescue but reported a significantly greater mortality of billbug adults on the infected grass.[71]

2. Secondary Metabolites and Nongraminivorous Insects

Several insects which have no natural association with grasses have been the subject of laboratory studies with the alkaloid groups produced in endophyte-infected plants. Some indole diterpenes and their derivatives are toxic to corn earworm (*Heliothis zea*).[47,66,127] Both the ergot and loline alkaloids are toxic to the large milkweed bug (*Oncopeltus fasciatus*).[119,134] *N*-formyl loline has the greatest activity against this insect,[134] with chronic exposure of 1 μg/g in water delaying development and reducing fecundity.[44] This alkaloid also has contact activity against the cat flea (*Ctenocephalides felis*), the American cockroach (*Periplaneta americana*), the adult face fly (*Musca autumnalis*), and the eggs of the tobacco budworm (*Heliothis virescens*).[44] Peramine has antifeedant activity against the cabbage stem flea beetle, *Phaedon cochleariae,* but not against the aphid, *Myzus persicae* (J.A. Pickett, personal communication).

C. NEUTRAL AND BENEFICIAL EFFECTS OF GRASS ENDOPHYTES ON INSECTS

Some species of insects are not affected by the presence of endophytes. The adult frit fly (*Oscinella frit*) is not deterred from laying eggs on endophyte-infected ryegrass.[84]

Another dipteran, the wheat sheath miner (*Ceredontha australis*), which mines the leaf sheath and stem of grasses, is also not affected by endophytes.[12] In New Zealand, the distribution of several homopteran leafhoppers, *Arawa novella, Deltocephalus hospes, Nesoclutha pallida,* and *Zygina zealandica,* was apparently unrelated to endophyte infection.[102] The tobacco hornworm, *Manduca sexta,* and the tobacco budworm, *Heliothis virescens,* showed no consistent differences in response to artificial diets containing plant material or plant extracts from endophyte-infected tall fescue.[73] Presumably, there are many other insects that exhibit no response to endophyte infection that are not reported in the literature.

No insect species is known to have the ability to exploit the plant–endophyte association, although there is one known example of an insect existing in a mutualistic relationship with a fungal endophyte. Larvae of the parasitic fly, *Phorbia phrenione,* feed exclusively on fungal stromata of *E. typhina,* and the sexual spores of the fungus are dispersed by the adult flies.[78] The adult fly deliberately drags its abdomen across the fungal stroma, picking up spermatia as it does so and acting as a "pollinator" for *E. typhina.*[25]

Three insect species, otherwise regarded as adversely affected by endophytes, have also shown positive responses to endophytes or their metabolites. Intermediate concentrations of some ergot alkaloids increased feeding by fall armyworm.[37] This was attributed to the nitrogen content of the alkaloids, which seems unlikely, as the minute quantities of nitrogen involved may be neither chemically available nor perceptible to the insect. Similarly, southern armyworm larvae feeding on endophyte-infected leaves (which had a higher nitrogen content than the endophyte-free leaves) exhibited higher mass gains and advanced development before mortality became a major factor.[3] Fecundity of the hairy chinch bug was also initially higher on endophyte-infected ryegrass than on ryegrass that was endophyte free.[86]

There are only two reports of insects preferring endophyte-infected grasses. Argentine stem weevil fed significantly more on ryegrass infected with a *Gliocladium*-like endophyte than on endophyte-free grass;[65] a species of leafhopper, *Exitianus exitiosus,* and an issid, *Bruchomorpha* sp., were reported to be more abundant on endophyte-infected tall fescue.[77] A more recent study, however, indicates that *E. exitiosus,* like other leafhoppers, prefers endophyte-free fescue.[91] These different results highlight the need for detailed studies of specific plant–insect interactions.

D. FACTORS AFFECTING INSECT RESPONSE TO ENDOPHYTES

Different insect stages respond differently to endophytes. Thus, Argentine stem weevil adults are only deterred by endophytes, whereas the larvae experience both deterrency and toxicity. Larval age also affects susceptibility to endophytes, particularly in the Lepidoptera. Fall armyworm,[68] the common armyworm (*Mythimna convecta*), and the common cutworm (*Agrotis infusa*)[55] are deterred by the presence of endophytes as early stage larvae but not as late stage larvae. Nymphs of the Russian wheat aphid are more susceptible to endophytes than the adults.[76]

The level of resistance to Argentine stem weevil afforded by *A. lolii*-infected ryegrass is reduced by the presence of the vesicular-arbuscular mycorrhizal fungus, *Glomus fasciculatum.*[11] Mycorrhizal infection alone has no effect on the weevils. Synergism between alkaloids produced by endophyte infection of grass can effect a higher mortality than would otherwise be caused by each compound alone.[134,137] Given that any insect response is likely to be determined by interactive cues from multiple compounds, scant attention has been paid to possible synergism between endophyte metabolites.

Environmental factors can affect the endophyte–host–insect interactions. Feeding preferences of *S. graminum* for endophyte-free perennial ryegrass are greater at 14 and 21°C than at 7 and 28°C, correlating with increased growth of ryegrass endophytes at more moderate temperatures.[22] Argentine stem weevil oviposition on endophyte-infected ryegrass can be similar to that on endophyte-free ryegrass in spring, when endophyte and alkaloid concentrations are low.[98] Nutrient status can also affect production of alkaloids by endophyte-infected grasses;[17] this is likely to affect interactions with insects.

E. INSECT RESPONSE TO TREE ENDOPHYTES

Although information is scarce, tree endophytes also elicit insect responses similar to those which occur in grass systems. In Greece, first year needles of *Pinus brutia* are heavily grazed by the pine processionary caterpillar unless the needles are infected with a pathogenic endophyte, *Elytroderma torres-juanii*.[46] Presence of the endophyte *Phomopsis oblonga* in elm bark disrupts reproduction of the elm bark beetle (*Scolytus multistriatus*) and causes populations to decline.[129] Laboratory studies show that the beetle prefers logs not infected by the endophyte, and few progeny are produced if beetles are confined to infected logs.[129]

Douglas fir needles are frequently infected in moist habitats by the endophyte *Rhabdocline parkeri,* which penetrates epidermal cells with increasing frequency as needles age. Larvae of three species of gall midges in the genus *Contarina* attack Douglas fir needles by chewing a hole in the needle and embedding themselves in needle tissue, forming a substantial gall. Galls can become infected with *R. parkeri* as the needle senesces, causing substantial mortality of the midge larvae.[28] Toxins produced by the fungi are thought to be responsible for the mortality[28]. Because the endophyte infection only proceeds as the needle is senescing and after gall formation by the midges, the endophyte does not in this case directly defend its host from attack by these insects. Rather, protection comes from a reduction in the population of midges, which subsequently must reduce the number of needles suffering damage. In the laboratory, *R. parkeri* has been shown to reduce growth and survival of the spruce budworm, *Choristoneura fumiferana*, significantly.[89] However, in another study, endophytic infection of balsam fir tree needles had no apparent effect on spruce budworm colonization.[72]

IV. EVOLUTIONARY ASPECTS OF ENDOPHYTE–INSECT INTERACTIONS

Endophytes and their host plants have almost certainly had a long evolutionary history. From origins in pathogenic parasitic fungi, typified today by members of the genus *Claviceps* and the choke-forming fungi of the genus *Epichloe*, the mutualistic associations which characterize the relationship between true endophytes and their hosts have emerged.[31,32] Plant and fungi coexist in a biotrophic relationship, the plant providing the nutrients needed for growth of the fungi and a means of transmission and the fungi producing secondary compounds which defend its food supply, the plant, from herbivory. Much has been written on the evolution of mutualism in endophyte–host relationships,[27,31,32,117,131] and it is not our intention to review the literature on this subject. Nevertheless, there are some aspects with respect to insect herbivores which are worth considering.

A. DOES THE PLANT HAVE A SAY IN ITS OWN DEFENSE?

Although the endophyte is primarily responsible for the defense of the grass, the role of the host need not be passive. While most of the secondary compounds so far identified in the association are produced by the endophyte, different quantities of these metabolites may be produced when an endophyte is artificially infected into a new host.[54,120] This may be due to the fitness of the individual plant and its ability to sustain fungal growth, or it may reflect a capability within the plant genotype to at least partially influence its own defense. Whether fortuitous or not, the resulting variable levels of alkaloids produced may have important implications for preventing insects from overcoming the defenses afforded by the plant–endophyte interaction.

The distribution of the alkaloids within the plant also suggests that the plant is actively influencing the use of the secondary metabolites. Peramine is translocated away from the endophyte that produces it within the plant and occurs in higher concentrations in younger ryegrass leaves than it does in older leaves.[75] Similarly, ergovaline levels are high in developing inflorescences and, subsequently, in seed.[110] There is no published information on the reasons for this, but it is possible that production is influenced by plant hormone levels.

B. ORIGINS OF SECONDARY METABOLITE PRODUCTION

The production of secondary compounds by fungal pathogens is widespread in nature, but the interaction of these compounds with insects is not well documented. The widespread occurrence of ergot alkaloids and indole diterpene mycotoxins in diverse fungi suggests that the evolution of a functional role for these metabolites preceded the occurrence of the plant–endophyte association. Secondary metabolite production is often regulated with the life stage of the fungus,[24,124] and in *Aspergillus*, *Claviceps*, and *Penicillium*, production of the indole diterpenes and ergot alkaloids coincides with the production of the sclerotia (or spores), which these metabolites protect from predation.[66,69,127] This suggests a possible evolutionary origin for the indole diterpene and ergot metabolites in endophyte systems. With the transition from parasitism to symbiote postulated to have occurred in the origin of endophytic fungi,[32] independent transmission of the fungus was lost, and protection of the sclerotia or spores was replaced by a protection for the fungus itself, through protection of the host plant. Epiphytic *Balansia* can effectively protect their host from herbivory, even though the fungal mycelia are highly localized and removed from where feeding occurs.[36,125] For seed-transmitted endophytes, this requires that secondary metabolite production be deregulated from the life cycle of the fungus and expressed constitutively in the symbiotic organism.

C. OPTIMAL DEFENSE THEORY AND ENDOPHYTE–PLANT INTERACTIONS

It is interesting to consider the endophyte as mediator of plant–insect interactions in the context of accepted theories on the evolution of plant defense against herbivores. In the theory of optimal defense,[109] a basic premise is that organisms evolve and allocate defenses in a way that maximizes individual fitness. Certainly, the mutualism between endophytes and their grass hosts appears to have maximized the fitness of both partners in the relationship. Endophytes not only defend themselves and their hosts against both mammalian and insect herbivores, but they also ameliorate against various abiotic stresses and may provide some protection against plant pathogens.[80] The predominance of endophyte-infected grasses in naturally occurring ryegrass pastures and in old pastures in Europe attests to the fitness of the endophyte–host relationship.[82]

An extension of this premise states that defenses are allocated within an organism according to the risk and value of a particular tissue. It is probably no coincidence then that endophytes in grasses are concentrated in the basal leaf sheath area. Grasses can survive and even benefit from grazing of leaf material, but damage to the base of the plant has more serious consequences. Nevertheless, extension of the mycelium throughout the aerial parts of the plant and translocation of secondary compounds to the roots ensure that protection from herbivores is not confined to the base of the plant alone. Also worth noting is the relatively high concentration of some alkaloids in the seed and developing inflorescence of endophyte-infected grasses. This may protect the seed from predation in the field, although there is no published information to show this actually happens. More importantly, these metabolites may protect the developing seedling from insect attack. For instance, peramine, stored in the infected ryegrass seed, is translocated into the developing seedling in sufficient quantities to provide protection from Argentine stem weevil until *de novo* synthesis of this compound begins.[10] Similar seedling resistance occurs in endophyte-infected annual ryegrass (*L. multiflorum*), even though peramine production by the endophyte is insufficient to protect the mature plant.[50] If the annual ryegrass–endophyte relationship represents an earlier evolution-ary step than that in perennial ryegrass, it would suggest that the original purpose of the defense afforded by the endophyte was to protect the seed and seedling; this has since evolved into a more complete protection against herbivory. Evolution, however, may have some way to go to perfect the system, as the viability of endophyte in seed declines rapidly after one year.

According to the second basic principle of optimal defense, defenses are costly for the plant to maintain because they utilize energy and nutrients that might otherwise be contributing to plant growth. At what cost to the plant, then, are the endophytes maintained? The answer, it seems, is surprisingly little, particularly considering the apparent effectiveness of the defense provided. Thus, even though the biotrophic relationship between endophyte and host must impose some demands on the host plant, in the absence of herbivory, the endophyte can still enhance the growth of its host, particularly in stressful situations such as drought.[5,118] This appears to contradict the belief that fitness in the well-defended individual is only a product of its ability to protect itself, and in the absence of enemies, its defensive mechanisms put it at a disadvantage in relation to the poorly defended individual. Furthermore, the endophyte–host plant relationship has elements of both the "quantitative" and "qualitative" defense mechanisms described by Feeny.[52] The grasses, as individuals, are unapparent plants, being relatively ephemeral and difficult for a herbivore to discover, but they often occur collectively in nature with a corresponding increase in apparency. For the most part, the grass–endophyte defense relationship can be considered to be a qualitative one, characteristic of unapparent plants, whereby low alkaloid levels produced at little cost to the plant mediate resistance to nonadapted herbivores. The exception to this is the loline alkaloids, which, because they can occur at very high concentrations, could be considered to be a quantitative defense mecha-nism. However, although as a general rule endophyte-infected plants have comparable growth rates to endophyte-free plants in the absence of herbivory, we cannot assume that this is true in all situations. To our knowledge there is no documented research that has specifically compared the growth of plants that produce high levels of loline alkaloids with those that do not. It may be that such plants endure a disadvantage in growth by the diversion of nutrients into the production of this quantitative defense. In any case, it must be remembered that all the research has been conducted on commercial grass cultivars artificially bred and selected for their superior growth in resource-rich environments. In native grassland habitats, fitness of the endophyte-infected individual may only be maintained through herbivore resistance.

D. INSECTS, ENDOPHYTES, AND EVOLUTION

Undoubtedly, insects have been a compelling force behind natural selection for endophyte-infected plants, and a study of the ecology of native grasses and grasslands where endophyte-infected plants are dominant would provide valuable insights into the selection for endophytism. In New Zealand, the selection pressure for endophytism has been clearly demonstrated where endophyte-free pastures rapidly become dominated by endophyte-infected plants as a result of Argentine stem weevil attack.[107]

There is now considerable evidence that plants and endophytes have coevolved. It is perhaps surprising, then, that so many insect graminivores are sensitive, and few, if any, are able to exploit an abundant host plant resource, relatively free of competition. There may be several reasons for this.

1. The insects concerned all have a choice. None are obligate grass feeders, and most utilize other endophyte-free Poaceae such as the cereals as alternative hosts. Even within their grass hosts, endophyte-free individuals are not uncommon. The adult-stage insects are highly mobile, aiding their ability to make appropriate host selections. Argentine stem weevil[13] and sod webworms[56] demonstrate such an ability when they select endophyte-free plants for oviposition. The mobility and dispersive abilities of insects

may be important factors in determining their responses to plant defense mechanisms. Two dipteran species that are not affected by endophyte, *Oscinella frit* and *Ceredontha australis*, have relatively short-lived adult stages, with mating and oviposition taking place soon after adult emergence. Such circumstances leave the insect little time to choose suitable oviposition sites, so, for these species, natural selection may have favored individuals with an ability to overcome the toxic factors produced in the endophyte-grass association.

2. Both toxic and deterrent factors are involved. In Argentine stem weevil at least, these are not mediated through the same secondary compounds, making it harder for the insect to overcome the resistance factors. In the laboratory, Argentine stem weevil insensitivity to the feeding deterrent peramine has been demonstrated (A. J. Popay, unpublished data). If such insensitivity occurs in the field, the presence of toxic compounds ensures that the insensitive individuals have reduced survival. The diversity of the secondary compounds produced in endophyte–plant interactions must also reduce the chances of the insect overcoming the defense mechanisms.

   The presence of deterrent factors has advantages for both the insect and the plant. For the plant, antifeedants prevent damage occurring, whereas a total reliance on toxic factors assumes that some consumption of the plant must take place. For the insect, deterrency aids in host selection. The adaptive significance of the latter is particularly well demonstrated for Argentine stem weevil. Here the adult is highly sensitive to deterrent factors in ryegrass that reduce oviposition and in turn protect the less mobile offspring from toxic factors likely to reduce survival. However, although such a system is beneficial to both host plant and insect, it is unlikely to be the result of a long coevolution, as Argentine stem weevils are thought not to have been exposed to endophytes in their native habitat of South America, until the relatively recent introduction of ryegrass (S. L. Goldson, personal communication). Sensitivity of Argentine stem weevils to endophytes may therefore be simply due to chance and to a rapid selection for individuals that avoid endophyte-infected plants. The idea is plausible if nonavoidance is accompanied by acute toxicity and very low survival.

3. There is spatial and temporal variation in the production of alkaloids due to environmental factors and the interaction between plant genotype and the endophyte. Thus, within any grassland environment at any one time, variability in plant genotype, nutrient status, and amount of grazing will cause variations in alkaloid production in individual plants. The effect of seasonal changes in temperature and moisture will produce a similar effect, so that over a period of time, an insect is not exposed to a uniform level of alkaloids.

V. ECONOMIC IMPORTANCE OF ENDOPHYTES

Increasing concerns worldwide about the use of chemical pesticides have resulted in a concerted effort by scientists to find satisfactory alternatives. Insect–host plant resistance is an attractive option, providing a self-replicating and environmentally-sustainable method of pest control. The use of endophytes to mediate host–plant resistance offers an opportunity to control a wide range of insect pests in grassland systems. Two such systems involving tall fescue and perennial ryegrass are vitally important for livestock production, with some 14 million hectares of tall fescue in the United States and 15 million hectares of perennial ryegrass in Australia and New Zealand utilized for this purpose. In addition, throughout the world these grasses are used extensively in turf culture. In the United States alone, the turf grass industry is estimated to be worth 25 billion dollars annually. Aesthetic standards demand very low levels of insect damage to turf grasses, and chemical inputs have been high.

Infection of both turf and pasture grasses with endophytes is widespread, and cultivars with high proportions of infected plants are commercially produced, due to their advantages over uninfected plants in terms of better growth and persistence. The superior growth of endophyte-infected fescue can in part be attributed to its adverse effects on pests such as sod webworms, armyworms, billbugs, chinch bugs, and aphids, with increasing evidence to suggest that the endophytes also mediate against root feeding pests, including the Japanese beetle and some species of nematode. Some of these pests are also important vectors of cereal diseases and utilize grass as a secondary host. However, such advantages have posed a dilemma for farmers and scientists, alike as the superior growth and persistence of *A. coenophialum*-infected tall fescue is often accompanied by severe mammalian toxicity.[7] In New Zealand, *A. lolii*-infected perennial ryegrass is specifically used to control Argentine stem weevil, despite the wide-spread occurrence of ryegrass staggers toxicity and associated stock management problems. Mammalian toxicity is also important in the turf grass industry when livestock graze pastures used for seed production.

Three major advances in endophyte research have improved the prospects for solving the dichotomy of insect resistance and mammalian toxicity:

1. The demonstration that mammalian toxicity and insect resistance could be mediated through different secondary metabolites of the endophyte,[65,112]
2. The ability to inoculate selected endophyte strains into new host grasses and, in particular, into pre-existing commercial cultivars,[81]
3. The discovery of a wide variety of endophyte strains that produce different combinations of alkaloids.[29,30]

These advances have allowed the selection of endophyte strains that, for example, produce peramine, the feeding deterrent to Argentine stem weevil but not the mammalian toxin, lolitrem B. Preliminary results using these endophytes in commercial ryegrass cultivars indicate that they retain a high level of resistance to Argentine stem weevil but do not cause ryegrass staggers.[54] However, caution must prevail in the use of such endophytes, as reliance on a single compound to provide insect resistance has inherent problems. Insensitivity to peramine may already exist in the population, as indicated earlier, and the successive removal of insect toxins from the host must inevitably result in the loss of resistance. Herein lies the challenge for agriculturalists as they attempt to manipulate endophyte–grass systems to retain favorable attributes such as enhanced growth and insect resistance without mammalian toxicity.

VI. CONCLUSION

While the documented evidence provides ample testimony to the defensive properties of the endophyte–plant relationship to insects, the information is tantalizingly incomplete, and there is still much to be learned. Finding other insect species that are sensitive to endophytic fungi will serve to delineate the extent of this phenomenon but will do little to add to the current state of knowledge. For those insects known to be affected we need to know more about the exact nature of the chemical, behavioral, and physiological responses involved.

The chemistry of the natural environment in which the insect encounters the endophyte has been only partially examined. Even at the most basic level, our knowledge of the primary and secondary metabolism of endophytes and their interaction with their hosts and, in turn, the herbivore is clearly rudimentary. Too often we test the chemical responses of insects in isolation, without acknowledging that the insect in its natural habitat is faced with an array of chemical cues both stimulatory and deterrent. Though it is apparent that both antifeedant and antibiotic responses are involved in the insect–endophyte–host interactions, the chemical basis for these is not well understood.

The endophyte–host–herbivore interaction may provide valuable insights into insect host selection. The fungi produce apparently asymptomatic infections in their host, so visual cues should not be a significant factor in host selection. Host-finding behavior will affect the ability of the insect to discriminate between infected and noninfected plants and may depend on the insect's mobility and potential to disperse. Finally, while chemical cues may determine host selection for adult feeding, we know little of their effect in determining ovipositional behavior.

A deeper understanding of these issues will advance our knowledge of plant–insect interactions. More significantly, unless we acquire such an understanding, attempts by agriculturalists to manipulate the endophyte–grass association to their advantage will ultimately not succeed.

REFERENCES

1. **Ahmad, S., Govindarajan, S., Funk, C. R., and Johnson-Cicalese, J. M.,** Fatality of house crickets on perennial ryegrass infected with a fungal endophyte, *Entomol. Exp. Appl.,* 39, 183, 1985.
2. **Ahmad, S., Johnson-Cicalese, J. M., Dickson, W. K., and Funk, C. R.,** Endophyte-enhanced resistance in perennial ryegrass to the bluegrass billbug, *Sphenophorus parvulus, Entomol. Exp. Appl.,* 41, 3, 1986.
3. **Ahmad, S., Govindarajan, S., Johnson-Cicalese, J. M., and Funk, C. R.,** Association of a fungal endophyte in perennial ryegrass with antibiosis to larvae of the southern armyworm, *Spodoptera eridania, Entomol. Exp. Appl.,* 43, 287, 1987.
4. **Asay, K. H., Minnick, T. R., Garner, G. B., and Harmon, B. W.,** Use of crickets in a bioassay of forage quality in tall fescue, *Crop Sci.,* 15, 585, 1975.
5. **Bacon, C. W. and Siegel, M. R.,** Endophyte parasitism of tall fescue, *J. Prod. Agric.,* 1, 45, 1988.
6. **Bacon, C. W., Porter, J. K., Robbins, J. D., and Luttrell, E. S.,** *Epichloe typhina* from toxic tall fescue grasses, *Appl. Environ. Microbiol.,* 34, 576, 1977.
7. **Bacon, C. W., Lyons, P. C., Porter, J. K., and Robbins J. D.,** Ergot toxicity from endophyte infected grasses: a review, *Agron. J.,* 78, 106, 1986.
8. **Ball, O. J.-P. and Prestidge, R. A.,** The effect of the endophytic fungus *Acremonium lolii* on adult black beetle (*Heteronychus arator*) feeding, *Proc. 45th N. Z. Plant Prot. Conf.,* 45, 201, 1992.
9. **Ball, O. J.-P. and Prestidge, R. A.,** The use of the endophytic fungus *Acremonium lolii* as a biological control agent of black beetle, *Heteronychus arator* (Coleoptera: Scarabaeidae), in *Proc. 6th Australasian Conf. Grassland Invert. Ecol.,* in press, 1993.
10. **Ball, O. J-P., Prestidge, R. A., and Sprosen, J. M.,** Effect of plant age and endophyte viability on peramine and lolitrem B concentration in perennial ryegrass seedlings, in *Proc. 2nd Int. Symp. Acremonium/Grass Interactions.* Hume, D. E., Latch, G. C. M., and Easton, H. S., Eds., AgResearch Grasslands, Palmerston North, New Zealand, 1993, 63.
11. **Barker, G. M.,** Mycorrhizal infection influences *Acremonium*-induced resistance to Argentine stem weevil in ryegrass, *Proc. 40th N. Z. Weed Pest Control Conf.,* 40, 199, 1987.
12. **Barker, G. M., Pottinger, R. P., Addison, R. J., and Oliver, E. H. A.,** Pest status of *Ceredontha* spp. and other shoot flies in Waikato pasture, *Proc. 37th N. Z. Weed Pest Control Conf.,* 37, 96, 1984.
13. **Barker, G. M., Pottinger, R. P., Addison, P. J., and Prestidge, R. A.,** Effect of *Lolium* endophyte fungus infection on behaviour of adult Argentine stem weevil, *N.Z. J. Agric. Res.,* 27, 271, 1984.
14. **Barker, G. M., Pottinger, R. P., and Addison, P. J.,** Effect of *Lolium* endophyte fungus infection on survival of larval Argentine stem weevil, *N.Z. J. Agric. Res.,* 27, 279, 1984.
15. **Barker, G. M., Pottinger, R. P., and Addison, P.J.,** Flight behaviour of *Listronotus bonariensis* (Coleoptera: Curculionidae) in the Waikato, New Zealand. *Environ. Entomol.,* 18, 996, 1989.
16. **Barker, G. M., Pottinger, R. P., and Addison, P. J.,** Population dynamics of the Argentine stem weevil (*Listronotus bonariensis*) in pastures in the Waikato, New Zealand, *Agric. Ecosyst. Environ.,* 26, 79, 1989.
17. **Beleskey, D. P., Studemann, J. A., Plattner, R. D., and Wilkinson, S. R.,** Ergopeptine alkaloids in grazed tall fescue, *Agron. J.,* 80, 209, 1988.
18. **Bell, N. L. and Prestidge, R. A.,** The effects of the endophytic fungus *Acremonium starii* on feeding and oviposition of the Argentine stem weevil, *Proc. 44th N. Z. Weed Pest Control Conf.,* 44, 181, 1991.
19. **Berde, B. and Schild, H. O.,** *Ergot Alkaloids and Related Compounds,* Berde, B. and Schild, H. O., Eds., Springer-Verlag, New York, 1978.

20. **Bernard, E. C., Cole, A. M., Oliver, J. B., and Gwinn, K. D.,** Survival and fecundity of *Folsomia candida* (Collembola) fed tall fescue tissues or ergot peptide-amended yeast, in *Proc. Int. Symp. Acremonium/Grass Interactions*, Quisenberry, S. S. and Joost, R. E., Eds., Louisiana Agric. Exp. Stn, Baton Rouge, LA, 1990, 125.

21. **Betina, V.,** Indole derived tremorgenic toxins, in *Mycotoxins Production, Isolation, Separation and Purification*, Betina, V., Ed., Developments in Food Science, 8, Elsevier, Amsterdam, 1984, 415.

22. **Breen, J. P.,** Temperature and seasonal effects on expression of *Acremonium* endophyte-enhanced resistance to *Schizaphis graminum* (Homoptera: Aphididae), *Environ. Entomol.,* 21, 68, 1992.

23. **Buckley, R. J., Halisky, P. M., and Breen, J. P.,** Variation in feeding deterrence of the corn leaf aphid related to *Acremonium* endophytes in grasses, *Phytopathology,* 81, 120, 1991.

24. **Bu'Lock, J. D.,** Secondary metabolism in fungi and its relationship to growth and development, in *The Filamentous Fungi I*, Smith, J. E. and Berry, D. R., Eds., Edward Arnold, London, 1975, 35.

25. **Bultman, T. L., Bowdish, T. I., Welch, A. M, and White, J. F., Jr.,** The mechanism of insect-mediated fertilization of *Epichloe typhina*, in *Proc. 2nd Int. Symp. Acremonium/Grass Interactions*, Hume, D. E., Latch, G. C. M., and Easton, H. S., Eds., AgResearch Grasslands, Palmerston North, New Zealand, 1993, 3.

26. **Carroll, G. C.,** The biology of endophytism in plants with particular reference to woody perennials, in *Microbiology of the Phyllophane*, Fokkema, N. J. and van den Heuvel, J., Eds., Cambridge University Press, Cambridge, 1986, 205.

27. **Carroll, G. C.,** Fungal endophytes in stems and leaves: from latent pathogen to mutualistic symbiont, *Ecology,* 69, 2, 1988.

28. **Carroll, G. C. and Carroll, F. E.,** Studies on the incidence of coniferous needle endophytes in the Pacific Northwest, *Can. J. Bot.,* 56, 3034, 1978.

29. **Christensen, M. J., Latch, G. C. M., and Tapper, B. A.,** Variation within isolates of *Acremonium* endophytes from perennial ryegrass, *Mycol. Res.,* 95, 918, 1991.

30. **Christensen, M. J., Leuchtmann, A., Rowan, D. D., and Tapper, B. A.,** Taxonomy of *Acremonium* endophytes of tall fescue (*Festuca arundinacea*), meadow fescue (*F. pratensis*) and perennial ryegrass (*Lolium perenne*), *Mycol. Res.,* 1993, in press.

31. **Clay, K.,** Fungal endophytes of grasses: a defensive mutualism between plants and fungi, *Ecology,* 69, 10, 1988.

32. **Clay, K.,** Clavicipitaceous endophytes of grasses: coevolution and the change from parasitism to mutualism, in *Coevolution of Fungi with Plants and Animals*, Hawksworth, D. and Pirozynski, K., Eds., Academic Press, London, 1988, 79.

33. **Clay, K.,** Clavicipitaceous endophytes of grasses: their potential as biocontrol agents, *Mycol. Res.,* 92, 1, 1989.

34. **Clay, K.,** Fungal endophytes, grasses and herbivores: ecological interactions among three trophic levels, in *Multitrophic Level Interactions among Microorganisms, Plants and Insects*, Barbosa, P., Kirschik, L., and Jones, E., Eds., 1992, in press.

35. **Clay, K., Hardy, T. N., and Hammond, A. M., Jr.,** Fungal endophytes of grasses and their effects on an insect herbivore, *Oecologia,* 66, 1, 1985.

36. **Clay, K., Hardy, T. N., and Hammond, A. M., Jr.,** Fungal endophytes of *Cyperus* and their effect on an insect herbivore, *Am. J. Bot.* 72, 1284, 1985.

37. **Clay, K. and Cheplick, G. P.,** Effect of ergot alkaloids from fungal endophyte-infected grasses on fall armyworm (*Spodoptera frugiperda*), *J. Chem. Ecol.,* 15, 169, 1989.

38. **Clement, S. L., Pike, K. S., Kaiser, W. J., and Wilson, A. D.,** Resistance of endophyte-infected plants of tall fescue and perennial ryegrass to the Russian wheat aphid (Homoptera: Aphididae), *J. Kan. Entomol. Soc.,* 63, 646, 1990.

39. **Clement, S. L., Lester, D. G., Wilson, A. D., and Pike, K. S.,** Behaviour and performance of *Diuraphis noxia* (Homoptera: Aphididae) on fungal endophyte-infected and uninfected perennial ryegrass, *J. Econ. Entomol.,* 85, 583, 1992.

40. **Cole, R. J., Kirksey, J. W., and Wells, J. M.,** A new tremorgenic metabolite from *Penicillium paxilli, Can. J. Microbiol.,* 20, 1159, 1974.

41. **Cole, R. J., Dorner, J. W., Lansden, J. A., Cox, R. H., Pape, C., Cunfer, B., Nicholson, S. S., and Bedell, D. M.,** Paspallum staggers: isolation and identification of tremorgenic metabolites from sclerotia of *Claviceps paspali, J. Agric. Food Chem.,* 25, 1197, 1977.

42. **Cole, A. M., Pless, C. D., and Gwinn, K. D.,** Survival of *Drosophila melanogaster* (Diptera: Drosophilidae) on diets containing roots or leaves of *Acremonium*-infected or non-infected tall fescue, in *Proc. 1st Int. Symp. Acremonium/Grass Interactions*, Quisenberry, S. S. and Joost, R. E., Eds., Louisiana Agric. Exp. Stn., Baton Rouge, LA, 1990, 128.

43. **Cubit, J. D.,** Interactions of seasonally changing physical factors and grazing affecting high intertidal communities on a rocky shore, Ph.D. dissertation, University of Oregon, 1974.

44. **Dahlmann, D. L., Eichenseer, H., and Siegel, M. R.,** Chemical perspectives on endophyte–grass interactions and their implications for herbivory, in *Multi-trophic Level Interactions among Microorganisms, Plants and Insects*, Barbosa, P., Kirschik, L., and Jones, E., Eds., John Wiley & Sons, New York, 1992, in press.

45. **Davis, N. D., Cole, R. J., Dorner, J. W., Weete, J. D., Backman, P. A., Clark, E. M., King, C. C., Schmidt, S. P., and Diener, U. L.,** Steroid metabolites of *Acremonium coenophialum*, an endophyte of tall fescue, *J. Agric. Food Chem.,* 34, 105, 1986.

46. **Diamandis, S.,** *Elytroderma torres-juanii* Diamandis and Minter. A serious attack on *Pinus brutia* L. in Greece, in *Current Research on Conifer Needle Diseases,* Millar, C. S., Ed., Aberdeen University Press, Aberdeen, Scotland, 1981, 9.

47. **Dowd, P. F., Cole, R. J., and Vesonder, R. F.,** Toxicity of selected tremorgenic mycotoxins and related compounds to *Spodoptera frugiperda* and *Heliothis zea, J. Antibiot.,* 41, 1868, 1988.

48. **Dymock, J. J., Rowan, D. D., and McGee, I. R.,** Effects of endophyte-produced mycotoxins on Argentine stem weevil and the cutworm, *Graphania mutans, in Proc. 5th Australasian Conf. Grassland Invert. Ecol.,* P. P. Stahle, Ed., D & D Printing Ltd., Victoria, Australia, 1989, 35.

49. **Dymock, J. J., Prestidge, R. A., and Rowan, D. D.,** The effects of lolitrem B on Argentine stem weevil, *Proc. 42nd N. Z. Weed Pest Control Conf.,* 42, 73, 1989.

50. **Dymock, J. J., Latch, G. C. M., and Tapper, B. A.,** Novel combinations of endophytes in ryegrasses and fescues and their effects on Argentine stem weevil (*Listronotus bonariensis*) feeding, in *Proc. 5th Australasian Conf. Grassland Invert. Ecol.,* Stahle, P. P., Ed., D & D Printing Pty. Ltd., Victoria, Australia, 1989, 28.

51. **Eichenseer, H., Dahlman, D. L., and Bush, L. P.,** Influence of endophyte infection, plant age and harvest interval on *Rhopalosiphum padi* survival and its relation to quantity of N-formyl and N-acetyl loline in tall fescue, *Entomol. Exp. Appl.,* 60, 29, 1991.

52. **Feeny, P. P.,** Plant apparency and chemical defense, *Rec. Adv. Phytochem.,* 10, 1, 1976.

53. **Fletcher, L. R. and Harvey, I. C.,** An association of a *Lolium* endophyte with ryegrass staggers, *N.Z. Vet. J.,* 29, 185, 1981.

54. **Fletcher, L. R., Popay, A. J., and Tapper, B. A.,** Evaluation of several lolitrem-free endophyte/perennial ryegrass combinations, *N.Z. Grassland Assoc.,* 53, 215, 1991.

55. **Frost, W. E., Quigley, P. E., and Cunningham, P. J.,** Research into the effect of *Acremonium lolii* on the feeding behaviour and nutrition of principal Australian pests of perennial ryegrass, in *Proc. Int. Symp. Acremonium/Grass Interactions,* Quisenberry, S. S. and Joost, R. E., Eds., Louisiana Agric. Exp. Stn., Baton Rouge, LA, 1990, 144.

56. **Funk, C. R., Halisky, P. M., Johnson, M. C., Siegel, M. R., Stewart, A. V., Ahmad, S, Hurley, R. H., and Harvey, I. C.,** An endophytic fungus and resistance to sod webworms: association in *Lolium perenne* L., *Biotechnol.,* 1, 189, 1983.

57. **Gallagher, R. T. and Hawkes, A. D.,** The potent tremorgenic neurotoxins lolitrem B and aflatrem: a comparison of the tremor response in mice, *Experientia,* 42, 823, 1986.

58. **Gallagher, R. T. and Prestidge, R. A.,** Structure-activity studies on indole diterpenes, including lolitrems and related indoles and tremorgens, in *Proc. Int. Symp. Acremonium/Grass Interactions.* Quisenberry, S. S. and Joost, R. E., Eds., Louisiana Agric. Exp. Stn., LA, 1990, 80.

59. **Gallagher, R. T., White, E. P., and Mortimer, P. H.,** Ryegrass staggers: isolation of potent neurotoxins lolitrem A and lolitrem B from staggers-producing pastures, *N.Z. Vet. J., 29,* 189, 1981.

60. **Gallagher, R. T., Hawkes, A. D., Steyn, P. S., and Vleggaar, R.,** Tremorgenic neurotoxins from perennial ryegrass causing ryegrass staggers disorder of livestock: structure elucidation of lolitrem B., *J. Chem. Soc. Chem. Commun.,* 614, 1984.

61. **Gallagher, R. T., Hawkes, A. D., and Stewart, J. M.,** Rapid determination of the neurotoxin lolitrem B in perennial ryegrass by high-performance liquid chromatography with fluorescence detection, *J. Chromatogr.,* 321, 217, 1985.

62. **Gallagher, R. T., Smith, G. S., and Sprosen, J. M.,** The distribution and accumulation of lolitrem B neurotoxin in developing perennial ryegrass plants, *Proc. 4th Anim. Sci. Congr. Asian-Australasian Assoc. Anim. Prod. Soc.,* Hamilton, New Zealand, 404, 1987.

63. **Garthwaite, I., Miles, C. O., and Towers, N. R.,** Immunological detection of the indole diterpenoid tremorgenic mycotoxins, in *Proc. 2nd Int. Symp. Acremonium/Grass Interactions,* Hume, D. E., Latch, G. C. M., and Easton, H. S., Eds., AgResearch Grasslands, Palmerston North, New Zealand, 1993, 88.

64. **Gaynor, D. L. and Hunt, W. F.,** The relationship between nitrogen supply, endophytic fungus, and Argentine stem weevil resistance in ryegrasses, *Proc. N. Z. Grassland Assoc.,* 44, 257, 1983.

65. **Gaynor, D. L., Rowan, D. D., Latch, G. C. M., and Pilkington, S.,** Preliminary results on the biochemical relationship between adult Argentine stem weevil and two endophytes in ryegrasses, in *Proc. 36th Weed Pest Control Conf.,* 36, 220, 1983.

66. **Gloer, J. B., Rinderknecht, B. L., Wicklow, D. T., and Dowd, P. F.,** Nominine: a new insecticidal indole diterpene from the sclerotia of *Aspergilllus nomius, J. Org. Chem.,* 54, 2530, 1989.

67. **Hardy, T., Clay, K., and Hammond, Jr A. M.,** Fall armyworm (Lepidoptera: Noctuidae): a laboratory bioassay and larval preference study for the fungal endophyte of perennial ryegrass, *J. Econ. Entomol.,* 78, 571, 1985.

68. **Hardy, T., Clay, K., and Hammond, A. M., Jr.,** Leaf age and related factors affecting endophyte-mediated resistance to fall armyworm (Lepidoptera: Noctuidae) in tall fescue, *Environ. Entomol.,* 15, 1083, 1986.

69. **Ibba, M., Taylor, S. J. C., Weedon, C. M., and Mantle, P. G.** Submerged fermentation of *Penicillium paxilii* biosynthesing paxilline, a process inhibited by calcium-induced sporulation, *J. Gen. Microbiol.,* 133, 3109, 1987.

70. **Huizing, H. J., Kloek, W., and Den Nijs, A. P. M.,** Endofyten in grassen induceren de vorming van alkaloiden, *Prophyta,* 42, 24, 1988.

71. **Johnson-Cicalese, J. M. and Funk, C. R.,** Host range and effect of endophyte on four species of billbug (*Sphenophorus* spp.) found on New Jersey turfs, *Agron. Abstr.,* 152, 1988.

72. **Johnson, J. A. and Whitney, N. J.,** A study of fungal endophytes of needles of balsam fir (*Abies balsamea*) and red spruce (*Picea rubens*) in New Brunswick, Canada, using culture and electron microscope techniques, *Can. J. Bot.,* 67, 3513, 1989.

73. **Johnson, M. C., Dahlman, D. L., Siegel, M. R., Bush, L. P., Latch, G. C. M., Potter, D. A., and Varney, D. R.,** Insect feeding deterrents in endophyte-infected tall fescue, *Appl. Environ. Microbiol.,* 49, 568, 1985.

74. **Kanda, K., Koga, H., Hirai, Y., Hasegawa, K., Uematu, T., and Tukiboshi, T.,** Resistance of *Acremonium* endophyte-infected perennial ryegrass and tall fescue to bluegrass webworm, *Parapediasia teterella,* abstr. trans., Proc. Pythopathol. Soc. Japan, Morioka, Iwate, May, 1992.

75. **Keogh, R. G. and Tapper B. A.,** *Acremonium lolii,* lolitrem B and peramine concentrations within vegetative tillers of perennial ryegrass, in *Proc. 2nd Int. Symp. Acremonium/Grass Interactions,* Hume, D. E., Latch, G. C. M., and Easton, H. S., Eds., AgResearch Grasslands, Palmerston North, New Zealand, 1993, 81.

76. **Kindler, S. D., Breen, J. P., and Springer, T. L.,** Reproduction and damage by Russian wheat aphid (Homoptera: Aphididae) as influenced by fungal endophytes and cool-season turfgrasses, *J. Econ. Entomol.,* 84, 685, 1991.

77. **Kirfman, G. W., Brandenburg, R. L., and Garner, G. B.,** Relationship between insect abundance and endophyte infestation level in tall fescue in Missouri, *J. Kan. Entomol. Soc.,* 59, 552, 1986.

78. **Kohlmeyer, J. and Kohlmeyer, E.,** Distribution of *Epichloe typhina* (Ascomycetes) and its parasitic fly, *Mycologia,* 66, 77, 1974.

79. **Koshino, H., Togiya, S., Joshihara, T., and Sakamura, S.,** Four fungitoxic C-18 hydroxy unsaturated fatty acids from stomata of *Epichloe typhina, Tetrahedron Lett.,* 28, 73, 1987.

80. **Latch, G. C. M.,** Physiological interactions of endophytic fungi and their hosts. Biotic stress tolerance imparted to grasses by endophytes, in *Acremonium/Grass Interactions,* Joost, R., and Quisenberry, S., Eds., *Agric. Ecosyst. Environ. (Special Issue),* 44, 143, 1993.

81. **Latch, G. C. M. and Christensen, M. J.,** Artificial infections of grasses with endophytes, *Ann. Appl. Biol.,* 107, 17, 1985.

82. **Latch, G. C. M., Potter, L. R., and Tyler, B. F.,** Incidence of endophytes in seeds from collections of *Lolium* and *Festuca* species, *Ann. Appl. Biol.,* 111, 59, 1987.

83. **Latch, G. C. M., Christensen, M. J., and Gaynor, D. L.,** Aphid detection of endophyte infection in tall fescue, *N.Z. J. Agric. Res.,* 28, 129, 1985.

84. **Lewis, G. C. and Clements, R. O.,** A survey of ryegrass endophyte (*Acremonium loliae*) in the U. K. and its apparent ineffectuality on a seedling pest, *J. Agric. Sci.,* 107, 633, 1986.

85. **Lyons, P. C., Plattner, R. D., and Bacon, C. W.,** Occurrence of peptide and clavine ergot alkaloids in tall fescue grass, *Science,* 232, 487, 1986.

86. **Mathias, J. K., Ratcliffe, R. H., and Hellman, J. L.,** Association of an endophytic fungus in perennial ryegrass and resistance to the hairy chinch bug (Hemiptera: Lygaeidae), *J. Econ. Entomol.,* 83, 1640, 1990.

87. **Miles, C. O., Munday, S. C., Wilkins, A. L., Ede, R. M., Hawkes, A. D., Embling, P. P., and Towers, N. R.,** Large scale isolation of lolitrem B, structure determination of some minor lolitrems, and tremorgenic activities of lolitrem B and paxilline in sheep, in *Proc. 2nd Int. Symp. Acremonium/Grass Interactions,* Hume, D. E., Latch, G. C. M., and Easton, H. S., Eds., AgResearch Grasslands, Palmerston North, New Zealand, 85, 1993.

88. **Miles, C. O., Wilkins, A. L., Gallagher, R. T., Hawkes, A. D., Munday, S. C., and Towers, N. R.,** Synthesis and tremorgenicity of paxitriols and lolitriol: possible biosynthetic precursors of lolitrem B, *J. Agric. Food Chem.,* 40, 234, 1992.

89. **Miller, J. D.,** Toxic metabolites of epiphytic and endophytic fungi of conifer needles, in *Microbiology of the Phyllosphere,* Fokkema, N. J. and Van den Heuvel, J., Eds., Cambridge University Press, Cambridge, England, 1986, 221.

90. **Mortimer, P. H. and di Menna, M. E.,** Ryegrass staggers: further substantiation of a *Lolium* endophyte aetiology and the discovery of weevil resistance of ryegrass pastures infected with *Lolium* endophyte, *Proc. N.Z. Grassland Assoc.,* 44, 240, 1983.

91. **Muegge, M. A., Quisenberry, S. S., Bates, G. E., and Joost, R. E.,** Influence of *Acremonium* infection and pesticide use on seasonal abundance of leafhoppers and froghoppers (Homoptera: Cicadellidae; Cercopidae) in tall fescue, *Environ. Entomol.,* 20, 1531, 1991.

92. **Nozawa, K., Horie, Y., Udagawa, S., Kawai, K., and Yamazaki, M.,** Isolation of a new tremorgenic indoloditerpene, 1'-O-acetylpaxilline, from *Emericella striata* and distribution of paxilline in *Emericella* spp., *Chem. Pharm. Bull.,* 37, 1387, 1989.

93. **Oliver, J. B., Pless, C. D., and Gwinn, K. D.,** Effect of endophyte, *Acremonium coenophialum,* in 'Kentucky 31' tall fescue, *Festuca arundinaceae,* on survival of *Popilla japonica,* in *Proc. Int. Symp. Acremonium/Grass Interactions,* Quisenberry, S. S. and Joost, R. E., Eds., Louisiana Agric. Exp. Stn., Baton Rouge, LA, 1990, 173.

94. **Penn, J., Garthwaite, I., Christensen, M. J., Johnson, C. M., and Towers, N. R.,** The importance of paxilline in screening for potentially tremorgenic *Acremonium* isolates, in *Proc. 2nd Int. Symp. Acremonium/ Grass Interactions,* Hume, D. E., Latch, G. C. M., and Easton, H. S., Eds., AgResearch Grasslands, Palmerston North, New Zealand, 1993, 88.

95. **Patterson, C. G., Potter, D. A., and Fannin, F. F.,** Feeding deterrency of alkaloids from endophyte-infected grasses to Japanese beetle grubs, *Entomol. Exp. Appl.,* 61, 28585, 1991.

96. **Pearson, W. D.,** The pasture mealy bug, *Balanococcus poae* (Maskell), in Canterbury: a preliminary report, *Proc. 5th Australasian Conf. Grassland Invert. Ecol.,* Stahle, P. P., Ed., D. & D. Printing Pty., Ltd., Victoria, Australia, 1989, 297.

97. **Plowman, T. C., Leuchtmann, A., Blaney, C., and Clay, K.,** Significance of the fungus *Balansia cyperi* infecting medicinal species of *Cyperus* (Cyperaceae) from Amazonia, *Econ. Bot.,* 44, 1990.

98. **Popay, A. J. and Mainland, R. L.,** Seasonal damage by Argentine stem weevil to perennial ryegrass pastures with different levels of *Acremonium lolii, Proc. 44th N. Z. Weed Pest Control Conf.,* 171, 1991.

99. **Popay, A. J., Prestidge, R. A., Rowan, D. D., and Dymock, J. J.,** The role of *Acremonium lolii* mycotoxins in insect resistance of perennial ryegrass (*Lolium perenne*), in *Proc. Int. Symp. Acremonium/ Grass Interactions,* Quisenberry, S. S. and Joost, R. E., Eds., Louisiana Agric. Exp. Stn., Baton Rouge, LA, 1990, 44.

100. **Popay, A. J., Mainland, R. A., and Sanders, C. J.,** The effects of endophytes in fescue grass on growth and survival of third instar grass grub larvae, in *Proc. 2nd Int. Symp. Acremonium/Grass Interactions,* Hume, D. E., Latch, G. C. M., and Easton, H. S., Eds., AgResearch Grasslands, Palmerston North, New Zealand, 1993, 174.

101. **Potter, D. A., Patterson, C. G., and Redmond, C. T.,** Influence of turfgrass species and tall fescue endophyte on feeding ecology of Japanese beetle and southern masked chafer grubs (Coleoptera: Scarabaeidae), *J. Econ. Entomol.,* 85, 900, 1992.

102. **Prestidge, R. A.,** Preliminary observations on the grassland leafhopper fauna of the central North Island Volcanic Plateau, *N.Z. Entomol.,* 12, 54, 1989.

103. **Prestidge, R. A.,** Susceptibility of Italian ryegrasses *Lolium multiflorum* Lam. to Argentine stem weevil *Listronotus bonariensis* Kuschel feeding and oviposition, *N.Z. J. Agric. Res.,* 34, 119, 1991.

104. **Prestidge, R. A. and Gallagher, R. T.,** Lolitrem B — a stem weevil toxin isolated from *Acremonium*-infected ryegrass, *Proc. 38th N.Z. Weed Pest Control Conf.,* 38, 38, 1985.

105. **Prestidge, R. A. and Gallagher, R. T.,** Endophyte fungus confers resistance to ryegrass: Argentine stem weevil studies, *Ecol. Entomol.,* 13, 429, 1988.

106. **Prestidge, R. A., Pottinger, R. P., and Barker, G. M.,** An association of *Lolium* endophyte with ryegrass resistance to Argentine stem weevil, *Proc. 35th N. Z. Weed Pest Control Conf.,* 35, 119, 1982.

107. **Prestidge, R. A., di Menna, M. E., van der Zijpp, S., and Badan, D.,** Ryegrass content, *Acremonium* endophyte and Argentine stem weevil in pastures in the volcanic plateau, *Proc. 38th N.Z. Weed Pest Control Conf.,* 41, 1985.

108. **Riedell, W. E. . Kieckhefer, R. W., Petroski, R. J., and Powell, R. G.,** Naturally occurring and synthetic loline alkaloid derivatives: feeding deterrent activity and toxicity against insects, *J. Entomol. Soc.,* 26, 122, 1991.

109. **Rhoades, D. F.,** Evolution of plant chemical defense against herbivores, in *Herbivores: Their Interaction with Secondary Plant Metabolites,* Rosenthal, G. A. and Janzen, D. H., Eds., Academic Press, New York, 1979, 4.

110. **Rottinghaus, G. E., Garner, G. B., Cornell C. N., and Ellis, J. L.,** An HPLC method for quantitating ergovaline in endophyte infected tall fescue: seasonal variation of ergovaline levels in stems, leaves and seed heads, *J. Agric. Food Chem.,* 39, 112, 1991.

111. **Rottinghaus, G. E., Rowan, D. D., and Tapper, B. A.,** Alkaloids in pasture grasses, in *Modern Methods of Plant Analysis,* Vol. 15. Alkaloids, Linsken, H. F. and Jackson, J. F., Eds., Springer-Verlag, Heidelberg, Germany, 1993, in press.

112. **Rowan, D. D. and Gaynor, D. L.,** Isolation of feeding deterrents against Argentine stem weevil from ryegrass infected with the endophyte *Acremonium loliae, J. Chem. Ecol.,* 12, 647, 1986.

113. **Rowan, D. D., Dymock, J. J., and Brimble, M. A.,** Effect of fungal metabolite peramine and analogs on feeding and development of Argentine stem weevil (*Listronotus bonariensis*), *J. Chem. Ecol.,* 16, 1683, 1990.

114. **Rowan, D. D., Hunt, M. B., and Gaynor, D. L.,** Peramine, a novel insect feeding deterrent from ryegrass infected with the endophyte *Acremonium loliae, J. Chem. Soc. Chem. Commun.,* 935, 1986.

115. **Saha, D. C., Johnson-Cicalese, J. M., Halisky, P. M., Van Heemstra, M. I., and Funk, C. R.,** Occurrence and significance of endophytic fungi in the fine fescues, *Plant Dis.,* 71, 1021, 1987.
116. **Schmidt, D.,** La quenouille rend-elle le fourrage toxique?, *Rev. Suisse Agric.,* 18, 329, 1986.
117. **Siegel, M. R. and Schardl, C. L.,** Fungal endophytes of grasses: detrimental and beneficial associations, in *Microbial Ecology of Leaves,* Andrews, J. H. and Hirono, S. S., Eds., Springer-Verlag, New York, 1992, in press.
118. **Siegel, M. R., Latch, G. C. M., and Johnson, M. C.,** Fungal endophytes of grasses, *Annu. Rev. Phytopathol.,* 25, 293, 1987.
119. **Siegel, M. R., Dahlman, D. L., and Bush, L. P.,** The role of endophytic fungi in grasses: new approaches to biological control of pests, in *Integrated Pest Management for Turfgrass and Ornamentals.* Leslie, A. R. and Metcalfe, R. L., Eds., U.S. Environmental Protection Agency, Office of Pesticide Programs, Washington, D.C., 1989, 169.
120. **Siegel, M. R., Latch, G. C. M., Bush, L. P., Fannin, N. F., Rowan, D. D., Tapper, B. A., Bacon, C. W., and Johnson, M. C.,** Fungal endophyte-infected grasses: alkaloid accumulation and aphid response, *J. Chem. Ecol.,* 16, 3301, 1990.
121. **Springer, T. L. and Kindler, S. D.,** Endophyte-enhanced resistance to the Russian wheat aphid and the incidence of endophytes in fescue species, in *Proc. Int. Symp. Acremonium/Grass Interactions,* Quisenberry, S. S. and Joost, R. E., Eds., Louisiana Agric. Exp. Stn., Baton Rouge, LA, 1990, 194.
122. **Stewart, A. V.,** Perennial ryegrass seedling resistance to Argentine stem weevil, *N. Z. J. Agric. Res.,* 28, 403, 1985.
123. **Steyn, P. S. and Vleggaar, R.,** Application of biosynthetic techniques in the structural studies of mycotoxins, in *Modern Methods in the Analysis and Structural Elucidation of Mycotoxins,* Cole, R. J., Ed., Academic Press, New York, 1986, 177.
124. **Stone, M. J. and Williams, D. H.,** On the evolution of functional secondary metabolites (natural products), *Mol. Microbiol.,* 6, 29, 1992.
125. **Stovall, M. E. and Clay, K.,** Adverse effects on fall armyworm feeding on fungus-free leaves of fungus-infected plants, *Ecol. Entomol.,* 16, 519, 1991.
126. **Tapper, B. A., Rowan, D. D., and Latch, G. C. M.,** Detection and measurement of the alkaloid peramine in endophyte-infected grasses, *J. Chromatogr.,* 463, 133, 1989.
127. **TePaske, M. R., Gloer, J. B., Wicklow, D. T., and Dowd, P. K.,** Aflavazole: a new antiinsectant carbazole metabolite from the sclerotia of *Aspergillus flavus, J. Org. Chem.,* 55, 5299, 1990.
128. **van Heeswijck, R. and McDonald, G.,** *Acremonium* endophytes in perennial ryegrass and other pasture grasses in Australia and New Zealand, *Aust. J. Agric. Res.,* 43, 1683, 1992.
129. **Webber, J.,** A natural control of Dutch elm disease, *Nature,* 292, 449, 1981.
130. **Weedon, C. M. and Mantle, P. G.,** Paxilline biosynthesis by *Acremonium loliae*: a step towards defining the origin of lolitrem neurotoxins, *Phytochemistry,* 26, 969, 1987.
131. **White, J. F., Jr.,** Endophyte–host associations in forage grasses. XI. A proposal concerning origins and evolution, *Mycologia,* 80, 442, 1988.
132. **Yates, S. G. and Powell, R. G.,** Analysis of ergopeptine alkaloids in endophyte-infected tall fescue, *J. Agric. Food Chem.,* 36, 337, 1988.
133. **Yates, S. G., Plattner, R. D., and Garner, G. B.,** Detection of ergopeptine alkaloids in endophyte infected, toxic Ky-31 tall fescue by mass spectrometry, *J. Agric. Food Chem.,* 33, 719, 1985.
134. **Yates, S. G. . Fenster, J. C., and Bartelt, R. J.,** Assay of tall fescue seed extracts, fractions, and alkaloids using the large milkweed bug, *J. Agric. Food Chem.,* 37, 354, 1989.
135. **Yoshihara, T., Togiya, S., Koshino, H., and Sakamura, S.,** Three fungitoxic cyclopentanoid sesquiterpenes from stromata of *Epichloe typhina, Tetrahedron Lett.,* 26, 5551, 1985.

Chapter 5

THE COST OF PLANT CHEMICAL DEFENSE AGAINST HERBIVORY: A BIOCHEMICAL PERSPECTIVE

Jonathan Gershenzon

TABLE OF CONTENTS

0-8493-4125-6/94/$0.00+$.50
© 1994 by CRC Press Inc.

I. INTRODUCTION

Plants synthesize an enormous variety of secondary metabolites that are toxic and deterrent to herbivores.[224,428] These substances are usually assumed to serve as defenses against herbivory. However, the benefits of secondary metabolites in reducing herbivore damage may be at least partly offset by their metabolic cost.[78,87,162,169,300,412,414,469] In many plant species, significant quantities of energy and nutrients appear to be devoted to the manufacture, storage, and maintenance of defensive compounds.

The costs of chemical defenses can be expressed in evolutionary as well as metabolic terms by computing the loss in fitness resulting from the commitment of resources to defensive purposes. It is generally acknowledged that plants have limited amounts of energy and nutrients to apportion among growth, reproduction, and defense.[29] Thus, allocation to defense necessarily reduces allocations to other purposes,[78,300,412,470] and, as a consequence, investment in defense can only be justified if it is likely that the value of the resources saved from herbivory will exceed the expected reductions in growth rate or reproductive output. The evolutionary cost of defense can theoretically be measured by comparing the fitness of defended plants to that of undefended plants in the absence of herbivores.[412,471]

In practice, the cost of plant chemical defense is difficult to assess directly in either metabolic or evolutionary terms. Nevertheless, it is widely believed that costs impose significant constraints on the types and amounts of defenses produced.[78,87,162,169,300,412,414] Hence, defense costs have remained a frequent topic of discussion in plant–herbivore research for many years, and cost has been accorded a prominent role in several hypotheses seeking to explain intraspecific and interspecific patterns of plant defense.[87,162,412,414] The costs of chemical defense have been examined theoretically with the aid of quantitative models,[87,150,209,471,473] and several attempts have been made to estimate costs empirically by calculating the proportion of a limiting resource, such as nitrogen or fixed carbon, that is allocated to specific defense compounds.[74,78,209,324,357,423,541]

The actual existence of chemical defense costs in plants can be inferred from two major lines of evidence. First, investment in defense has often been shown to reduce the resources available for growth or reproduction. Inverse correlations between growth rate and the concentration of defensive substances or reproductive output and the amount of defenses have been reported in numerous plant species,[18,33,40,56,73,86,125,167,211,223,273,345] although this is not always the case.[53,56,60,434] Second, defenses appear to be reduced or eliminated when not needed. Plants growing in environments where herbivore pressure is low have been found to have reduced defenses compared to conspecifics growing in areas of greater herbivore pressure.[73,88,126,145,237,261,565]

Chemical defense may not always incur a measurable fitness cost. Simms and Rausher found that the constitutive resistance of *Ipomoea purpurea* to several species of insect herbivores had no detectable cost.[470,471] These authors and others have also criticized much of the evidence commonly adduced in support of the costliness of plant defense.[60,169] For example, Simms and Rauscher point out that, within a species, the diminished defenses sometimes seen in populations found in areas of low herbivore density could result from many other selective forces besides reductions in herbivore pressure.[471] Similarly, the inverse correlations reported between defense and growth rate or defense and reproductive output could result from spurious comparisons among plants of very different genetic or environmental backgrounds, rather than from direct trade-offs between defense and other plant functions.[60] Thus, allocation to defense may not always represent a significant drain on plant resources. Fitness costs may only be evident at certain developmental stages[18,56] or under particular environmental conditions, such as stress or intense competition.

If chemical defense against herbivory imposes significant fitness costs on plants, then the metabolic processes involved in deploying defensive compounds are worth careful study. This chapter focuses on the metabolic or physiological costs ("direct costs"[209]) of chemical defense. Plants that rely on secondary metabolites for protection against herbivores must be able to 1) synthesize these substances from primary metabolic intermediates, 2) build and maintain suitable storage structures for them, 3) transport them from the site of synthesis to the site of storage, if necessary, and 4) maintain an effective level of protection by making up for any losses that occur in storage. Each of these tasks may require a substantial expenditure of energy and nutrients. In addition, all activities must be carried out in a way that minimizes the risk of self-poisoning, since defense compounds are often general metabolic toxins that have the potential to be as harmful to the plant as to their intended herbivore targets.[354]

Although these various components of defense costs are often mentioned in the literature on plant–herbivore interactions, their overall contributions to the chemical defense budget have not been evaluated critically in light of recent progress in the biochemistry of secondary plant metabolites. In the last few years, our knowledge of secondary product biosynthesis has increased dramatically. New information has become available on the sequences of pathway intermediates, the characteristics of some key enzymes, and the control mechanisms that regulate metabolic flux. In addition, our understanding of the transport, turnover, and storage of plant secondary metabolites has improved substantially. This information permits some interesting new insights into the costs associated with plant chemical defense. To be sure, we are still a long way from being able to calculate these costs rigorously. However, it is now possible to identify several emerging general properties of plant secondary metabolism and suggest how these can be expected to influence defense costs. This review examines the costs of synthesizing, transporting, storing, and maintaining antiherbivore chemical defenses in the context of recent biochemical advances. Examples are chosen from a range of different compound classes, with an emphasis placed on recent research on monoterpene metabolism. Throughout, it is assumed that the majority of terpenoid, phenolic, and nitrogenous secondary metabolites in plants serve as defensive substances, even though direct evidence for this function is often lacking.

II. THE COST OF SYNTHESIS

The most widely discussed costs of plant chemical defense are those associated with the synthesis of defensive substances. These costs are not inherently different from the costs of producing other cellular metabolites. Supplies of certain substrates and cofactors are required, as are enzymes, nucleic acids, and other appropriate macromolecular machinery. The principal pathways involved in the formation of defensive secondary metabolites in plants originate from just a few key intermediates of primary metabolism, including acetyl coenzyme A, mevalonic acid, and shikimic acid, and the amino acids phenylalanine, tyrosine, tryptophan, lysine, and ornithine. Among the essential cofactors are ATP, NADPH, and S-adenosylmethionine, all of which need to be continually regenerated by respiration or photosynthesis.

A. RAW MATERIALS: SUBSTRATES AND COFACTORS

A number of attempts have been made to quantify the raw materials costs for the production of plant defense compounds[74,78,209,304,357,541] based on methodology originally developed for calculating growth yield and growth efficiency in microorganisms.[351,391] In this approach, cost is determined by computing the quantity of carbohydrate needed to provide the necessary carbon skeletons and cofactors for biosynthesis. Glucose is usually chosen as the starting material, because it can be respired readily to give ATP and redox equivalents. Carbohydrates, such as glucose, appear to be a suitable "currency" for estimating biosynthetic costs, as they constitute the usual storage and transport forms of carbon and energy in plants, and the supply of fixed carbon is often limiting to growth.[74] Similar methods, in which costs are defined in terms of ATP equivalents, have also been employed.[12,324]

The glucose requirement for synthesizing a given metabolite can be calculated in several different ways. If the biosynthetic pathway is known in sufficient detail, direct inspection allows one to add up the number of moles of glucose needed for both the carbon skeleton and the formation of the required cofactors.[74,78,357,391,541] Alternatively, the molecular formula can be used to estimate the number of carbon atoms and reducing equivalents necessary for biosynthesis,[351,541] ignoring any glucose required for the synthesis of ATP, *S*-adenosylmethionine, or the reduction of molecular oxygen.

Using the "pathway inspection" method (see Appendix), the raw materials costs for the synthesis of a variety of secondary metabolites representing most of the major classes of plant defenses have been computed. The results are presented in Table 1 as the grams of glucose required per gram of defense compound (structures given in Figure 1). All calculations are based on the most recent available biosynthetic information, and, consequently, they are often quite different from cost estimates made for similar or identical compounds in prior publications.[74,304,357,391,541] The costs of some primary metabolites are listed for comparison in Table 2.

The overall range of glucose costs is large, with the most expensive compound, nicotine (3.62 g glucose/g), being almost three times as costly as the cheapest compound, tellimagrandin II, a hydrolyzable tannin (1.28 g glucose/g). Of the major classes of secondary metabolites, the terpenoids are among the most expensive to produce per gram (average cost = 3.11 g glucose/g) due to their high level of chemical reduction. Accordingly, the costliest terpenes are the ones with the lowest proportions of oxygen. Compare the glucose requirements for caryophyllene ($C_{15}H_{24}$, 3.54 g glucose/g) or menthone ($C_{10}H_{18}O$, 3.37) with those for papyriferic acid ($C_{35}H_{56}O_8$, 2.72) or aucubin ($C_{15}H_{22}O_9$, 2.39).

In contrast to the terpenoids in Table 1, the phenolic compounds listed are not especially expensive (average cost = 2.11), except for the furanocoumarin psoralen (3.39). The biosynthesis of the furan ring of psoralen involves the attachment of a C_5 terpenoid side chain to the basic coumarin skeleton with the accompanying loss of three carbon atoms. The high cost of 5,6,7,8-tetramethoxyflavone (2.37) relative to the other flavonoids listed is due to the presence of the four methoxyl groups in this molecule, each of which is formed by the addition of a methyl group from the cofactor *S*-adenosylmethionine. The regeneration of *S*-adenosylmethionine via homocysteine and methionine is quite expensive, requiring ATP and a tetrahydrofolate derivative. In fact, the cost of a single carbon atom from this cofactor is nearly twice as much as one derived directly from glucose.[12] Yet, the methylation of oxygen, nitrogen, or sulfur atoms by *S*-adenosylmethionine is a common feature of the biosynthesis of many plant defense compounds,[396] possibly because it is a convenient way to increase hydrophobicity.

Note that the biosynthetic costs of tannins are not very high, with hydrolyzable tannins being particularly inexpensive. Similar results were obtained when tannin costs were computed on the basis of ATP equivalents.[324] Tannin synthesis was erroneously depicted as costly, relative to terpenes and nicotine, in an otherwise thoughtful evaluation of defense costs recently published by Skogsmyr and Fagerstrom, due to an invalid comparison between costs expressed on a per gram basis and costs expressed on a per mole basis.[473] (In this chapter, I

TABLE 1
Raw Materials Costs for the Formation of Various Antiherbivore Defense Compounds

Chemical class	Compound (and corresponding number in Fig. 1)	Cost per gram (g glc/g)	Concentration in plant	Species and reference	Organ	Cost per gram of plant tissue (mg glc/g)
Terpenoids						
Monoterpenes	Camphor (1)	3.10	0.6% DW[a]	*Tanacetum vulgare*[183]	Mature leaves	19
Monoterpenes	Menthone (2)	3.37	0.9% DW	*Mentha x piperita*[183]	Mature leaves	30
Iridoid monoterpenes	Aucubin (3)	2.39	3.0% DW	*Plantago lanceolata*[53]	Mature leaves	72
Sesquiterpenes	Caryophyllene (4)	3.54	0.5% DW	*Hymenaea courbaril*[306,307]	Leaves	19
Sesquiterpenes	Germacrone (5)	3.34	0.5% DW	*Ledum groenlandicum*[408]	Leaves	17
Diterpenes	Pachydictyol A (6)	3.35	0.7% DW	*Pachydictyon coriaceum*[239]	Fronds	23
Diterpenes	Abietic acid (7)	3.34	0.1% DW	*Larix laricina*[381]	Needles	3
Triterpenes	Cucurbitacin B (8)	2.87	0.3% FW	*Cucurbita ecuadorensis*[358]	Roots	8
Triterpenes	Papyriferic acid (9)	2.72	11.3% DW	*Betula resinifera*[409]	Juvenile twig internodes	307
Phenolics						
Phenol glycosides	Salicin (10)	1.91	3.0% DW	*Populus trichocarpa*[435]	Leaves	57
Hydroxycinnamic acid esters	Chlorogenic acid (11)	1.59	3.5% FW	*Salix integra*[347]	Leaves	56
Simple coumarins	Daphnetin (12)	2.02	0.9% FW	*Daphne mezereum*[572]	Leaves	17
Furanocoumarins	Psoralen (13)	3.39	0.1% DW	*Ruta graveolens*[573]	Leaves	4
Hydrolyzable tannins	Tellimagrandin II (14)	1.28	0.4% FW	*Liquidambar formosana*[233]	Leaves	5
Condensed tannins	Linear procyanidin polymer (15)	2.08	1.2% DW	*Betula alleghieniensis*[17]	Leaves	25
Flavonoids	Apigenin (16)	2.06	5.0% DW	*Isocoma acradenia*[82]	Leaves	103
Flavonoids	5,6,7,8-Tetramethoxyflavone (17)	2.37	0.3% DW	*Godmania aesculifolia*[485]	Leaves	7
Isoflavonoids	Daidzein (18)	2.25	0.1% DW	*Phaseolus mungo*[467]	Seeds	2
Alkaloids						
Benzylisoquinoline	Thebaine (19)	3.01	0.3% DW	*Papaver somniferum*[250]	Capsules and upper stem	9
Monoterpene-indole	Ajmalicine (20)	3.47	0.5% DW	*Catharanthus roseus*[132]	Roots	17
Quinolizidine	Lupanine (21)	3.23	0.2% DW	*Lupinus polyphyllus*[544,549]	Leaves	7
Pyridine/pyrrolidine	Nicotine (22)	3.62	0.5% DW	*Nicotiana sylvestris*[16]	Leaves	18
Purine	Caffeine (23)	2.89	2.5% DW	*Camellia sinensis*[382]	Leaves	72

TABLE 1 (Continued)
Raw Materials Costs for the Formation of Various Antiherbivore Defense Compounds

Chemical class	Compound (and corresponding number in Fig. 1)	Cost per gram (g glc[a]/g)	Concentration in plant	Species and reference	Organ	Cost per gram of plant tissue (mg glc/g)
Other nitrogenous defenses						
Cyanogenic glycosides	Prunasin (24)	2.11	1.5% DW	*Amelanchier alnifolia*[336]	Leaves	32
Cyanogenic glycosides	Linamarin (25)	1.96	0.4% DW	*Hevea brasiliensis*[325]	Seedling leaves	8
Glucosinolates	Methyl glucosinolate (26)	1.70	1.2% DW	*Cleome serrulata*[332]	Leaves	20
Glucosinolates	3,4-Dihydroxy-benzylglucosinolate (27)	1.80	0.7% DW	*Bretschneidera sinensis*[51]	Leaves	13
Nonprotein amino acids	Mimosine (28)	2.83	1.5% DW	*Leucaena leucocephala*[344]	Mature leaves	42
Nonprotein amino acids	Canavanine (29)	2.43	0.2% DW	*Medicago sativa*[360]	Seedlings	5
Proteinase inhibitors	Inhibitor I from tomato	2.71	0.01% FW	*Lycopersicon esculentum*[200]	Leaves 24 h after wounding	0.27
Proteinase inhibitors	Inhibitor II from tomato	2.59	0.004% FW	*Lycopersicon esculentum*[201]	Leaves 24 h after wounding	0.10

Note: Costs were calculated as the quantity of glucose required to make all the starting materials and cofactors necessary for biosynthesis. (For additional information, see Appendix.) Chemical structures are given in Figure 1. Costs per gram of plant tissue were computed using the concentrations listed. Concentration data from leaves were used whenever possible to facilitate comparisons among compounds.

[a] Abbreviations: glc = glucose; DW = dry weight, FW = fresh weight.

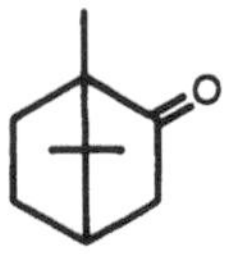

camphor (**1**)

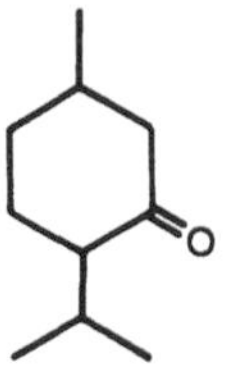

menthone (**2**)

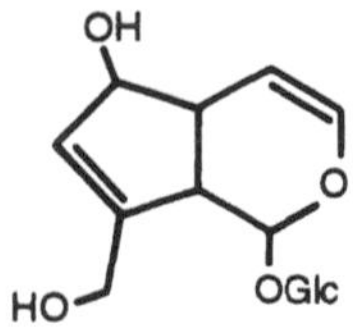

aucubin (**3**)

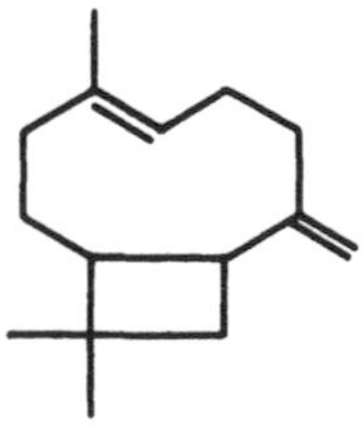

caryophyllene (**4**)

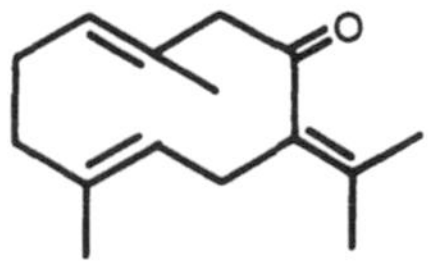

germacrone (**5**)

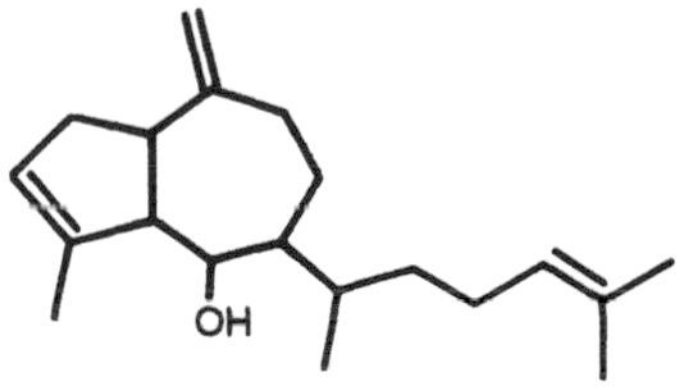

pachydictyol (**6**)

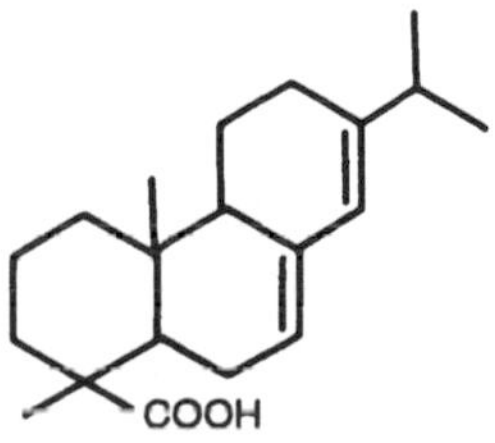

abietic acid (**7**)

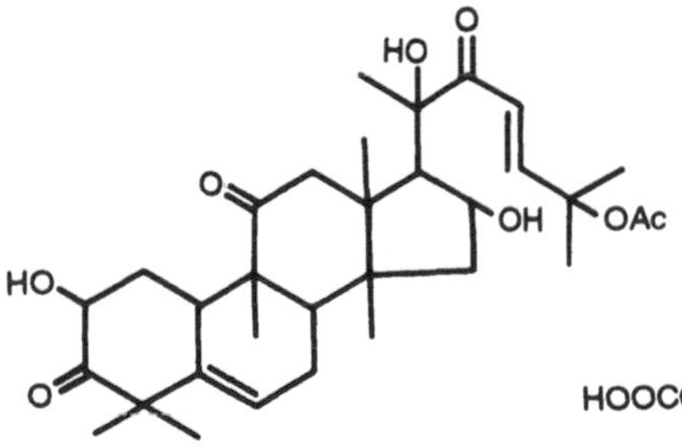

cucurbitacin B (**8**)

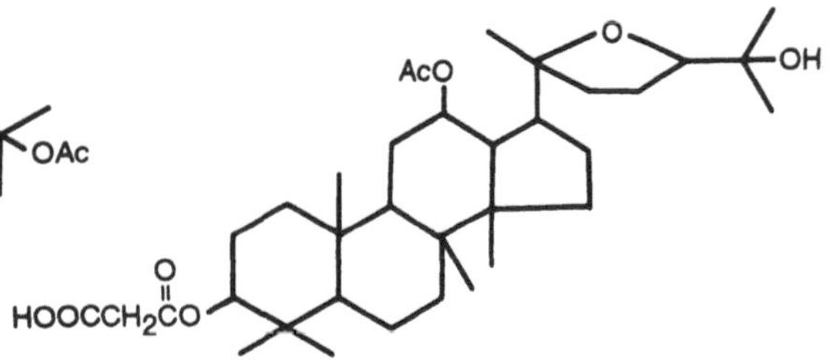

papyriferic acid (**9**)

FIGURE 1A. Structures of antiherbivore defense compounds listed in Table 1.

have calculated costs on a per gram basis because the effectiveness of defense compounds is generally assumed to depend on their fraction by weight of plant tissue rather than their fraction by mole.) The cheapest phenolic compounds in Table 1 include salicin (a phenolic

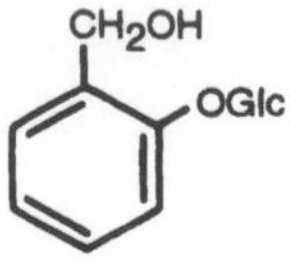

salicin (**10**)

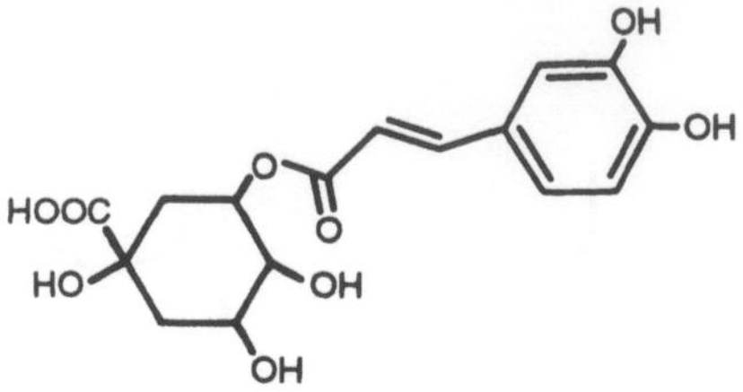

chlorogenic acid (**11**)

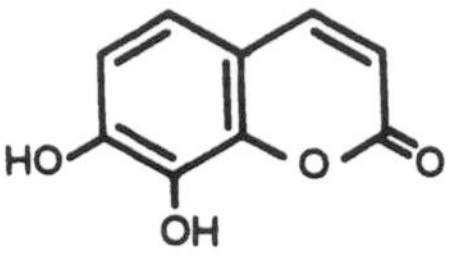

daphnetin (**12**)

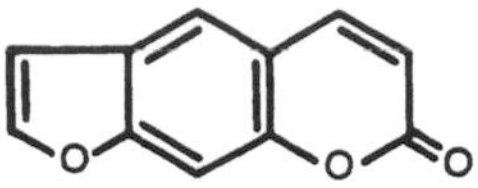

psoralen (**13**)

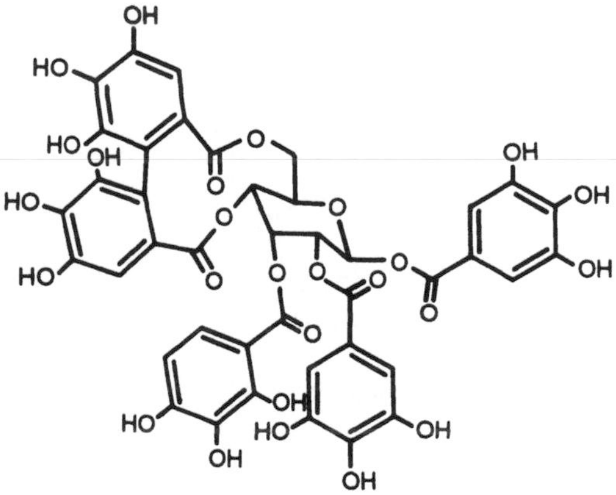

tellimagrandin II (**14**)

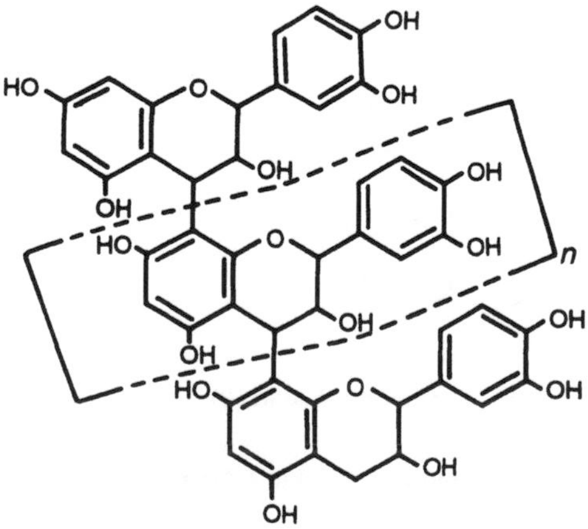

linear procyanidin polymer (**15**)
(*n*=1,2,3,4,5...)

FIGURE 1B.

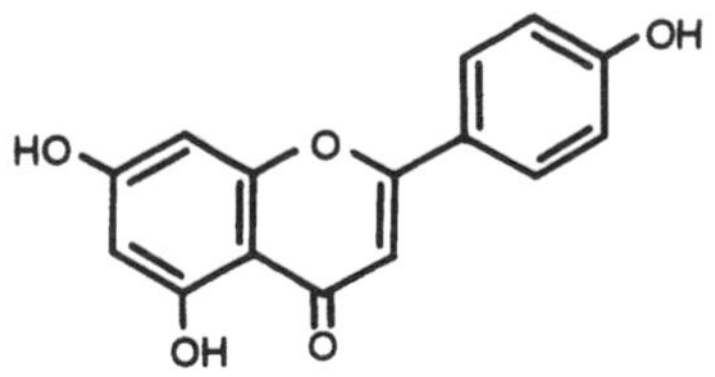

apigenin (**16**)

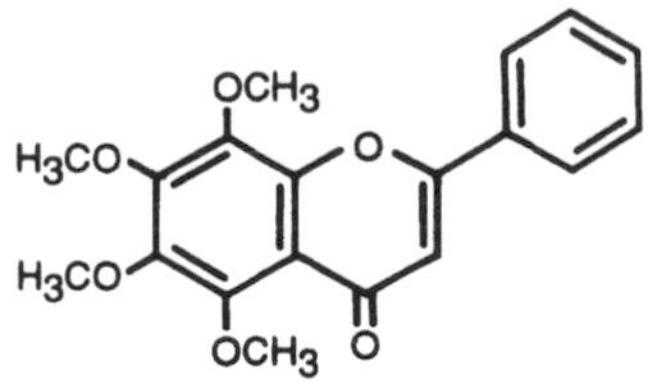

5,6,7,8 - tetramethoxyflavone (**17**)

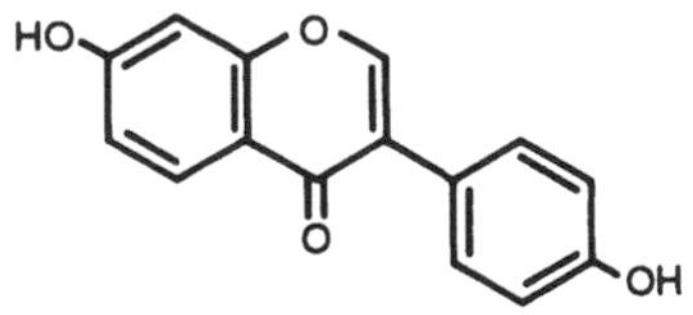

daidzein (**18**)

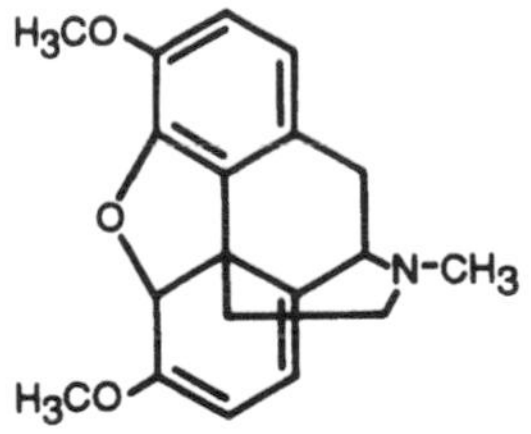

thebaine (**19**)

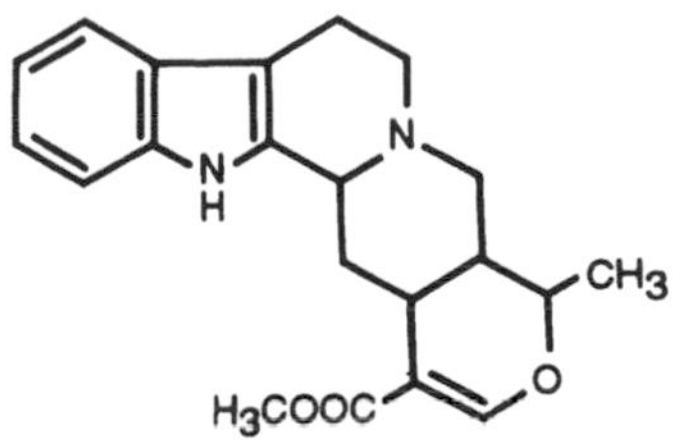

ajmalicine (**20**)

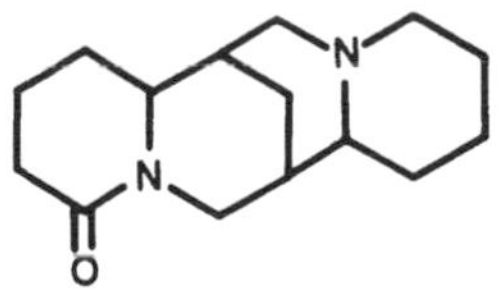

lupanine (**21**)

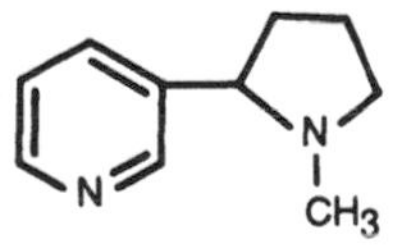

nicotine (**22**)

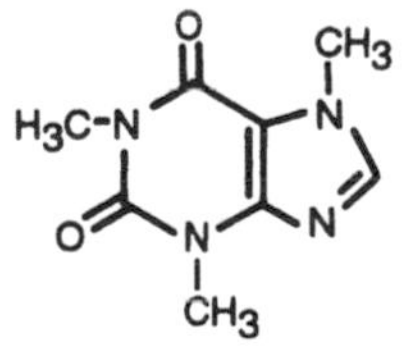

caffeine (**23**)

FIGURE 1C.

glycoside), chlorogenic acid (a quinate ester of caffeic acid), and tellimagrandin II (a hydrolyzable tannin), all of which contain a highly oxygenated carbohydrate moiety.

Alkaloids constitute the most expensive class of defense substances in Table 1, with the average biosynthetic cost of the compounds listed being 3.24 g glucose/g. However, these high costs were computed on the assumption that nitrogen was supplied as nitrate rather than as ammonium (see Appendix). The use of ammonium in biosynthesis results in a considerable savings of reducing power, equivalent to 4 NADPH molecules (= 0.342 glucose units) per nitrogen atom. For example, the costs of constructing nicotine (2.86) and caffeine (1.61) from

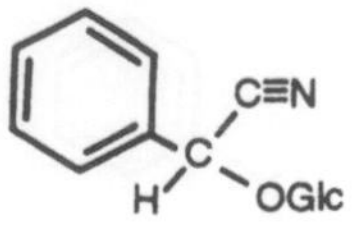

prunasin (**24**)

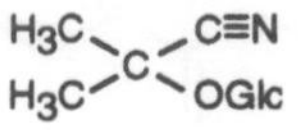

linamarin (**25**)

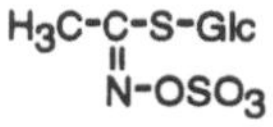

methyl glucosinolate (**26**)

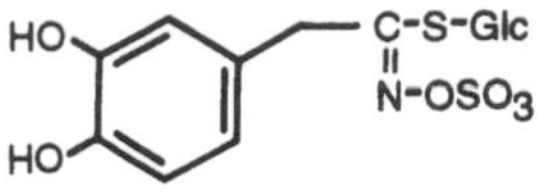

3,4-dihydroxybenzyl glucosinolate (**27**)

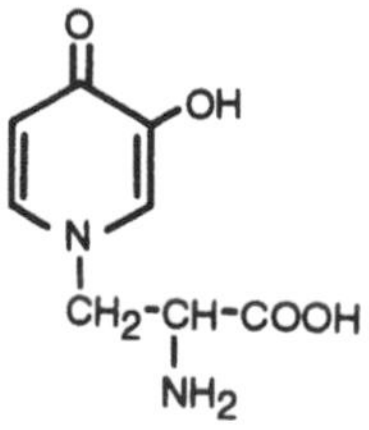

mimosine (**28**)

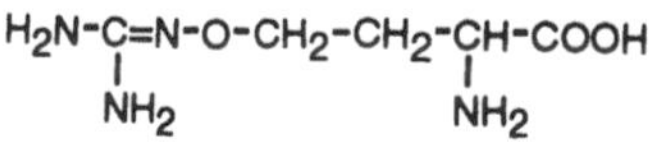

canavanine (**29**)

FIGURE 1D.

ammonium nitrogen are much lower than their construction costs from nitrate nitrogen (nicotine = 3.62, caffeine = 2.89). One of the most expensive alkaloids, ajmalicine, owes its high cost to the fact that its formation incorporates a monoterpene moiety.

The biosynthetic costs of the other nitrogen-containing defense compounds compiled in Table 1 are not especially high (average cost = 2.27 g glucose/g). The glucose requirements for producing these substances generally reflect those of their amino acid progenitors, with defenses derived from the more expensive amino acids, such as phenylalanine or lysine (Table 2), usually being more costly to synthesize than defenses originating from the cheaper amino acids, such as aspartate. Interestingly, most of the nitrogenous defenses listed are not at all expensive to make from their parent amino acids, a trend previously observed by other workers.[78,541] In fact, the glucose costs of the cyanogenic glycosides and glucosinolates in Table 1 are actually slightly lower than those of their precursor amino acids (Table 2). Thus, given the ready availability of the appropriate amino acids, the production of nitrogenous defensive compounds is not energetically expensive. However, since plant growth is commonly limited by low nitrogen availability,[75] the formation of nitrogen-containing defenses may frequently be restricted by the availability of nitrogen itself.[209] As a consequence, the real cost of nitrogen-based defenses might be more accurately assessed if nitrogen, rather than glucose, were used as the standard currency for evaluating cost.[74] When nitrogen is a limiting resource, its use in defense must come at the expense of growth or reproduction, making the synthesis of nitrogen-containing defenses a costly proposition. Under such conditions, plants may deploy nitrogenous defenses only if these compounds have additional functions besides defense or if the nitrogen in these substances can eventually be recovered for other uses (see Section VI).

TABLE 2
Raw Materials Costs for the Biosynthesis of Some Important
Primary Plant Metabolites

Compound	Cost (g glc/g)	Compound	Cost (g glc/g)
Carbohydrates		Amino acids	
Glucose	1.00	Alanine	1.76
Fructose	1.03	Arginine	2.79
Sucrose	1.09	Asparagine	1.77
Starch	1.11	Aspartate	1.23
Cellulose	1.11	Cysteine	2.39
Organic acids		Glutamate	1.56
Pyruvate	0.87	Glutamine	2.02
Malate	0.68	Glycine	1.52
Oxaloacetate	0.61	Histidine	2.58
Citrate	0.77	Isoleucine	2.39
Lipids		Leucine	2.31
Palmitic acid	3.01	Lysine	2.48
Linoleic acid	3.18	Methionine	2.82
Nucleotides		Phenylalanine	2.22
Adenosine monophosphate	1.80	Proline	2.31
Guanosine monophosphate	1.74	Serine	1.44
Cytosine monophosphate	1.54	Threonine	1.62
Uridine monophosphate	1.27	Tryptophan	2.58
Proteins		Tyrosine	1.93
Rubisco (large and small subunits)	2.41	Valine	2.11

Note: Calculations were performed as for Table 1 (see Appendix), and were based on established pathways.[197,498,523] The cost of Rubisco was computed using published amino acid sequences for the large subunit of maize[353] and the small subunit of pea.[30]

The availability of an ample supply of nitrogen does not by itself guarantee the production of nitrogenous defense compounds. For example, young, rapidly growing plants usually have a high nitrogen content but typically produce only low concentrations of many types of defenses. This is attributed to the fact that, in rapidly growing plants, the synthesis of defense compounds incurs high indirect costs or "opportunity costs" because it causes substantial reductions in growth rate[87,150,209] and could give less defended but more rapidly growing competitors a chance to preempt the space needed to acquire light, water, and other resources for future growth.[209,318] Rapid growth rates may also increase defense costs by progressively diluting the concentration of defensive metabolites present, thus necessitating a higher rate of synthesis to maintain a given concentration.

Besides growth rate, several additional factors could have a major influence on the biosynthetic costs of defense compounds. In computing the glucose requirements in Table 1, NADPH was assumed to be derived from glucose via the pentose phosphate pathway, and ATP was assumed to arise from glycolysis and the citric acid cycle coupled to electron transport. However, in photosynthetic cells, both of these cofactors can be produced in light-driven electron transport reactions, making their glucose costs substantially lower than in nonphotosynthetic cells.[79,351] Therefore, plant defense compounds synthesized in green tissues, such as diterpenes in *Nicotiana tabacum*,[280] isoflavonoids in *Glycine max*,[44] and quinolizidine alkaloids in *Lupinus* species,[545] may be considerably cheaper to construct than the estimates above would indicate.

The net cost of making chemical defenses depends not only on the amount of resources required for each gram of antiherbivore compound produced but also on the actual quantity of defensive substances present in the plant. For each compound listed in Table 1, its concentration in a representative species is also given. These were used to calculate costs per unit weight of plant tissue (see last column of Table 1). Costs expressed in this fashion vary over three orders of magnitude, a much larger range than that for costs per gram of compound alone. Thus, the overall outlay to plant defense may be influenced much more heavily by changes in the concentrations of defensive compounds than by differences among particular compounds in their costs per gram, a pattern first noted by Gulmon and Mooney.[209] For example, although the iridoid glycoside aucubin has a lower estimated cost per gram than the monoterpene menthone (2.39 versus 3.37), the actual aucubin concentration in *Plantago lanceolata* leaves is approximately 3.0% of dry weight compared to 0.9% for menthone in *Mentha x piperita* leaves. Therefore, the net cost of aucubin production in *P. lanceolata* (72 mg glucose/g leaf tissue) is over twice as high as that of menthone in *M. piperita* (30 mg glucose/g leaf tissue).

It is instructive to compare the raw materials costs of defensive compounds (Table 1) with those for various primary metabolites (Table 2). The production of defenses may require either a larger or smaller investment of resources than the production of primary plant metabolites, depending on the type of primary constituent under consideration. Carbohydrates, organic acids, and nucleotides have lower biosynthetic costs than nearly all the defense compounds listed, but acyl lipids, with their high degree of chemical reduction, are as expensive to make as most terpenoids and alkaloids. As previously noted, there is not much difference between the costs of the amino acids and those of most nitrogenous defense compounds.

B. BIOSYNTHETIC MACHINERY

The production of plant defenses obviously requires more than just a supply of raw materials. A variety of different types of macromolecular machinery is needed to transform metabolic intermediates into defensive products. Chief among these are the enzymes that catalyze the reactions of the biosynthetic pathway. Other essential supporting equipment includes the nucleic acids and ribosomes from which the enzymes are derived and the cellular and subcellular membranes that must be present to keep pH, redox potential, and ionic strength in a range where the enzymes can function properly.

1. Enzymes

The net cost of the enzymes required in the formation of plant defenses will vary with the overall length of the biosynthetic pathway. From the examples given in Table 1, a biosynthetic route from primary precursor to secondary product can be as short as 5 steps (e.g., phenylalanine to apigenin) or at least as long as 13 steps (e.g., mevalonate to papyriferic acid). Long and short pathways are found in all major groups of plant defense compounds. A full accounting of enzyme costs must also include the enzymes present in glucosinolate- or cyanogenic glycoside-accumulating plants that, upon cell disruption, mediate the hydrolysis of these glycosides to toxic end products.[331,397]

The metabolic cost of an individual enzyme varies with its molecular weight, its amino acid composition, its rate of turnover, and its catalytic efficiency (which determines the concentration required for a sufficient rate of catalysis).[295] Unfortunately, for the enzymes of plant secondary metabolism, not enough information is available about these properties to permit meaningful conclusions to be drawn. Measurements of protein turnover would be especially useful for cost comparisons, but there is little reliable data on the turnover rate of most plant enzymes.[110,390,520] In animals, the half-lives of individual proteins vary from less than an hour to several weeks or even months and thus may have a significant impact on cost.[236,295]

The potentially high costs of protein turnover have been rationalized on the grounds that this process benefits an organism by allowing the constant adjustment of metabolic processes to varying conditions.[295,236] Consistent with this belief, regulatory proteins have often been found to have especially rapid turnover rates.[236,248] At this point, we can only speculate about

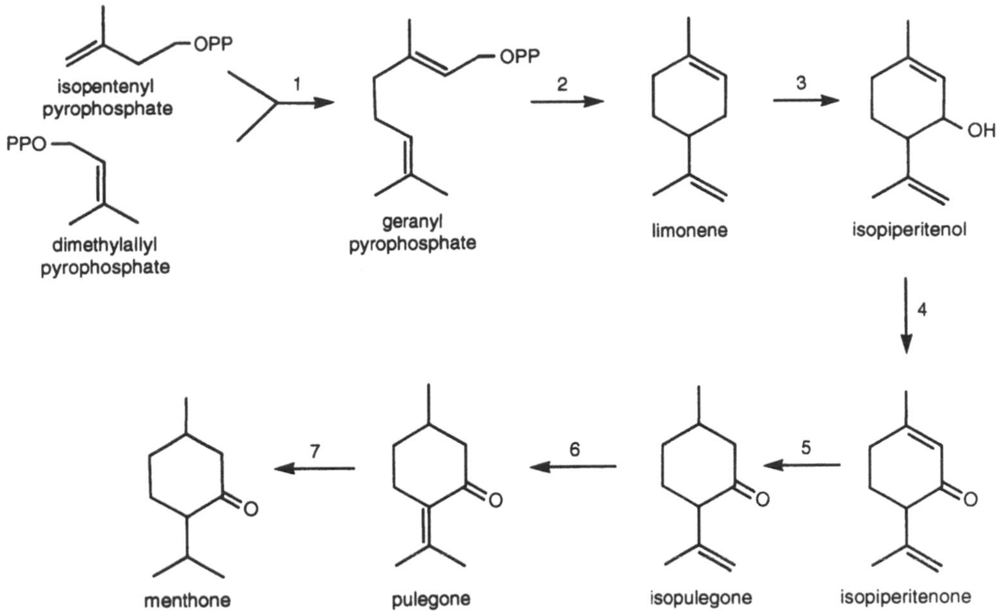

FIGURE 2. Pathway of menthone biosynthesis from isopentenyl pyrophosphate and dimethylallyl pyrophosphate in peppermint (*Mentha x piperita*). Enzyme names are given in Figure 3.

the turnover rates of most of the regulatory enzymes of plant secondary metabolism. However, phenylalanine ammonia-lyase and flavanone synthase, two possible rate-limiting enzymes of flavonoid biosynthesis, were calculated to have biological half-lives of just 5 to 10 h in cultured parsley cells,[215] considerably shorter than the half-lives reported for many other plant proteins.[520] In addition, tryptophan decarboxylase, which catalyzes the conversion of trypto-phan to tryptamine, a rate-controlling step in indole alkaloid biosynthesis, was found to have a biological half-life of about 21 h in cell suspension cultures of *Catharanthus roseus*.[380] Furthermore, it is interesting to note that a key enzyme of terpene biosynthesis, 3-hydroxy-3-methylglutaryl coenzyme A reductase (HMGR), has been shown to possess a PEST amino acid sequence in its structure,[15,66,81,370] a sequence with a high content of proline (P), glutamine (E), serine (S), and threonine (T) residues that appears to target proteins for rapid degrada-tion.[248] HMGR is an important regulatory enzyme of isoprenoid synthesis in animals;[138,441] it has been postulated to play a similar role in plants.[76,185,186]

Clearly, more information about enzyme turnover is needed before accurate estimates can be made of the costs of the biosynthetic machinery involved in making plant defenses. However, in the absence of such data, it could be helpful to consider what is known about changes in the activity of the enzymes of plant defense biosynthesis. If enzymes are only present in the plant for a short period of time, there should be less opportunity for them to undergo turnover.

As part of an ongoing investigation of the regulation of monoterpene metabolism in peppermint (*Mentha x piperita*), we recently examined the formation of monoterpenes in relation to leaf development.[187] Monoterpene biosynthesis was found to be restricted to a brief period during the first 2 weeks of leaf ontogeny. Assays of the seven enzymes of the monoterpene pathway between dimethylallyl pyrophosphate and menthone (Figure 2) showed that all of these activities were quite high during the first 2 weeks of leaf development but declined to very low levels thereafter (Figure 3). Therefore, if changes in enzyme activity are assumed to be reflective of changes in the levels of enzyme protein, then the enzymes of

THE ENZYMES OF MONOTERPENE BIOSYNTHESIS ARE ACTIVE FOR ONLY A BRIEF PERIOD DURING LEAF DEVELOPMENT

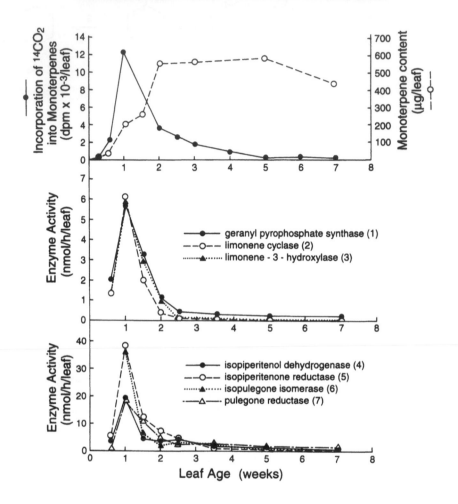

FIGURE 3. Changes in the monoterpene content, the rate of monoterpene biosynthesis, and the activities of monoterpene biosynthetic enzymes during leaf development in peppermint (*Mentha x piperita*). Leaves are fully expanded when 2 weeks old. To depict the activities of all the enzymes on the same scale, the activity of limonene-3-hydroxylase is represented as 100 × the actual value and the activity of isopiperitenol dehydrogenase as 1/10 the actual value. The pathway of monoterpene biosynthesis from isopentenyl pyrophosphate and dimethylallyl pyrophosphate to menthone is given in Figure 2. (Gershenzon, J. and Croteau, R., unpublished results.)

monoterpene biosynthesis appear to be present for only a brief period in leaf ontogeny, thereby reducing the likelihood of turnover with its attendant costs of replacement. Similar patterns are encountered for many other plant defense compounds. For example, enzymes involved in the biosynthesis of flavone glycosides in *Secale cereale*,[453] indole alkaloids in *Catharanthus roseus*,[111] and cyanogenic glycosides in *Sorghum bicolor*[218] are also only active for short periods in plant development, therefore reducing the potential costs of protein turnover. Such fluctuations in enzyme activity may, in fact, have important roles in regulating the rates of biosynthesis of these compounds.

In many plants, the production of defensive compounds is induced by herbivore or pathogen attack.[225,412,500] The enzymes participating in the biosynthesis of induced defenses have also been found to be active for only brief periods of time following bouts of herbivory or infection and are often not present in unattacked plants,[105,137,180,398,484] thus eliminating the opportunity for protein turnover. For example, in *Abies grandis*, the enzymatic machinery necessary for producing induced monoterpenes was not detectable in uninjured plants by immunoblotting but appeared several days after wounding.[192] The monoterpenes of *A. grandis* are important constituents of a viscous oleoresin that provides a defense against bark beetle attack.[188,322] Induced defenses are often touted as being more economical than constitutive defenses because the costs of synthesis, storage, and maintenance are incurred only intermittently (when the plant is under attack). As the pattern of biosynthetic enzyme occurrence indicates, expenditures for the synthesis of induced defenses are reduced not only by savings in raw materials but also by decreased production and upkeep costs for the enzymes themselves. However, plants that employ induced systems of defense must set aside a sufficient store of carbon and nutrients to rapidly synthesize protective compounds when needed.

2. Messenger RNA

Another important component of cellular biosynthetic machinery is the messenger RNA that encodes biosynthetic enzymes. Generally speaking, mRNA costs will be significantly less than enzyme costs since biosynthesis requires much smaller amounts of mRNA than enzymes. Each molecule of mRNA usually directs the formation of thousands of protein molecules. The cost of each species of mRNA is a function of its length, turnover rate, and the number of copies needed for an adequate rate of translation. Unfortunately, here again little information is available that is pertinent to plant secondary metabolism. The length of an mRNA molecule depends on the size of the corresponding protein and the proportion of introns to exons in the genomic sequence. Although the noncoding intron sequences are excised prior to translation, the size of the initially formed message is contingent upon the length of the introns as well as the exons. To date, reports on the structures of the genes encoding proteinase inhibitors[84,312] and various enzymes of terpenoid,[149,370,384] phenolic,[182,247,297,330,521] and alkaloid biosynthesis[54] indicate that their exon–intron architecture is diverse and, as a whole, not different from that of other plant genes. Hence, there is no indication that the length of the mRNAs involved in plant defense varies significantly from that of other plant mRNAs.

The rate of turnover could make a significant contribution to the cost of mRNA. However, the turnover rates of most plant mRNAs do not appear to vary as much as the turnover rates of their corresponding proteins. Most of the nuclear genes studied in plants have been shown to be regulated transcriptionally,[301] and, as a consequence, changes in mRNA levels seem to arise largely from changes in the rate of transcription rather than from differences in mRNA stability. Given the primacy of transcriptional control, it may be expected that each species of mRNA will be present in the plant only when the synthesis of its encoded protein is required and that the rate of mRNA turnover will be high to facilitate rapid adjustment to changing developmental and environmental conditions. Both predictions are borne out by investigations on the formation of induced defenses. The mRNAs for several proteinase inhibitors[199] and for enzymes of terpenoid and phenolic phytoalexin biosynthesis[129,176,329,524] have been shown to be present in plants for only brief periods of time immediately following simulated herbivory or pathogen infection and to possess half-lives of 12 h or less. Thus, for mRNAs associated with induced plant defense, as for their corresponding proteins, maintenance and upkeep costs are minimized by the transient need for these components of the biosynthetic apparatus. It will be interesting to see if this conclusion is also applicable to the machinery for making constitutive defenses.

3. Other Regulatory Apparatus

Until about 20 years ago, the formation of most plant secondary metabolites was not thought to be regulated closely, since these substances were assumed to be largely waste products. However, it is now clear that the biosynthesis of defensive secondary metabolites is usually restricted to particular tissues, organs, and stages of plant development[333] and thus

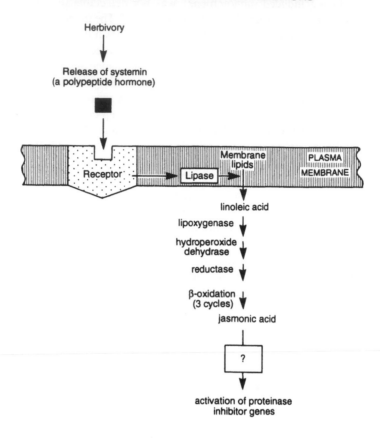

ELABORATE SIGNALING PATHWAY REGULATES THE
BIOSYNTHESIS OF PROTEINASE INHIBITORS

Herbivory

Release of systemin
(a polypeptide hormone)

Membrane lipids | PLASMA
Receptor | Lipase | MEMBRANE

linoleic acid

lipoxygenase

hydroperoxide dehydrase

reductase

β-oxidation (3 cycles)

jasmonic acid

?

activation of proteinase
inhibitor genes

FIGURE 4. Proposed signaling pathway for the induction of proteinase inhibitor biosynthesis in herbivore-damaged tomato leaves. This pathway provides an example of the complex mechanisms involved in regulating the formation of chemical defenses in plants.[160]

seems in most cases to be under tight control. This section surveys some of the ways in which the formation of plant defenses is regulated in order to illustrate that control may be mediated by complex and costly mechanisms.

Within the last few years, a substantial number of genes have been isolated that code for enzymes of secondary metabolism. Many of these exist as part of multigene families consisting of up to ten different genes, each specifying the same or nearly the same protein.[84,149,182,228,296,297,308,312,330] The various members of a multigene family may be activated in different ways in different tissues in response to environmental or developmental signals. For example, potato tubers contain multiple genes for 3-hydroxy-3-methylglutaryl coenzyme A reductase (HMGR), a possible regulatory enzyme of terpenoid biosynthesis.[80,559] Wounding activates three distinct HMGR genes in this tissue, but fungal infection activates only two of these genes while suppressing the expression of the third.[80]

The presence of regulatory genes may introduce an additional layer of control. Upon elicitation by specific environmental or developmental cues, regulatory genes express proteins called transcription factors that in turn modulate the expression of structural genes by binding to specific sequences in the promoter region. In secondary metabolism, regulatory genes have so far been identified only in connection with flavonoid biosynthesis.[131,182] For example, in maize, the products of the R gene family stimulate the transcription of genes encoding chalcone synthase, dihydroflavonol-4-reductase, and UDP-glucose-flavonoid-3-oxy-glucosyltransferase.

The expression of antiherbivore defense genes in plants may be triggered by elaborate signal transduction pathways. The best studied example of this phenomenon is the wound-induced production of proteinase inhibitors in the leaves of tomato, potato, and other members of the Solanaceae.[438,440] Damage to the leaves of one of these species can induce the synthesis of proteinase inhibitors, not only at wound sites, but also in distal, undamaged leaves. According to the latest model,[160] tissue damage is thought to initiate the signaling pathway by stimulating the release of systemin, a polypeptide hormone (Figure 4).[388] Systemin binds to receptors on leaf cell plasma membranes, enhancing the activity of certain lipases that catalyze the release of linoleic acid from membrane lipids. Free linoleic acid is then converted via several intermediary steps to jasmonic acid, which in turn activates the expression of the proteinase inhibitor genes.[160] Other substances, such as abscisic acid and the oligouronide fragments of damaged plant cell walls, are believed to participate in this signaling pathway as well.[161,389,439] Complex signal transduction networks are also associated with the induction of phytoalexin biosynthesis in several plant species.[128,210,217,447]

The production of plant defenses may also be regulated by the compartmentation of pathways (or pathway segments) within specific cellular organelles.[11,185,334] Compartmentation allows the independent control of different metabolic sequences at separate sites within the cell. A good example of subcellular compartmentation in secondary metabolism is furnished by the biosynthesis of protoberberine-type benzylisoquinoline alkaloids in certain species of Annonaceae, Berberidaceae, Menispermaceae, and Ranunculaceae. In these plants, four sequential enzymes of the protoberberine alkaloid pathway are found exclusively in a specific class of cellular vesicles.[8,566,569] These organelles, which are probably derived from the smooth endoplasmic reticulum, seem to function exclusively in the synthesis of protoberberine alkaloids.

Multienzyme complexes, made up of enzymes catalyzing consecutive steps of a metabolic pathway, represent another form of subcellular compartmentation. These noncovalent aggregates are believed to help regulate cellular metabolism by channeling substrate through a reaction sequence without allowing intermediates to diffuse away, thus effectively segregating competing pathways from one another.[476] Multienzyme complexes have been described in a number of different branches of plant secondary metabolism.[185,256,257,478,479] For instance, in sorghum, several enzymes of cyanogenic glycoside biosynthesis form complexes that are associated with microsomal membranes.[89,368] Multienzyme complexes are easily disrupted during conventional protein extraction procedures and so may be much more widespread than present evidence would indicate.[478]

It is difficult to estimate the costs associated with subcellular compartmentation or any of the other regulatory mechanisms discussed in this section. Nevertheless, the elaborate nature of these mechanisms illustrates that plants may invest a significant amount of resources in regulating the biosynthesis of defense compounds. Of course, this investment could ultimately lead to a reduction in the overall allocation to defense if the regulatory machinery is adjusted so that defenses are produced only when and where they are most needed.

Several other types of cellular machinery are necessary for biosynthetic processes, including DNA, ribosomes, and various types of cellular membranes. Unfortunately, little is known of the costs of these components,[390] and it is not easy to separate those costs attributable to the manufacture of defense compounds from those due to other activities. Membranes may require the expenditure of a considerable amount of energy because both the acyl lipids and sterols needed for construction have very high raw materials costs (Table 2). In addition, substantial resources are needed for the maintenance of ion gradients and for the active transport of materials across the membrane.[12,390] By controlling pH and ion concentrations within the narrow ranges necessary for enzyme activity, membranes constitute a vital part of cellular biosynthetic machinery.

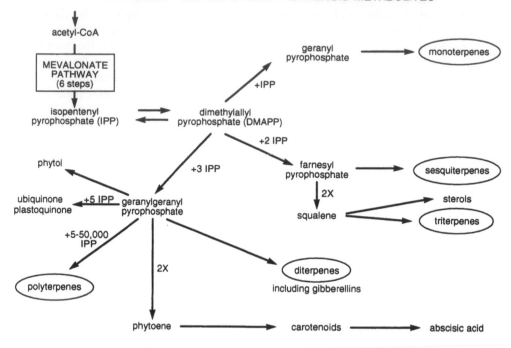

FIGURE 5. Outline of terpenoid biosynthesis with the major groups of secondary metabolites circled. The high degree of overlap among pathways for making primary and secondary metabolites illustrates how in theory plants could reduce their biosynthetic costs by sharing the use of a number of enzymes. However, plant cells actually possess several distinct terpenoid pathways located in different organelles.

4. Is the Biosynthetic Machinery Necessary for Making Plant Defenses also Used for Synthesizing Primary Metabolites?

The early steps in the biosynthesis of plant defense compounds often overlap with those involved in the formation of certain primary metabolites. Most shikimate pathway-derived products, including the amino acids phenylalanine, tyrosine, and tryptophan, as well as a variety of aromatic defense compounds, share at least seven biosynthetic steps from the intermediate 3-deoxyarabinoheptulosonic acid-7-phosphate (DAHP) to chorismic acid. The biosynthesis of all terpenoid metabolites begins with the seven basic reactions of the mevalonate pathway, from acetyl-CoA to dimethylallyl pyrophosphate (Figure 5). Terpenoid metabolites include a long list of primary compounds (steroids, carotenoids, dolichols, etc.), in addition to an assortment of defensive substances.

The use of the same enzymes (and their corresponding mRNAs) for making both primary and secondary metabolites could effect a considerable savings in the outlay for biosynthetic machinery. However, recent findings suggest that, rather than a single set of shikimate or terpenoid pathway enzymes, plant cells possess multiple sets distributed among several different subcellular locations. Detailed studies of several shikimate pathway enzymes provide strong evidence for the existence of two separate pathways composed of immunologically distinct isozymes with different allosteric properties.[255,256,371,395] One pathway, located in the plastids, is assumed to supply aromatic amino acids for protein synthesis, while a second pathway, located in the cytoplasm, is thought to produce precursors for lignin and secondary products, as well as the aromatic amino acids.[255,256,263,395]

With respect to terpenoid biosynthesis, the subcellular locations of certain pathway enzymes are still somewhat controversial.[202,291] However, many experimental results point to the presence of separate pathways located in the plastids, the Golgi vesicles, and the cytosol.[14,241,291,327,402,403,455,456,496,543] Among these compartments, there is no clear separation between the synthesis of primary and secondary metabolites. Chloroplasts and other types of plastids are apparently responsible for making carotenoids, for forming the prenyl side chain of chlorophyll, and for carrying out at least the early steps of monoterpene and diterpene formation.[137,195,202,291,302] Elsewhere, the Golgi seem to be the site of plastoquinone and ubiquinone production,[496] while sterols, sesquiterpenes, and triterpenes are synthesized in the cytosolic compartment, which also includes the endoplasmic reticulum.[31,291] Therefore, even when defensive metabolites are formed from the same basic biosynthetic reactions as certain primary metabolites, plants do not necessarily economize by using a single set of biosynthetic enzymes for making both types of products. Perhaps the need to separate different metabolic sequences physically to facilitate their independent control (see Section II.B.3) is of overriding importance.

The possibility of "enzyme sharing" must also be considered for individual steps in the biosynthesis of plant defenses. In theory, enzymes catalyzing certain general types of reactions, such as the oxidation of a secondary alcohol group to a ketone, could be borrowed from primary metabolic pathways. However, plants do not appear to employ this strategy for reducing biosynthetic costs. Most enzymes of secondary metabolism that catalyze general reaction types are able to use only a limited range of substrates, contrary to what was once assumed;[64,333] thus, they do not seem to function in more than one pathway. High substrate specificities are exhibited by hydroxylases,[277,278] dehydrogenases,[113,288,289] double-bond reductases,[99,106,112] glucosidases,[244,397] glucosyltransferases,[383,454,490,571] O-methyltransferases,[142,234,258,396] N-methyltransferases,[396,518] and sulfotransferases.[22] For example, the hydroxylase that converts limonene to *trans*-isopiperitenol during the biosynthesis of monoterpenes in peppermint (*Mentha x piperita*) is highly selective with regard to its substrate (Figure 6).[278] Out of 13 other structurally-related monoterpenes tested, only *p*-menth-1-ene was hydroxylated with a rate approaching that of limonene.

A particularly remarkable case of substrate specificity concerns the O-methyltransferases that participate in the formation of polymethylated flavonol glycosides in *Chrysosplenium americanum*.[238,284] The methylation of these substances occurs in a stepwise fashion, one methyl group at a time, under the catalysis of a series of distinct, position-specific enzymes, each of which possesses a high substrate specificity (Figure 7). For instance, the 3-O-methyltransferase in this series requires the unmethylated flavonol quercetin as its substrate, so the 3 position is always methylated first. The 7-O-methyltransferase only accepts 3-methylquercetin as a substrate; thus, the 7 position is methylated next, and so forth. The net result is that up to five O-methyl substituents are added in precise order.

The high degree of substrate specificity in the enzymes of plant secondary metabolism suggests that metabolic grids are not common in the formation of defensive compounds. Metabolic grids or metabolic matrices are multidimensional networks of parallel reaction paths that occur when a reaction sequence is catalyzed by several consecutive enzymes, each of which tolerates considerable structural variation in its substrate. Within a grid, a given intermediate may be converted to an end product by several different routes. For example, in the presence of methyltransferases and glucosyltransferases of low substrate specificity, a flavonoid could be first methylated and then glucosylated, or first glucosylated and then methylated. Although metabolic grids were regularly depicted in older textbooks and reviews,[64,181,196,216,418] in most cases, their existence has not withstood careful experimental scrutiny (however, see Reference 479).

5. The Cost of Making Mixtures

A striking feature of plant defense chemistry is that many species produce complex mixtures of closely related defensive metabolites rather than just one or two individual compounds. Mixtures are thought to possess several significant advantages over the equivalent amount of a single defensive substance: 1) the components of a mixture may act

SUBSTRATE SPECIFICITY OF ENZYMES OF SECONDARY METABOLISM:
A MONOTERPENE HYDROXYLASE

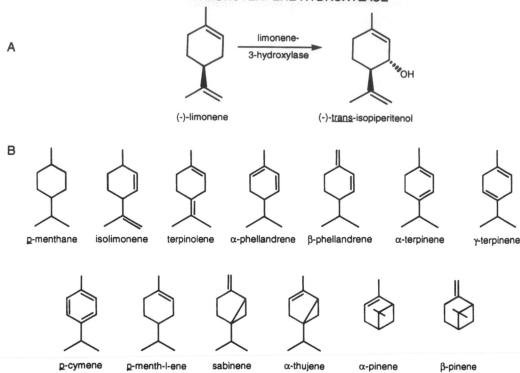

FIGURE 6. Substrate specificity of an enzyme of secondary metabolism. A) Limonene-3-hydroxylase converts limonene to *trans*-isopiperitenol, an important intermediate in monoterpene formation in peppermint (*Mentha x piperita*). B) Although this enzyme catalyzes a general type of reaction, it exhibits great selectivity for its natural substrate, limonene. Of the 13 structurally-related monoterpenes tested, none were hydroxylated at a significant rate, except for *p*-menth-1-ene, which was hydroxylated at 37% of the rate of limonene. [278]

synergistically to provide greater toxicity or deterrency at a lower total concentration;[35,354] 2) mixtures may afford more effective protection against a wider range of herbivores;[63,238] 3) the rate at which herbivores can evolve resistance to plant defenses may be slowed by the presence of mixtures;[394] and 4) mixtures may serve to maintain defensive products in a liquid rather than in a crystalline state, in which they would seem to be less effective repellents.

One of the possible disadvantages of producing mixtures is the cost of any extra biosynthetic machinery required. However, additional enzymes may not always be necessary to generate a mixture of defensive metabolites. Several monoterpene cyclases, enzymes that catalyze the cyclization of geranyl pyrophosphate to various monoterpene skeletal types, have been shown to yield multiple olefinic products.[7,93,101,179,222,323,532] For example, (–)-pinene cyclase from garden sage (*Salvia officinalis*) synthesizes a mixture of five monoterpenes, (–)-camphene, (–)-α-pinene, (–)-β-pinene, myrcene, and (–)-limonene, from geranyl pyrophosphate (Figure 8).[101] In sesquiterpene and cyanogenic glycoside biosynthesis,[108,372] enzymes (or enzyme complexes) have also been reported to generate more than one end product. The existence of multiple product enzymes may be a direct result of natural selection for this ability or simply an unintended consequence of the particular reaction mechanism employed. Nevertheless, regardless of how they originated, multiproduct enzymes are likely to lower the potential costs of producing mixtures of defensive metabolites.

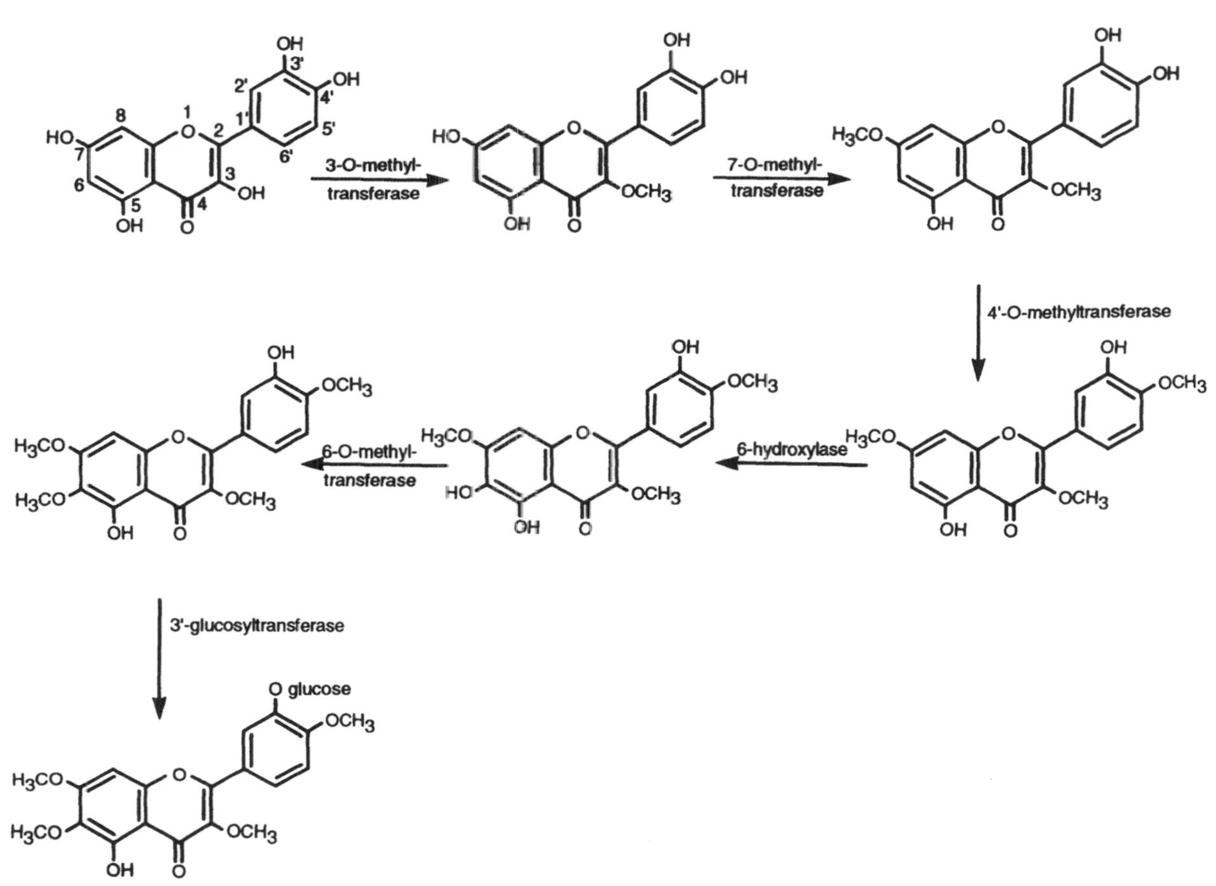

FIGURE 7. Substrate specificity of enzymes of secondary metabolism. The final steps in the biosynthesis of the flavonol, 3,6,7,4'-tetramethylquercetin-3'-O-glucoside, in *Chrysosplenium americanum* involve the transfer of four methyl groups and a glucose residue to the basic quercetin skeleton. These reaction types are sometimes thought to be catalyzed by general enzymes that accept a broad range of substrates. However, in *C. americanum*, the enzymes that take part in these transformations have high specificities for both substrate and site of reaction. Hence, the reactions proceed in stepwise fashion and in precise order.[258,284]

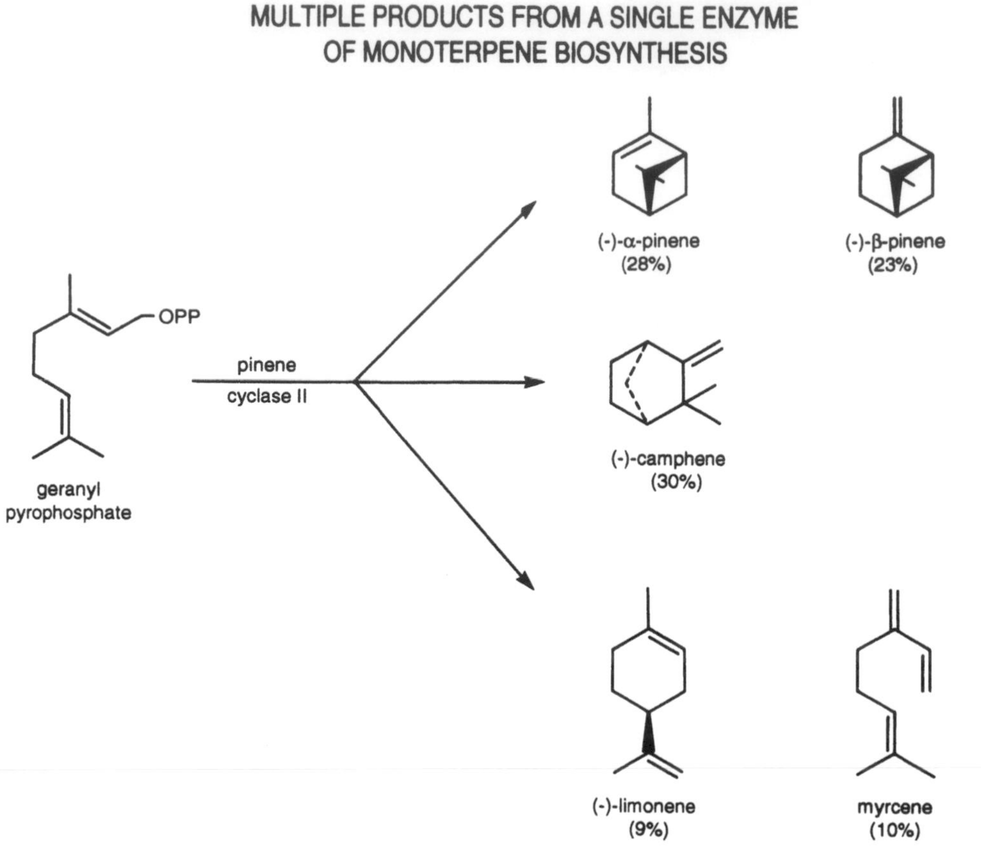

MULTIPLE PRODUCTS FROM A SINGLE ENZYME
OF MONOTERPENE BIOSYNTHESIS

FIGURE 8. Products of (–)-pinene cyclase, a monoterpene cyclase from *Salvia officinalis* that synthesizes a mixture of monoterpene olefins from the substrate geranyl pyrophosphate. The numbers in parentheses represent the proportion of each product in the total mixture.[101,179]

III. THE COST OF STORAGE

In considering the metabolic costs of plant chemical defense, most attention has focused on the process of biosynthesis. However, substantial expenses could also be associated with the storage, transport, and maintenance of defenses. For instance, in comparing cyanogenic and acyanogenic phenotypes of *Trifolium repens*, Kakes found that plants producing cyanogenic glycosides made only half as many flowers as plants lacking cyanogenic glycosides, but the calculated cost of synthesizing cyanogenic glycosides was much lower than the caloric value of the additional flowers.[273] Thus, other factors may contribute significantly to the overall cost of plant defense compounds. In the balance of this chapter, some of these are discussed, beginning with the cost of storage.

Antiherbivore defense compounds are not evenly distributed in plant tissue but instead accumulate in specific organs, tissues, cells, and organelles. The principal cellular and subcellular storage sites include glandular trichomes,[282,286,422,564] resin ducts,[43,151,319] secretory cavities,[36,305,437] surface wax,[1,13,556,573] laticifers,[170,377,406] vacuoles,[229,249,346,528] and cell walls.[487] The storage of defense metabolites may play an important role in enhancing their effectiveness against herbivores. For example, many defenses are found in high concentration in epidermal and subepidermal cell layers or are situated directly on the plant surface as resinous or waxy coatings or in glandular hairs.[48,117,282,340,354,442,555] These substances may present a formidable barrier to smaller herbivores, such as insects. Defensive compounds located in ducts, canals, or laticifers may be quite effective against small herbivores as well, because the contents of these structures are stored under pressure.[140] Thus, when feeding activity severs a duct or canal, the contents flow toward the cut surface, creating a high concentration of defensive substance at the site of attack.

The storage of defenses in specific locales may also serve to limit their toxicity to the plant itself. Because of the basic biochemical similarities between plant and animal cells, many defenses are as poisonous to the plant that produces them as they are to attacking herbivores.[168,354] For example, terpenes inhibit root and shoot growth,[164] furanocoumarins crosslink DNA,[36] tannins bind proteins,[214] quinolizidine alkaloids inhibit the translation of mRNA,[298] hydrogen cyanide blocks cellular respiration,[475] and certain photosensitizers generate toxic oxygen species.[38,133,511] The most common way for plants to avoid these harmful consequences is to sequester noxious defense substances away from sensitive metabolic processes.

The metabolic costs of storing chemical defenses are not easy to measure, and so have largely been ignored by students of plant–herbivore interactions. In the remainder of this section, several processes involved in defense storage are described, and speculations are made about the costs that might be associated with them. Special emphasis is placed on how storage is accomplished while avoiding autotoxicity.

A. CONSTRUCTION OF STORAGE SITES

Antiherbivore defense compounds are often sequestered in elaborately fashioned compartments. Many lipophilic defenses, for instance, accumulate in different types of multicellular secretory structures, such as glandular trichomes, resin ducts, and secretory cavities (Figure 9).[151,188,282] The formation of these complex structures requires a substantial commitment of resources. Glandular trichomes, resin ducts, and secretory cavities all contain a large intercellular space in which defense compounds accumulate. This space is lined with a layer of specialized epithelial cells that are active in the biosynthesis and secretion of the stored product.[151,152,335,451] Adjacent cells often possess heavily thickened or suberized walls, which seem to function in supporting and strengthening the secretory structure and in regulating the apoplastic movement of water, nutrients, or defense compounds. Thus, secretory structures are made up of a large number of variously differentiated cells (Figure 9). Since most of these are not photosynthetically active, except in certain glandular trichomes, the construction of secretory structures must be regarded as an expensive proposition. Such construction may impose considerable nitrogen costs in particular, as fertilization with nitrogen has been reported to significantly increase the size and number of resin ducts in the needles of *Pinus sylvestris*.[45]

Water soluble defense compounds are typically deposited in cell vacuoles.[346,528] The production and maintenance of these storage compartments is also presumed to be a costly venture. Large amounts of energy and nutrients are needed to synthesize the tonoplast, to take up the required osmoticum, and to maintain the electrochemical potential of the membrane.[404] In addition, compared to a cell without a vacuole, a vacuolated cell requires both a larger plasmalemma and a larger and thicker cell wall to surround the cell and resist the increased turgor pressure.[404] Clearly, these costs cannot all be charged to defense purposes, since vacuoles have myriad other functions. However, under certain conditions, antiherbivore defenses may make up a significant proportion of the contents of this organelle. For example, in the epidermal vacuoles of young sorghum leaves, the concentration of the cyanogenic glycoside dhurrin is approximately 200 mM,[90] nearly half of the total osmotic strength of a typical plant vacuole.[404] Thus, for certain water soluble defensive metabolites, the resources used in vacuole construction and upkeep may represent a major proportion of total defense costs.

Several classes of antiherbivore defenses that are stored in the vacuole as glycosides are readily converted to active (or more active) forms by the catalysis of specific glycosidases when the cell is ruptured. These substances, which include cyanogenic glycosides,[91,397,459] glucosinolates,[77,331] the glycosides of hydroxamic acids,[378] ranunculin, and other compounds,[374]

SECRETORY STRUCTURES CONTAIN A VARIETY OF SPECIALIZED CELL TYPES

A. GLANDULAR TRICHOME

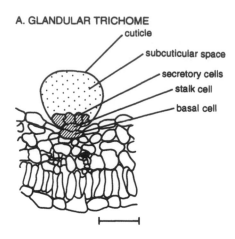

B. SECRETORY CAVITY

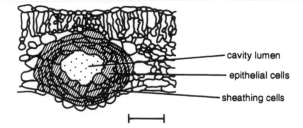

FIGURE 9. Sketches of two types of secretory structures, illustrating the location of the secreted, lipophilic defense product (dotted region) and the number of specialized cells associated with each structure (hatching). A) Transverse section of the leaf of M*entha x piperita* (peppermint) showing a peltate glandular trichome. This structure accumulates a defensive secretion containing mostly monoterpenes, which are synthesized by the secretory cells and deposited in the subcuticular space. The stalk cell supports the trichome, and the basal cell anchors it to the epidermis. In other species, the cuticular covering may rupture, allowing the secreted defenses to volatilize or exude onto the plant surface forming a resinous coating. Length of the bar is 50 μm. (Drawn from a photo by J. Gershenzon and M. Maffei). B) Transverse section of the leaf of *Tagetes erecta* (African marigold) showing a secretory cavity. This structure accumulates a defensive secretion containing mainly indole and two monoterpenes.[437] The secretion is synthesized by the epithelial cells, the first layer of cells surrounding the cavity lumen. The sheathing cells function to strengthen the cavity wall. Length of the bar is 100 μm. Adapted from reference 436.

are not normally hydrolyzed in the intact plant because the glycoside and the specific glucosidase are separated spatially to avoid autotoxicity. Special expenditures of resources may be necessary to ensure that this compartmentation is maintained. In the case of cyanogenic compounds, the glycosides and their corresponding glycosidases are normally sequestered in separate cells or organelles,[91,173] so no additional costs may be required to keep these substances apart. However, for glucosinolates, both the glycosides and their corresponding glycosidases (called β-thioglucosidases or myrosinases) seem to be situated in vacuoles of the same cells[204,243,251] necessitating some form of additional compartmentation to keep them apart.

B. CHEMICAL MODIFICATION OF DEFENSES FOR STORAGE

As we have already noted, the site in which plant chemical defenses are stored is strongly dependent on their polarity. Lipophilic defenses usually accumulate in secretory structures or on the surface, while hydrophilic substances are sequestered in the cell wall or in vacuoles. Therefore, one might imagine that some of the chemical transformations involved in the synthesis of defense compounds have been selected for their ability to increase the lipophilicity or hydrophilicity of particular substances, according to the storage site to be utilized. For example, methylation of the free hydroxyl groups of a flavonoid (Figures 7, 10) may be an adaptation for deposition in a lipophilic compartment. Highly methoxylated flavonoids are nearly always found on the plant surface, in leaf wax or bud excretions,[556] or in association with secretory structures, especially glandular trichomes. Glycosylation, on the other hand, may be a method to make metabolites sufficiently water-soluble for vacuolar storage (Figure 10). Most plant glycosides, including phenylpropanoids, flavonoids, iridoids, triterpenes, steroidal alkaloids, and cyanogens, usually occur in the vacuole.[254,346] Another chemical modification that increases water solubility is the conversion of pyrrolizidine alkaloids from their free base forms to the corresponding N-oxides. The free bases of pyrrolizidine alkaloids are weakly lipophilic molecules that are less suited for vacuolar storage than the polar, salt-like N-oxides, since the unprotonated form of the free base may readily diffuse through the tonoplast.[144,230] With our present state of knowledge, it is difficult to decide whether any of these chemical modifications is indeed a direct adaptation for storage or simply a way to increase defensive potency (which should probably be treated as a cost of synthesis). In either case, the quantities of enzymes and raw materials needed for methylation, glycosylation, and similar reactions could be substantial. For example, each of the four O-methylations that take place during the biosynthesis of 5,6,7,8-tetramethoxyflavone (Figures 1C, 10) requires 1 mole of S-adenosylmethionine, raising the raw materials cost of this compound from 2.00 to 2.37 g of glucose/g (Table 1), nearly a 20% increase.

The storage of defensive metabolites may also be chemically mediated without any actual structural modification. For example, *Coptis japonica* solubilizes large quantities of the benzylisoquinoline alkaloid berberine (a quaternary amine) in its vacuoles by employing high concentrations (>150 mM) of the dicarboxylic acid malate as a counter ion (Figure 11).[445] Berberine has no free hydroxyl groups, so its solubility in water cannot be increased by glycosylation. The use of such high concentrations of malate as a counterion is metabolically expensive, since it requires a large quantity of fixed carbon that becomes unavailable for other purposes. However, in theory, both malate and the sugars used in glycosylation are recoverable when the defense compounds that they have helped solubilize are no longer needed.

C. OTHER ADAPTATIONS TO AVOID AUTOTOXICITY

The compartmentation of defensive metabolites may break down on occasion, allowing phytotoxic compounds to penetrate sensitive areas of the cell. Plants appear to have several types of protection systems to mitigate the effects of such an occurrence. For instance, some species possess enzymes capable of degrading their own defensive metabolites. One of the best known of these activities is β-cyanoalanine synthase, which, in the presence of cysteine, converts hydrogen cyanide to β-cyanoalanine and hydrogen sulfide (Figure 12). β-Cyanoalanine synthase is present in all plants, presumably to detoxify the hydrogen cyanide produced during ethylene biosynthesis.[560] However, this enzyme occurs at much higher levels in species that accumulate cyanogenic glycosides[363] and so is thought to play a role in deactivating any hydrogen cyanide accidentally released from cyanogenic glycosides. Since the principal target of hydrogen cyanide in plants is cytochrome oxidase, which is part of the terminal complex of the mitochondrial electron transport chain, it is not surprising that the bulk of β-cyanoalanine synthase is located in the mitochondrion.[557] Besides β-cyanoalanine synthase, an additional cyanide-detoxifying enzyme, rhodanese, has been demonstrated in plants. Rhodanese is also claimed to occur in higher concentrations in cyanogenic glycoside-containing plants than in

THE STORAGE SITES OF FLAVONOIDS WITH
DIFFERENT CHEMICAL MODIFICATIONS

Chemical modification	Compound	Species and storage site
methoxylation		*Helichrysum nitens* leaf and stem surface
prenylation		*Lupinus albus* leaf surface
glycosylation		*Secale cereale* vacuoles of leaf mesophyll
glycosylation and malonylation		*Petroselinum crispum* vacuoles

FIGURE 10. Plant chemical defenses are stored in different locations depending on their polarity. Lipophilic defenses are typically stored in secretory structures or on the plant surface, while hydrophilic defenses usually accumulate in the cell wall or in vacuoles. This pattern is demonstrated here using flavonoids with different chemical modifications as examples. Some of these modifictions may in fact have been selected to facilitate storage at a particular site. The formation of malonic acid conjugates, like glucose conjugates, increases solubility in water, and a wide variety of secondary metabolites with malonyl conjugates has recently been discovered. Malonyl conjugates may be far more prevalent than currently suspected, because the malonyl moiety is readily hydrolyzed during plant extraction.[28] Abbreviations: Glur = glucuronic acid; Glc = glucose. References: *Helichrysum nitens*,[509] *Lupinus albus*,[227] *Secale cereale*,[10] *Petroselinum crispum*.[342]

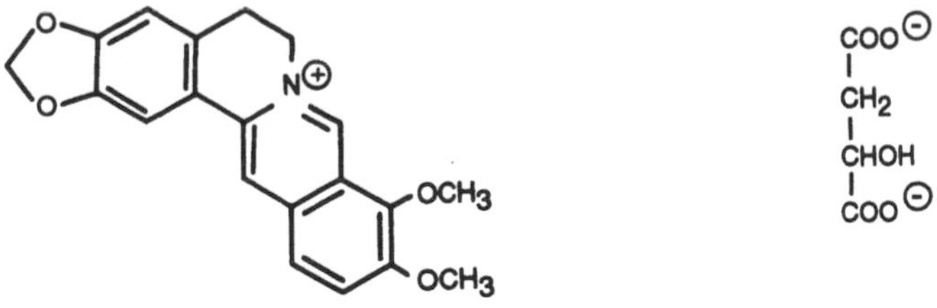

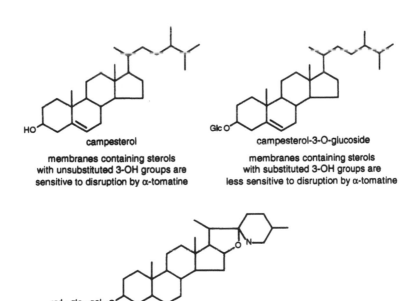

FIGURE 11. Large quantities of the alkaloid berberine are stored in the vacuoles of *Coptis japonica* using malate as a counter ion.[445]

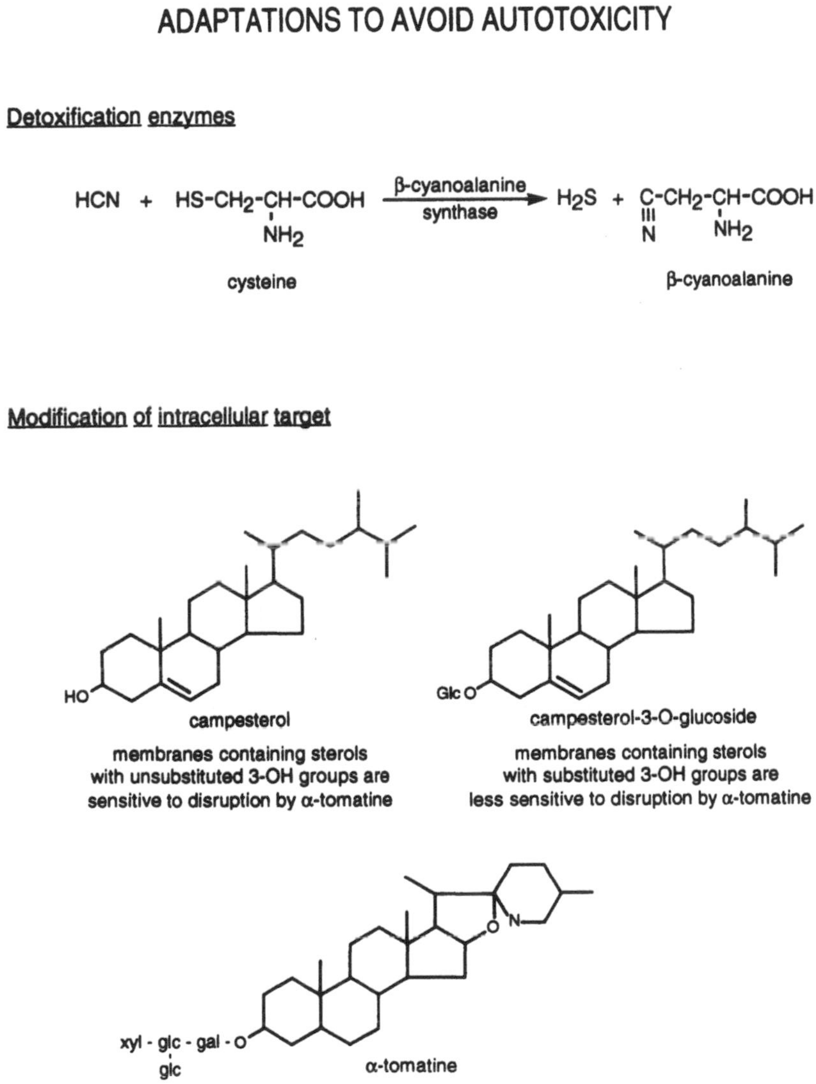

ADAPTATIONS TO AVOID AUTOTOXICITY

Detoxification enzymes

$$HCN \ + \ HS\text{-}CH_2\text{-}CH\text{-}COOH \xrightarrow[\text{synthase}]{\beta\text{-cyanoalanine}} H_2S \ + \ \underset{N}{\overset{\underset{\textstyle |||}{}}{C}}\text{-}CH_2\text{-}CH\text{-}COOH$$

cysteine   β-cyanoalanine

Modification of intracellular target

campesterol

membranes containing sterols
with unsubstituted 3-OH groups are
sensitive to disruption by α-tomatine

campesterol-3-O-glucoside

membranes containing sterols
with substituted 3-OH groups are
less sensitive to disruption by α-tomatine

xyl - glc - gal - O
glc   α-tomatine

FIGURE 12. Examples of mechanisms, other than compartmentation, by which plants avoid being poisoned by their own defense compounds.

plants that do not accumulate cyanogenic glycosides.[468] However, the importance of rhodanese in ridding plants of free hydrogen cyanide has been questioned.[274,363] Other enzymes capable of detoxifying defensive substances are known,[26,190,359,431] but the actual roles of these activities *in vivo* are still unclear. The synthesis of large amounts of enzymes solely for detoxification purposes seems costly, and many of these activities may function also (or instead) to degrade plant defenses that are no longer needed (see Section VI).

An additional protective mechanism is found in plants with light-activated antiherbivore defense compounds. Several classes of defenses, including furanocoumarins, polyacetylenes, benzylisoquinoline alkaloids, and quinones, show much greater toxicity to herbivores and pathogens in the presence of sunlight than they do in darkness.[37,38,134,511] Upon irradiation, these substances can react with a variety of essential biological molecules or interact with oxygen to form singlet oxygen and various toxic oxygen radicals.[134,511] While such behavior is thought to function principally in defense, it also poses a hazard to the plants themselves. All green plants possess carotenoids, tocopherols, ascorbic acid, and other antioxidants to quench free radicals and reactive oxygen species since, given the presence of the photosensitizing pigment chlorophyll, they are all susceptible to photo-oxidative damage.[310] Species that accumulate light-activated defenses might be expected to have increased quantities of such antioxidants. In fact, members of the Rutaceae and Apiaceae, two families that produce large amounts of phototoxic defenses, are notably rich in ascorbic acid (Rutaceae) and carotenoids (Apiaceae),[39] suggesting that these plants do require supplemental protection against photo-oxidative damage, protection that may be metabolically expensive.

Plants may also avoid the deleterious effects of their own antiherbivore defenses by altering potential physiological targets so that they are no longer sensitive to the types of defensive compounds being stored. Such modifications may entail reconfiguration of the active site of an enzyme[168] or changes to other cellular components, such as ribosomes[298] and membranes.[481] An interesting example concerns α-tomatine, a steroidal glycoalkaloid found in tomato and other species of the Solanaceae that disrupts cell membranes.[421] α-Tomatine is much more effective at disrupting membranes containing sterols with a free 3-β-OH group than membranes made up of sterols without a free 3-β-OH group (Figure 12).[481] The tolerance of tomato to endogenous tomatine results from the fact that its membranes contain a high level of sterols in which the 3-β-OH is substituted with sugar residues. Sterol glycosides were found to constitute over 80% of the total sterols of tomato leaves compared to 5 to 25% of total sterols in the leaves of an assortment of non-solanaceous species.[139] Target site modification, like other adaptations for avoiding autotoxicity, may exact certain costs. In this case, tomato must forgo the use of the sugar molecules needed to modify the 3-OH functions of its membrane sterols. This sacrifice is only temporary, however, as the proportion of glycosylated sterols decreases substantially on leaf senescence.[139]

In summary, the costs of storing antiherbivore defenses are in large part dictated by the need to avoid autotoxicity. As we have seen, plants employ a variety of mechanisms to accomodate the presence of these substances in their tissues, including 1) physical separation of defenses from living cytoplasm, 2) synthesis of detoxifying enzymes, 3) production of extra quantities of antioxidants as protection against phototoxins, and 4) modification of physiological targets so that they are no longer susceptible to stored toxins. While little effort has been made to assess the costs of any of these mechanisms precisely, all clearly require some investment of energy and nutrients.

IV. THE COST OF TRANSPORT

Most plant chemical defenses are stored in a different location than where they are made and so must be transported from the site of synthesis to the site of storage. The costs of such

UPTAKE OF ALKALOIDS INTO THE VACUOLE: POSSIBLE MECHANISMS

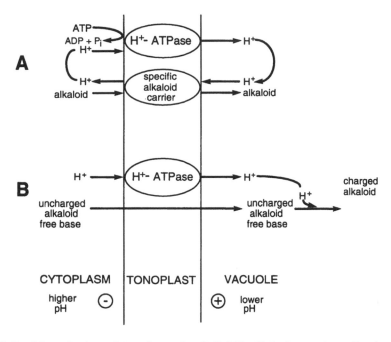

FIGURE 13. Possible mechanisms of vacuolar uptake of alkaloids. A) *Active, carrier-mediated transport.* H+-ATPases pump protons into the vacuole at the expense of ATP hydrolysis.[497] This results in a gradient across the tonoplast, with the vacuole being more electropositive and having a higher proton concentration (lower pH) than the cytoplasm. Alkaloids are then taken into the vacuole via specific carrier proteins which exchange them for protons (antiport transport). B) *Ion-trap mechanism.* Uncharged alkaloid free bases passively diffuse into the vacuole. Once inside, they become protonated, due to the lower pH, and are thus trapped because they cannot pass back through the tonoplast in charged form.[47,235,349,411,417] No carrier proteins are required for this mechanism, but H+-ATPases are needed to maintain the proton gradient.

transport have not been considered explicitly in previous analyses of defense costs. However, in the last few years, a substantial amount of research has been carried out on the uptake and translocation of certain plant secondary metabolites, which should help in assessing the metabolic costs of these processes. Two types of transport may be distinguished: intracellular and intercellular.

A. INTRACELLULAR TRANSPORT

Antiherbivore defenses are typically produced in the cytoplasm and then exported to the vacuole, cell wall, surface wax, or intercellular spaces of secretory structures. Of these destinations, most attention has focused on movement into the vacuole. The transport of defenses, such as alkaloids,[121,122,230,349,356,527,558] phenolics,[343,538] and triterpene glycosides,[260,299] into this organelle has been demonstrated to occur via active processes that are mediated by highly specific carriers. For example, the uptake of alkaloids into the isolated vacuoles of *Fumaria capreolata* was significantly stimulated by ATP.[122] These vacuoles readily accumulated the benzylisoquinoline alkaloids *S*-reticuline and *S*-scoulerine (both naturally synthesized by *F. capreolata*) but did not take up the unnatural enantiomers *R*-reticuline and *R*-scoulerine, nor did they accumulate eight other alkaloids of various structural classes.

The notion that secondary metabolites are transported into plant vacuoles in a selective, energy-dependent fashion is consistent with current models of the vacuolar uptake of ions, sugars, and amino acids (Figure 13A). The vacuolar membrane or tonoplast is believed to possess an H^+-ATPase that pumps protons into the vacuole at the expense of ATP hydrolysis.[497] This activity creates an electrochemical gradient across the tonoplast, with the vacuole side being more electropositive and having a higher proton concentration than the cytoplasmic side. Embedded in the tonoplast are also special protein channels and carriers to facilitate the transport of hydrophilic compounds that otherwise would only pass through the lipophilic membrane at very slow rates. These channels and carriers are very selective for specific classes of compounds. The uptake of anionic species occurs through highly selective channels and is driven directly by the electrical gradient. Positively charged and neutral substances, on the other hand, enter the vacuole via specific carriers, which exchange them for protons (Figure 13A). The energy for exchange is derived from the proton gradient set up by H^+-ATPase activity. Thus, the costs of actively transporting substances into the vacuole include the ATP needed to set up the proton gradient and the resources used in construction and maintenance of both the ATPases and carrier proteins. Since plants accumulate many primary metabolites in their vacuoles in addition to defenses, these expenses must be apportioned between both defense and nondefense budgets.

Not all alkaloid transport into vacuoles is mediated by specific carrier proteins. In a number of species, an ion-trap mechanism appears to operate, in which alkaloids passively diffuse through the membrane as uncharged free bases (Figure 13B).[47,235,349,411,417] (Uncharged compounds can penetrate membranes directly without the assistance of carrier proteins.) Once in the vacuole, uncharged alkaloids may become protonated as a result of the lower pH and are therefore "trapped" because, in their positively-charged form, they cannot pass back through the tonoplast. Due to the low concentration of the uncharged form of the alkaloid inside the vacuole, additional quantities continue to diffuse in and become protonated, allowing a high concentration of the charged form to accumulate. The ion-trap mechanism does not involve carrier proteins but still requires expenditures for both the H^+-ATPases, to generate the proton gradient, and the ATP. Alkaloid retention in vacuoles may also be brought about by complexation with various anionic species.[235,411]

In contrast to what is known about how defenses are taken up into vacuoles, our knowledge of transport to the cell wall, surface wax, or intracellular spaces is extremely meager. Ultrastructural studies suggest that many defense compounds are secreted into vesicles that move through the cytoplasm and fuse with the plasma membrane, releasing their contents outside the cell.[152,335] The nature of these processes and their resulting costs are still unclear.

B. INTERCELLULAR TRANSPORT

Defenses may also be transported over much greater distances within a plant between cells, tissues, and organs. A variety of different compounds exhibit such mobility, including monoterpenes,[103,107,480,526] triterpenes,[4,231,337] simple phenolic acids,[127] flavonoids,[42,71] alkaloids,[16,135,190,230,287,377,534,550,551,554] cyanogenic glycosides,[83,174,292,365] glucosinolates,[50,193] nonprotein amino acids,[430,570] and hydroxamic acid glucosides.[194] The direction of transport may vary depending on the organ and stage of development. For example, in mature *Lupinus albus*, quinolizidine alkaloids are exported from their site of synthesis in the leaves to stems, roots, pods, and seeds, but in germinating seedlings, they are transferred from the seed back to the developing organs of the young plant.[550]

1. The Vascular System

Most of the information available on the intercellular transport of defenses concerns long distance movements through the vascular system. Representatives of many classes of compounds have been detected in either the xylem or phloem,[16,71,127,194,230,337,377,385,534,550,551,554,570] with the bulk of reports pertaining to the phloem.

The costs of phloem transport include the energy and materials required to maintain the structure of the sieve elements and associated companion cells[191] and the expenses of loading and possibly unloading the transport stream. Sugars (principally sucrose) and amino acids are actively loaded into the phloem by means of specific carriers, with the driving force being a proton gradient produced by the pumping action of plasma membrane H^+-ATPases.[92,191,386] The energy cost of sucrose loading has been estimated at 1.1 to 1.4 molecule of ATP per sucrose molecule.[191] For defense compounds, it is unclear whether specific carrier proteins are needed, since materials can be taken into the phloem by simple diffusion and then swept along in the translocation stream by bulk flow. However, tracer experiments suggest that the transport of pyrrolizidine alkaloid N-oxides in the phloem takes place only in species that produce pyrrolizidine alkaloids[230] and so may involve specific carrier proteins. With or without specific carriers, energy must be expended to create the proton gradient needed to generate flow.

The xylem seems a less useful vehicle for translocating defense compounds than the phloem because, instead of being bidirectional, movement is limited to the acropetal direction only. However, the costs of xylem transport are also lower, since flux is driven by transpiration rather than energy-dependent sugar uptake. In addition, since the xylem is not composed of living cells, use of this tissue for translocating defenses may reduce the risk of autotoxicity (see Section IV.B.3). The expenses of utilizing either the xylem or the phloem must be shared among the variety of substances that circulate in these tissues.

Outside the vascular system, there have been few studies of intercellular transport of plant defenses. In a pioneering investigation, Wink and Mende examined the movement of quinolizidine alkaloids from the mesophyll to the epidermis of *Lupinus polyphyllus*.[548] In this species, alkaloids are synthesized principally in the green mesophyll cells of leaves and stems but accumulate to much higher levels in adjacent epidermal tissue. Experiments with isolated epidermal strips suggest that uptake into the epidermis proceeds in an active fashion and is mediated by specific carrier proteins in the plasma membrane. Thus, like vascular transport, cell to cell movements of defenses may also be brought about by selective, energy-requiring processes that exact a metabolic cost.

2. Chemical Modification for Transport

Not surprisingly, the antiherbivore defenses that have been reported to be transported intercellularly are all highly soluble in water. The vast majority are either glycosides[4,42,49,50,71,83,103,107,174,193,194,292,337,365,480,526] or alkaloids (or alkaloid N-oxides)[16,135,230,287,534,550,551,554] that exist predominantly in their charged forms at neutral pH. From an evolutionary perspective, the chemical transformations that make these substances water-soluble, such as glycosylation, malonation, or N-oxide formation (see Section III.B and Figure 10), can be thought of as specific adaptations for intercellular movement and thus should be considered part of the costs of transport. The major expense of these reactions probably lies in the production of the necessary enzymes, since the raw material requirements are minimal. Indeed, the formation of glycoside or malonate conjugates for transit may actually assist in the mobilization of stored carbon.

3. Adaptations to Avoid Autotoxicity

As discussed in Section III, plants resort to several different types of mechanisms to protect themselves from being intoxicated by their own antiherbivore defense compounds. Unfortunately, the act of transporting defenses could expose many additional cells to potential toxins, and so might make it necessary to extend some of these costly protective mechanisms to other parts of the plant. The movement of cyanogenic glycosides through the vascular system may pose a particularly formidable challenge in this regard. All plants that produce cyanogenic glycosides also synthesize specific glucosidases to hydrolyze the glycosides in the event the

TRANSPORT OF CYANOGENIC GLYCOSIDES AS DIGLYCOSIDES TO AVOID THE AUTOTOXIC RELEASE OF HYDROGEN CYANIDE

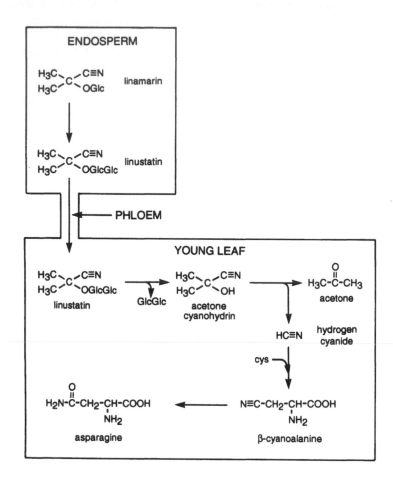

FIGURE 14. During the germination of rubber tree (*Hevea brasiliensis*) seedlings, the cyanogenic glycoside linamarin is transported from the endosperm to the young leaves through the phloem in the form of the diglycoside linustatin.[464,465] Conversion to linustatin is thought to be a specific adaptation for transport since, unlike linamarin, linustatin cannot be hydrolyzed by apoplastic glucosidases to release hydrogen cyanide. Once in the leaves, linustatin is believed to be cleaved by a specific diglucosidase to give acetone cyanohydrin, which dissociates to evolve hydrogen cyanide.[326] The hydrogen cyanide formed is immediately incorporated into β-cyanoalanine,[363,557] which is then converted to asparagine.[72]

plant is crushed, triggering the release of hydrogen cyanide.[91,397,459] In a number of species, glucosidases are compartmentalized in the apoplast as a means of separating them from the glycosides, which are typically stored in the vacuole.[173,462] Since phloem loading and unloading often occur via apoplastic routes,[92,191] translocation of cyanogenic glycosides in the phloem could well result in the evolution of hydrogen cyanide unless special precautions are taken. The rubber tree (*Hevea brasiliensis*) appears to have solved this problem by converting its cyanogenic glycoside, linamarin, to the diglycoside linustatin, a compound that is not susceptible to cleavage by the apoplastic glucosidase of this species (Figure 14).[464,465] Therefore, in germinating seedlings of the rubber tree, for example, linustatin can be transported safely from the endosperm via the phloem to the young leaves (where it is then degraded by the sequential action of a specific diglucosidase and β-cyanoalanine synthase to recover the nitrogen for primary metabolic processes; see Section VI.A.2). The synthesis of diglycosides as protected transport forms of cyanogenic glycosides seems to occur in cassava and other cyanogenic species as well.[174,365,466] The only costs of this ingenious protective mechanism may be the price of the specific enzymes needed (the glucosyltransferase and the diglucosidase), since the export of additional sugar residues to young leaves or other sink tissues should be beneficial to growth.

To summarize, recent research on the intracellular and intercellular transport of plant defense compounds has begun to shed a great deal of light on the mechanisms of movement and hence their costs. Investigations on the uptake of defenses into vacuoles, isolated cells, and the phloem have shown that uptake frequently results, not from simple diffusion, but from active transport involving very selective membrane carriers. Nevertheless, the metabolic costs of transporting antiherbivore defenses are undoubtedly lower than those of synthesis or storage, since much of the basic equipment is also used for other plant functions. Perhaps the major expenses of transport are construction of the specific membrane carrier proteins and the enzymes needed for making water-soluble transport derivatives.

V. THE COST OF MAINTENANCE

While in storage, antiherbivore defenses may be susceptible to losses caused by metabolic degradation, volatilization to the atmosphere, or the leaching action of rain or dew. If these losses are substantial, plants may find it necessary to synthesize additional amounts of defense compounds to retain an effective level of protection. The costs of such resynthesis depend on just how frequently defenses are lost and replaced. In this section, I survey the reported incidence of secondary metabolite losses caused by turnover, volatilization, and leaching, and consider whether or not missing quantities of these substances are actually replenished.

A. METABOLIC TURNOVER

The metabolic pools of many plant defense substances are thought to be in a state of rapid flux or turnover. Pulse-chase experiments with a variety of terpenoids, phenolics, alkaloids, and other nitrogenous compounds have shown that the radioactivity incorporated into these substances from administered precursors is frequently lost within hours or days of its initial assimilation.[2,9,20,21,23,25,52,55,65,96,102,109,123,146,153,154,171,172,190,203,240,253,262,275,313,315,316,338,339,348,364,367,379,443,457,461,482,483,489,503,506,512,513,533,563] In addition, large diurnal variations in concentration have sometimes been observed.[155,156,252,266,267,504,547,550] These results have been taken to indicate that certain defense compounds are subject to continual degradation and resynthesis.[26,28,87,304,419,460,461]

Clearly, a rapid rate of metabolic turnover could dramatically increase the expense of maintaining a given concentration of antiherbivore defense, and in fact turnover is commonly believed to be a major component of defense costs.[87,150,209,473] However, turnover by itself is not necessarily costly unless it leads to the actual degradation of defenses and their subsequent replacement. "Turnover" is often defined to include biosynthetic interconversions, conjugation reactions, and polymerization,[26,28] transformations that are not expensive as long as they produce other defensive metabolites or allow the original compound to be easily regenerated. In addition, degradation that is not accompanied by replacement, such as the one-time catabolism of defenses in germinating seeds or senescing organs, is not costly either but facilitates the recovery of resources no longer needed in defense (see Section VI.A). Unfortunately, in most of the studies cited as evidence for turnover, only the simple disappearance of defense compounds was measured and not their actual metabolic fate. Hence, we generally lack sufficient information for estimating the costs of this process.

1. Terpenoids

The most often discussed example of rapid metabolic turnover is the flux of monoterpenes observed in peppermint (*Mentha x piperita*) cuttings.[65,96,102] In this tissue, up to 90% of the

METABOLIC TURNOVER OF
Mentha x piperita MONOTERPENES

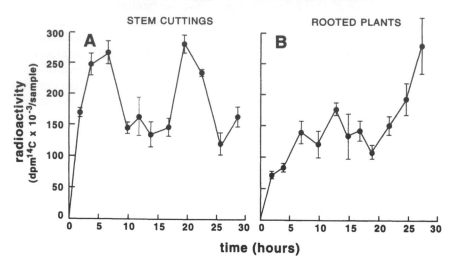

FIGURE 15. Time course of ^{14}C incorporation into leaf monoterpenes of *Mentha x piperita* (peppermint) stem cuttings (**A**) and rooted plants (**B**) after a pulse of [14]CO_2. Plants were exposed to [14]CO_2 for 5 min in a sealed plexiglass chamber under high intensity lights. After the pulse, samples were harvested periodically over the next 29 h and then subjected to simultaneous steam distillation-pentane extraction to isolate the monoterpenes. Isolated monoterpenes were quantified by gas-liquid chromatography and the incorporation of ^{14}C measured by liquid scintillation counting. For both cuttings and rooted plants, there were no significant alterations in leaf monoterpene content during the course of the experiment; however, dramatic changes were observed in the level of ^{14}C found in the monoterpene pool. In stem cuttings, monoterpenes exhibited pronounced metabolic turnover. There were two episodes of ^{14}C incorporation and loss, with peaks of incorporation (at 7 h and 20 h) differing significantly (Tukey's studentized range test, $P < 0.05$) from subsequent sampling intervals (10 to 17 h and 26 to 29 h, respectively). Rooted plants, on the other hand, showed only a steady increase in ^{14}C incorporation into monoterpenes over the time span of the experiment with no significant losses of radioactivity ($P > 0.05$). Each point represents the mean (bars indicate standard error) of three samples, each consisting of the apical tip of five separate plants. (From Mihaliak, C. A., Gershenzon, J., and Croteau, R., *Oecologia*, 87, 373, 1991. With permission.)

monoterpenes labeled with a pulse of radioactive CO_2, glucose, or mevalonic acid are lost within a 10 h chase period, even though the total quantity of monoterpenes present does not change appreciably. Several years ago, we repeated these experiments using both cuttings and rooted plants,[362] because of concern that the dynamics of monoterpene pools in detached shoots might be different from those of intact plants.

In separate experiments, peppermint stem cuttings and rooted plants were exposed to [14]CO_2 in a sealed plexiglass chamber under high intensity lights. After a 5 min pulse, the chamber air was flushed through a KOH trap, and leaf samples were removed periodically over the next 29 h. Monoterpenes were isolated by simultaneous steam distillation-pentane extraction, and the incorporation of ^{14}C was measured by liquid scintillation counting. The total amount of the monoterpenes in each sample was determined by gas-liquid chromatography.

In agreement with prior work, the monoterpenes of stem cuttings were found to undergo rapid metabolic turnover. Two episodes of ^{14}C incorporation and loss were observed during the 29 h course of the experiment (Figure 15A), while the total monoterpene concentration remained relatively constant. In contrast, the rooted plants exhibited a steady increase in the incorporation of ^{14}C into monoterpenes with no significant losses of radioactivity over the time span of the experiment (Figure 15B). These results suggested that rapid monoterpene turnover does not occur normally in peppermint leaves but is an artifact seen only in detached cuttings.

More recently, we repeated this experiment using only rooted peppermint plants but

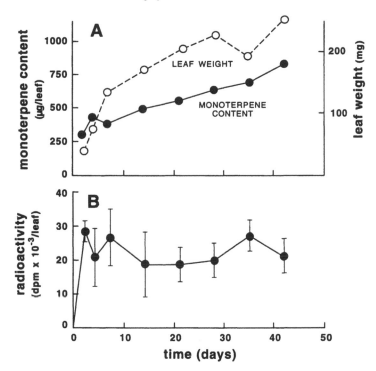

METABOLIC TURNOVER OF
Mentha x piperita MONOTERPENES

FIGURE 16. Long-term time course of ^{14}C incorporation into leaf monoterpenes of rooted *Mentha x piperita* (peppermint) after a pulse of [14]CO_2.[187] Experiment was performed in a similar fashion to that depicted in Figure 15. except that after the 5 min pulse of [14]CO_2, plants were sampled for 6 weeks. (A) Leaf weight (dashed line) and monoterpene content (solid line) steadily increased over the course of the experiment. (B) ^{14}C incorporation into monoterpenes did not change significantly (Tukey's studentized range test, $P > 0.05$) after the initial increase of the first 2 d, indicating the lack of any measurable turnover. Each point represents the mean of three samples, with bars indicating standard errors.

extending the sampling period to 6 weeks after the [14]CO_2 pulse.[187] Although leaf weight and monoterpene level steadily increased over this time (Figure 16A), there was no significant change in the radioactivity of extracted monoterpenes after the initial increase of the first 2 d (Figure 16B). Thus, the monoterpenes of intact peppermint plants do not appear to undergo any discernible turnover for at least 40 d after synthesis. Corroborating evidence for this metabolic stability comes from inspection of the changes in monoterpene content and in the rate of monoterpene biosynthesis measured during peppermint leaf development (Figure 3). In leaves more than 2 weeks old, the overall rate of monoterpene biosynthesis from CO_2 and the activities of the individual enzymes of the biosynthetic pathway drop to extremely low levels. Yet, monoterpene content does not change significantly, indicating that the rate of turnover must be negligible.

We have also expanded our work to other species and classes of terpenes by carrying out virtually identical experiments with rooted plants of *Salvia officinalis* (garden sage, Lamiaceae), *Tanacetum vulgare* (common tansy, Asteraceae), and *Pinus contorta* (lodgepole pine, Pinaceae) (Figure 17).[187] Samples were taken over a 10 to 14 d time span after the initial pulse of [14]CO_2. In all cases, the terpenes studied (monoterpenes in *S. officinalis*, *T. vulgare*, and *P. contorta*, sesquiterpenes in *S. officinalis*, and diterpene resin acids in *P. contorta*) did not show any significant loss of radioactivity and hence did not seem to be subject to rapid metabolic turnover.

METABOLIC TURNOVER

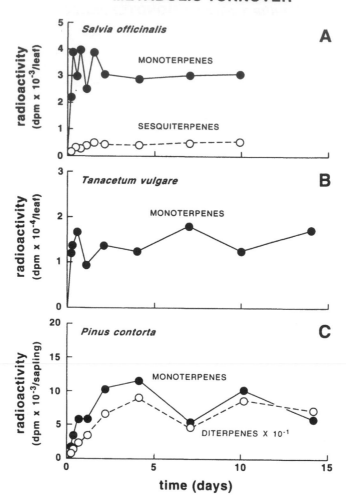

FIGURE 17. Time course of ^{14}C incorporation into foliage terpenes in rooted plants of A) *Salvia officinalis* (garden sage), B) *Tanacetum vulgare* (common tansy), and C) *Pinus contorta* (lodgepole pine) after a pulse of $^{14}CO_2$.[187] Experimental methods were as in Figures 15 and 16, except that the diterpene resin acids of *P. contorta* were isolated by soaking in *tert*-butyl methyl ether for 2 d at room temperature and analyzed by gas-liquid chromatography after methylation with diazomethane.[46,321] The level of ^{14}C did not decline significantly (Tukey's studentized range test, $P > 0.05$) over the course of these experiments for any of the terpene types in all three of the species studied. Hence, none of these substances seems to have undergone metabolic turnover. Each point represents the mean of at least three samples.

Our results are inconsistent with those of nearly all other investigations that have measured terpene turnover rates. In most previous pulse-chase labeling studies of monoterpenes,[20,21,65,96,102,171,240,379,457] sesquiterpenes,[20,96,102] and diterpenes,[55] rapid turnover was observed, with the compounds of interest having half-lives of 24 h or less, assuming complete mixing of metabolic pools and first-order kinetics. However, the vast majority of these experiments were carried out on detached stem cuttings, foliage, or flowers rather than on rooted plants and so must be viewed with suspicion.[109,362] In contrast, the handful of studies using intact plants either measured much lower rates of terpene turnover (half-lives of 5 to 100 d) or failed to detect any loss of label at all.[32,361,407,461,489]

In conclusion, if one accepts only the results of studies on intact plants, terpenoid defense compounds do not appear to be turned over readily. In view of the fact that most terpenoids are stored in the extracellular spaces of glandular trichomes, secretory cavities, or resin ducts, away from living cytoplasm, their metabolic stability is perhaps not surprising. However, many additional plant species and types of terpenoids must be carefully examined before this generalization can be considered sufficiently robust.

2. Phenolic Compounds

The metabolic turnover rates of a variety of plant phenolic compounds have also been measured. These investigations have employed mainly pulse-chase studies, but observations of diurnal variation have also been made, and several experiments have looked for changes in phenolic levels after treatment with L-α-aminooxy-β-phenylpropionic acid, an inhibitor of phenylpropanoid biosynthesis that blocks phenylalanine ammonia-lyase activity.[9,262] Unfortunately, as was true for terpenoids, a number of investigations have used detached organs rather than intact plants,[339,367,482,503] which may have led to erroneous conclusions.[109,362] In buckwheat (*Fagopyrum esculentum*), for example, flavonoid turnover rates are substantially higher in detached cotyledons or hypocotyls than in intact seedlings.[339]

Most research on phenolic turnover has been concerned with flavonoids. If we limit ourselves to studies performed with intact plants, the majority of flavonoids examined did not exhibit rapid metabolic turnover.[6,143,339,399,486] Prominent exceptions are the isoflavonoids of *Cicer arietinum* (chick pea). In the roots of this species, formononetin 7-*O*-glucoside-6″-*O*-malonate had a half-life of 110 to 160 h, while the pool of another isoflavonoid, biochanin A 7-*O*-glucoside-6″-*O*-malonate, turned over much more slowly, with a half-life of at least 300 h.[262] Formononetin 7-*O*-glucoside-6″-*O*-malonate provides one of the best documented examples of a plant secondary metabolite that undergoes turnover. Its metabolic fate is at least partly known (After sequential cleavage of the malonyl and glucose moieties, the resulting aglycone is degraded.), and the reported half-life has been confirmed by several different methods.[9,23,25,262] In addition, the turnover of this substance has been shown to occur at a time when there is no net change in plant isoflavonoid level, indicating that any losses are quickly replaced.[262] Other phenolic compounds exhibiting metabolic turnover include the phenolic glycosides in the leaves of *Salix* species,[504] coniferin in *Picea abies* (which may be an intermediate in lignin biosynthesis),[338] and several types of defenses, whose accumulation is induced by herbivory or fungal infection.[123,253]

3. Alkaloids and Other Nitrogenous Defense Compounds

There is more evidence for the turnover of alkaloids than for any other group of plant defense compounds. Rapid turnover rates have been reported for alkaloids of many different structural types,[419] including piperidine (coniine),[155] pyridine (nicotine[313,315,316,512,513,563] and ricinine[533]), quinolizidine (lupanine),[547,550] tropane (atropine and cocaine),[266,534] benzylisoquinoline (morphine),[153,154,156,364] phenylethylamine (hordenine),[172] indole (gramine),[190] purine (caffeine),[275,348,491,492] and steroid (tomatine)[146,443] alkaloids, with calculated half-lives ranging from 6 h to 1 week. In several pulse-chase experiments, the loss of label was found to be accompanied by a net increase in the total quantity of alkaloid (labeled and unlabeled),[512,533,563] suggesting that if alkaloids are actually being degraded, they must be replaced immediately. Interestingly, the turnover rate of a given alkaloid may vary with organ,[190] development,[146] or changes in certain environmental conditions. For example, lupines (*Lupinus succulentus*) fertilized with nitrate exhibited pronounced diurnal fluctuations in quinolizidine alkaloid concentration, while unfertilized lupines, which relied on symbionts for nitrogen fixation, showed no such diurnal variation.[267] In tobacco, there was significantly less loss of labeled nicotine in plants that had been "topped" (terminal inflorescence removed) than in undamaged controls.[563]

Most pulse-chase experiments describing rapid alkaloid turnover were performed on intact plants rather than on detached stem or leaf cuttings.[109,153,154,172,190,203,275,313,315,316,348,364,512,513,533,563] The use of intact plants is definitely preferred because cuttings can have an artificially high rate of alkaloid turnover.[109] Nevertheless, for our purposes, nearly all published studies describing alkaloid turnover are potentially flawed, since the labeled alkaloids themselves were administered to the plant rather than biosynthetic precursors.[146,153,154,172,190,203,275,313,315,316,348,364,444,491,492,512,513,533,563] In most cases, alkaloids were fed directly to the roots. While a variety of alkaloids are known to be synthesized in the roots and transported to the shoots,[16,120,189,230,377,534] alkaloids fed to the roots from the outside might be taken up and metabolized by cells in the periphery of the root that do not usually contact these compounds. Such cells may then degrade alkaloids via detoxification reactions or unusual, salvage-type pathways, or even form conjugates of radically-altered solubility.[24,28,177,317,419,546] These transformations will generate atypical labeled metabolites that could give misleading information on the rate and direction of natural metabolic turnover. To properly assess turnover, the best exogenous substrate for green plants is simply CO_2 fed at atmospheric concentration.[419] Curiously, the only alkaloid turnover study that used labeled CO_2 as a precursor, an investigation of the monoterpene-indole alkaloids in the shoots of *Catharanthus roseus*, did not detect any significant turnover in intact plants.[109] Alkaloid turnover was also not observed in the fruits of *Capsicum frutescens* (capsaicin),[221] the root of *Senecio erucifolius* (pyrrolizidine alkaloids),[444] and seedlings of *Cinchona ledgeriana* (quinoline alkaloids).[3] Therefore, at present, the evidence for rapid alkaloid turnover must be considered equivocal because the vast majority of pulse-labeling studies did not apply radioactive substrates in a physiologically appropriate manner. Even the reports of diurnal variation are not definitive in this regard, since most did not rule out the possibility that the observed fluctuations in leaves could be due to the transport of alkaloids to roots, seeds, or other organs, as was observed for the quinolizidine alkaloids of lupines.[547]

For other nitrogenous compounds, only meager information is available on metabolic turnover. The wound-induced proteinase inhibitors of tomato leaves apparently turn over at a very slow rate. A half-life of several weeks or longer was inferred for these substances by comparing the profiles of biosynthesis and accumulation after wounding.[212,439] Cyanogenic glycosides, on the other hand, display somewhat faster turnover rates. The half-life of dhurrin in etiolated *Sorghum vulgare* seedlings was indicated to be about 10 h,[52] while a recent, extensive investigation of dhurrin turnover in green *Sorghum bicolor* seedlings estimated the half-life of this substance at 250 to 400 h.[2] Since these experiments were performed while the net concentration of dhurrin was increasing, any losses of this cyanogenic glycoside were more than compensated.[2]

Taking the major classes of antiherbivore compounds together, it is difficult to determine the frequency with which defenses are actually lost to metabolic turnover. Few experiments have been designed to look for turnover in intact plants under physiologically realistic conditions.[2,3,109,123,143,212,262,339,362,399,407,483,486,489] While the majority of such designed experiments did not detect any significant turnover,[3,109,143,212,339,362,399,407,486] many more species and types of substances need to be surveyed before any meaningful generalizations can be made. The existence of metabolic turnover could lead to substantial increases in the cost of defense, but only if turned-over metabolites are ultimately degraded and then resynthesized. Unfortunately, in most studies, little effort has been expended toward deducing the precise metabolic path of turnover. As a consequence, there is currently almost no reliable evidence to indicate whether turnover contributes to defense costs. This is not to say that such a contribution could not be substantial. Using the most carefully determined half-life measurements available (formononetin 7-*O*-glucoside-6″-*O*-malonate in chick pea: 110 to 160 h;[262] dhurrin in sorghum: 250 to 400 h[2]), and assuming that turned-over metabolites are indeed degraded and replaced, defense pools might require total replenishment as often as once a week!

B. VOLATILIZATION, LEACHING, AND EXUDATION

The inherent volatility of many defenses, including monoterpenes, sesquiterpenes, and certain phenylpropanoid derivatives, suggests that such compounds could be lost continually from the aerial parts of plants. However, published rates of volatilization, ignoring those from flowers, are much too low to lead to significant losses of defensive compounds.[119,148,303,448,507,508,517,562] For example, in *Salvia mellifera* and various conifer species, less than 10% of the total monoterpenes would be volatilized in a typical 6 month growing period.[119,448,508,517]

Another possible route for defense loss is leaching due to rainfall, mist, fog, or dew. Plant leachates have been found to contain a variety of simple phenolic compounds, such as benzoic and cinnamic acids, coumarins, phenolic glycosides, and flavonoids.[5,115,116,311,515] Although claims have been made that these substances are leached in "appreciable amounts",[415] quantitative data are lacking. Simple phenolics have also been reported to be exuded from roots.[415,501,502] However, the quantities detected have been very small (a few micrograms per plant per day), and appear to represent only a tiny fraction of what remains in the plant. Therefore, it seems unlikely that volatilization, leaching, root exudation, and metabolic turnover cause significant losses of antiherbivore defenses, implying that the overall costs of maintenance are minimal.

VI. ADAPTATIONS FOR COST REDUCTION

As previous sections of this chapter attest, the synthesis, storage, and transport of antiherbivore defenses often require significant expenditures of resources. Therefore, plants should be under strong selection pressure to mitigate the costs of these activities. A number of adaptations that could reduce defense costs have already been mentioned, including the use of induced rather than constitutive defenses, the sharing of biosynthetic enzymes among multiple pathways, and the minimization of protein and mRNA turnover. In this section, two other cost-reduction tactics are discussed in more detail.

A. CATABOLISM OF DEFENSE COMPOUNDS THAT ARE NO LONGER NEEDED

Pools of antiherbivore defense compounds may constitute major repositories of energy and nutrients in plants. If these substances are metabolically degraded when they are no longer necessary for protection, the costs of defense might be substantially lowered, provided degradation is carried out in a way that permits most of the stored resources to be recovered in usable form. To this end, degradative reactions must be coupled directly to the synthesis of ATP and reduced pyridine nucleotides or must generate simple compounds, such as acetyl-CoA or citric acid cycle intermediates, that can serve as raw materials for the biosynthesis of new molecules or can undergo further oxidation. When nitrogen-containing defenses are broken down, the nitrogen atom must be transferred to other substances by transamination or released as ammonia.

Unfortunately, our knowledge of the catabolism of secondary compounds is rather limited, as was noted in Section V, due at least partially to various experimental constraints.[26] For instance, although secondary metabolites administered to intact plants or cell-free extracts are often broken down readily, the observed transformations may only represent detoxification or salvage-type reactions that do not ordinarily take place in the intact plant. To study the actual metabolic fate of secondary compounds *in vivo*, isotopically-labeled tracers must be fed only to cells (or the extracts of cells) in which these secondary compounds normally occur. Even with such precautions, exogenously-applied substances may be metabolized differently than endogenously-formed substances because of the difficulties in getting substances into the proper subcellular compartment. Another obstacle in studying catabolism is that intermediates are usually present in very low concentrations. Thus, it is not surprising that very few pathways of secondary product catabolism have been completely elucidated. Moreover, much of our fragmentary knowledge was obtained from experiments with undifferentiated cells in culture and so is of uncertain significance. In the next few paragraphs, some of the better documented examples of the catabolism of plant defense compounds are presented.

1. Terpenoids

The lipophilic terpenoid constituents of plant essential oils and resins are commonly believed to be metabolically inert. However, recent investigations of certain species of the Lamiaceae indicate that monoterpenes can be catabolized at late stages of leaf development. For example, in peppermint (*Mentha x piperita*), approximately 50 to 75% of the monoterpenes present in the mature leaves are degraded at the time of flowering.[94] The principal monoterpene in this species, the ketone menthone, is reduced to approximately equal amounts of the stereoisomeric alcohols menthol and neomenthol (Figure 18).[100,289] While the menthol produced remains in the leaves, the neomenthol becomes glucosylated and is transported to the rhizome.[107] Once in the rhizome, neomenthol glucoside is sequentially hydrolyzed and oxidized to reform menthone, which is then converted to menthone lactone by an unusual ring-opening reaction.[107] Finally, the menthone lactone produced is subjected to a modified β-oxidation sequence generating 6 units of acetyl-CoA.[98] An analogous process occurs in the mature leaves of garden sage (*Salvia officinalis*) where, during flowering, the monoterpene camphor is converted to a water-soluble transport derivative that is exported to the root and oxidatively degraded to acetyl-CoA.[103,104] The early steps of camphor catabolism have recently been clarified by feeding experiments conducted in cell cultures.[178]

Taken together, these investigations not only demonstrate that monoterpenes are mobilized and degraded prior to leaf senescence but also that the fixed carbon and reducing equivalents released by degradation are recycled into primary metabolism. Administration of labeled monoterpenes to peppermint rhizomes and sage roots revealed that the acetyl-CoA units derived from β-oxidation serve as substrates for the synthesis of sterols and acyl lipids,[98,103] and the reduced pyridine nucleotides formed take part in the biosynthesis of sugars and starch.[98]

Terpenoid defenses may be subject to catabolism in many other species as well, since a large variety of sesquiterpenes,[70,130,450] diterpenes,[165] triterpenes,[246,341,416] and other monoterpenes[136,220,437,488,495] are known to disappear from vegetative organs late in development. However, in none of these cases has volatilization, leaching, or transport to other parts of the plant yet been ruled out as an explanation for terpene loss.

2. Nitrogen-containing Compounds

A variety of different types of nitrogen-containing defense compounds appear to be catabolized in plants, particularly in germinating seedlings. One of the best described examples of this process is the breakdown of the nonprotein amino acid canavanine in the legume *Canavalia ensiformis* (Figure 19). Rosenthal and coworkers deciphered the basic route of canavanine degradation in this species and established that all four canavanine nitrogen atoms are recirculated into primary pathways (Figure 19).[424,425,429,432] The rest of the canavanine molecule is apparently utilized as well, since feeding [14C]canavanine to *C. ensiformis* seedlings resulted in the evolution of $^{14}CO_2$.[429] In addition to nonprotein amino acids, legume seeds often contain large concentrations of higher molecular weight nitrogen-containing defenses, such as proteinase inhibitors, amylase inhibitors, and lectins. During germination, these substances are catabolized in much the same fashion as other storage proteins.[352,446,542]

We have already seen (see Section IV.B.3) that, in the germinating seedlings of the rubber tree, the cyanogenic glycoside linamarin is transported from the seeds to the young leaves in the form of the diglycoside linustatin, a compound more resistant to hydrolysis than

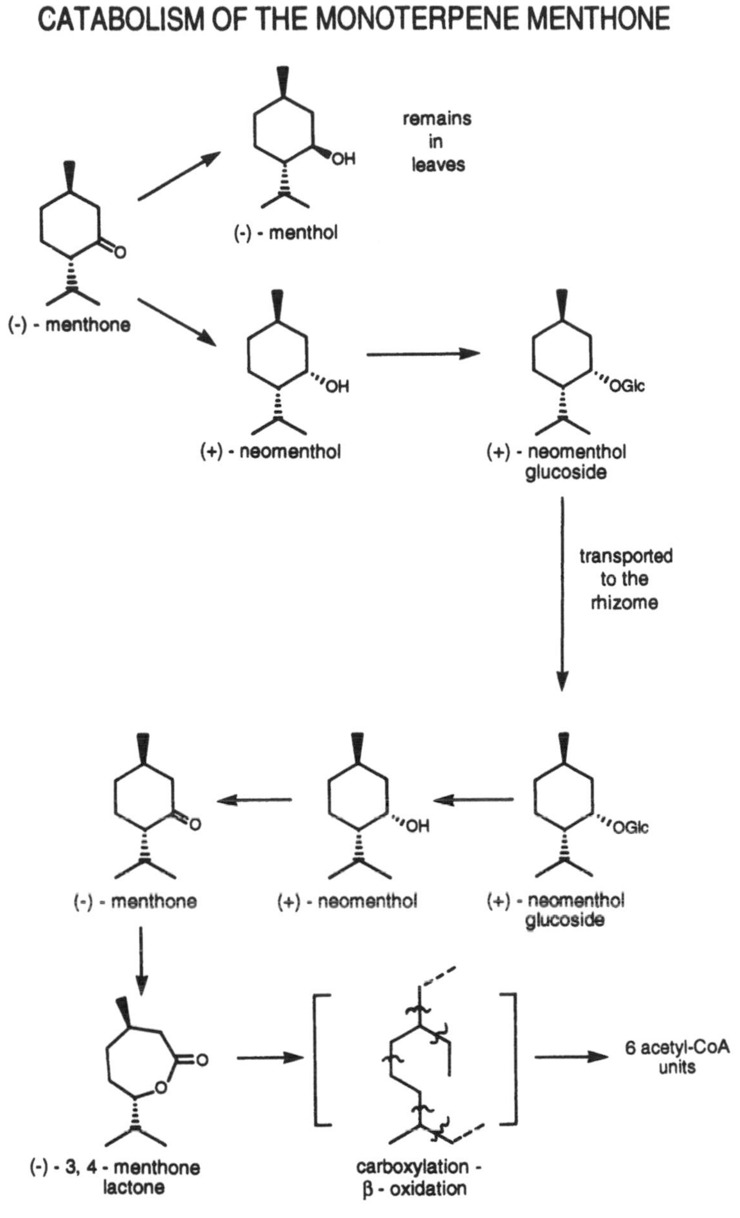

CATABOLISM OF THE MONOTERPENE MENTHONE

FIGURE 18. Menthone, the principal monoterpene of the mature leaves of peppermint (*Mentha x piperita*), is degraded at the time of flowering by the pathway shown.[94,98,100,107,289] The proposed scheme of β-oxidation is shown in brackets, with the dotted lines indicating likely sites of carboxylation and the wavy lines showing the bonds cleaved to form the acetyl-CoA units.[98] The acetyl-CoA produced is incorporated into sterols and acyl lipids, while the reduced pyridine nucleotides formed are used in carbohydrate formation.[98]

linamarin.[464,465] Once in the leaves, linustatin is broken down into noncyanogenic compounds without the detectable liberation of hydrogen cyanide.[326] This conversion has been postulated to proceed via the hydrolysis of linustatin, mediated by a specific diglucosidase, and the dissociation of the resulting acetone cyanohydrin (Figure 14). The hydrogen cyanide evolved is then immediately incorporated into β-cyanoalanine (by the action of β-cyanoalanine synthase, see Section III.C).[363,557] β-Cyanoalanine is subsequently converted into asparagine,[72] thereby recycling the nitrogen atom into primary metabolism. All these enzyme activities, except the

CATABOLISM OF THE NON-PROTEIN
AMINO ACID CANAVANINE

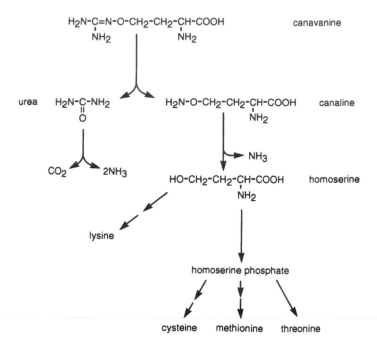

FIGURE 19. The metabolic degradation of canavanine in germinating jack bean (*Canavalia ensiformis*) seedlings is one of the best documented examples of the catabolism of a plant defense compound. This process allows all four nitrogen atoms of canavanine to be recycled into primary metabolism.[425,429,432]

asparagine synthase, have been demonstrated in cell-free extracts of rubber tree seedlings, although flux through the entire pathway has not yet been shown.[326,464]

There are also several good examples of catabolism among the alkaloids. Caffeine, for instance, is metabolically degraded in ripening coffee (*Coffea arabica*) fruits.[275,348,494] Extensive feeding studies have shown that the catabolism of caffeine begins with the loss of the three *N*-methyl carbon atoms (Figure 1C, **23**), followed by several oxidative, ring-opening transformations that result in the eventual formation of glyoxylate and urea. Hydrolysis of urea affords carbon dioxide and ammonia, which can be reincorporated into anabolic pathways via glutamate and glutamine.[498]

The literature contains many other reports concerning the degradation of nitrogenous defense compounds,[26,146,153,190,364,460,461,513,533,534] although in the majority of these cases definite proof is lacking that the reactions described actually occur in the intact plant. In addition, it has been observed that the levels of nitrogen-containing defenses frequently decline in germinating seeds, ripening fruits, senescing leaves, or other stages of development.[85,387,437,463,493,510,519,536,549,550] Some of these declines are undoubtedly attributable to leaching or transport, rather than to metabolic degradation. However, viewed in concert with previously-discussed examples, they suggest that the catabolism of nitrogenous defense compounds is a common pattern. The metabolic lability of nitrogen-containing defenses could be an adaptive consequence of the fact that shortages of this nutrient often limit plant growth.[75] In many environments, nitrogen may simply be too valuable to sequester in defense compounds unless it can later be reclaimed for growth or reproduction. Under these conditions, nitrogen-containing plant defenses may act as nitrogen storage reserves in addition to providing protection from herbivores.

Besides monoterpenes and nitrogen compounds, there are strong indications that other groups of plant defenses also undergo catabolism. Flavonoids, for instance, have been shown to be degraded metabolically in a number of cases,[23,25,26,28,262] by a process that may involve peroxidases.[26,28] In addition, both terpenoid and flavonoid phytoalexins seem to be broken down rapidly after synthesis, although this conclusion is based largely on studies performed with cell cultures rather than intact plants.[27,57,506,540]

3. The Costs and Benefits of Catabolism

As this brief survey suggests, several classes of defense compounds are frequently catabolized at particular stages of plant development. Moreover, where the biochemical details of this process have been investigated, it appears that much of the energy and nutrients originally expended in biosynthesis are salvaged, thereby reducing the overall costs of defense. The magnitude of this savings, as computed in terms of glucose equivalents (see Table 1, Appendix), can be quite substantial. For example, during the catabolism of menthone to acetyl-CoA units discussed above, nearly 80% of the original raw materials cost is recovered, assuming β-oxidation proceeds with complete recovery of reducing equivalents. Similarly, the breakdown of canavanine described reclaims about 85% of the raw materials originally invested in biosynthesis and, perhaps more importantly, recycles all four nitrogen atoms.

Of course, catabolism may impose costs of its own that must be balanced against the potential benefits of this process. For example, specific enzymes could be required to catalyze certain degradative reactions.[178,279,352,542] In addition, if the substance being catabolized is normally sequestered away from sensitive metabolic activities, special physiological adjustments may be necessary to avoid autotoxicity at the time of degradation.[464,465] Thus, it may be unreasonable to expect that plants can recoup the entire cost of synthesizing defensive substances. Nevertheless, catabolism can clearly make a sizable contribution toward balancing plant defense budgets. Additional research is urgently needed to determine how widespread this process is.

B. RECRUITMENT OF DEFENSE COMPOUNDS FOR ADDITIONAL FUNCTIONS

Many plant secondary metabolites that are assumed to serve as antiherbivore defenses have also been implicated in a large variety of other plant functions,[460] such as defense against pathogens,[58,226,283,474] attraction of pollinators and dispersal agents,[41,59] mediation of allelopathy,[401,415,505] reduction of transpiration,[117,118,355,413] interception of UV radiation,[67,68,276,320,525] protection against oxidative damage,[341,499] regulation of growth and development,[163,259,281,400] promotion of pollen germination,[366,561] storage of nutrients and energy (see Section VI.A), stimulation of nodule formation by N_2-fixing symbionts,[392,393,405] and stimulation of mycorrhizal growth.[472] If antiherbivore defenses actually take part in any of these additional functions, the costs of their synthesis, storage, transport, and maintenance will be shared among the functions performed, thereby reducing the expenditures necessary for defense. Unfortunately, many of these other roles are poorly documented. For example, the possible involvement of plant phenolics in regulating growth and development[163,259,400] is highly speculative at this point due to the lack of information on internal concentration and compartmentation.[281] Hence, at present it is difficult to decide whether defense compounds are commonly recruited for other purposes.

Nevertheless, plants that have multipurpose defense compounds and/or that catabolize defenses when they are no longer needed may enjoy substantial decreases in defense costs. In fact, the widespread adoption of these and other cost-reducing traits might explain why defense costs are sometimes undetectable when attempts are made to measure them empirically.[56,60,469-471] As Simms and Fritz point out, the cost of defense should be subject to continual reduction by the action of natural selection.[469]

VII. COST AND CARBON–NUTRIENT BALANCE

The amounts of antiherbivore defenses in plants are frequently altered by changing environmental conditions.[166,184,188,331,534,535] Many of the alterations observed can be rationalized by the carbon–nutrient balance hypothesis, which attributes phenotypic changes in the concentration of chemical defenses to variations in the relative availability of fixed carbon and nutrients for growth and defense.[62,516] The carbon–nutrient balance hypothesis is based on the observation that the rate of growth is more sensitive to nutrient deficiency than the rate of photosynthesis.[75] Thus, when nutrient supply is limited (high carbon to nutrient ratio), a plant's growth rate may slow dramatically, but photosynthesis continues apace, leading to a buildup of carbohydrate reserves in excess of those required for growth. This is thought to provide additional substrate for the biosynthesis of non-nitrogenous (carbon-based) defense compounds. On the other hand, when growth is carbon-limited rather than nutrient-limited (low carbon to nutrient ratio), such as under low light conditions, carbohydrate reserves are predicted to decline, leading to the reduced production of carbon-based defenses.

The alterations in secondary metabolite concentrations observed under different light and nutrient regimes often follow the predictions of the carbon-nutrient balance hypothesis.[158,410,433] However, a number of exceptions have been noted in recent years.[45,141,147,309,350,373] In addition, species grown with an augmented supply of carbon dioxide, and so presumably having higher carbon to nutrient ratios, do not show increases in carbon-based defenses.[157-159,269,270,328]

Implicit in the carbon–nutrient balance approach is the notion that carbon-based defensive substances are synthesized largely from excess carbohydrate and so have very low raw materials costs. However, while a high carbon to nutrient ratio may increase the availability of substrate for producing defense compounds, it may not necessarily lower the other costs of chemical defense, including those of biosynthetic machinery, storage, transport, or maintenance. If these costs are substantial, high carbon to nutrient ratios may not automatically result in increased concentrations of carbon-based defenses. Conversely, if the costs of biosynthetic machinery, storage, transport, and maintenance are low, decreases in carbon to nutrient ratios need not always cause declines in the concentrations of carbon-based defenses. Reichardt and coworkers have noted that the apparent failure of the carbon–nutrient balance model may also be due to the fact that increased concentrations of defenses can be masked by rapid metabolic turnover (see Section V.A).[410]

Perhaps a more fundamental weakness of the carbon–nutrient balance hypothesis is its assumption that the production of plant defenses is freely alterable by changes in the supply of carbohydrate reserves. Substrate levels appear to be just one of the factors controlling the biosynthesis of defense compounds. As we have seen in this chapter (see Section II.B.3), changes in gene expression, enzyme activity, and subcellular compartmentation may all contribute to the regulation of secondary metabolite formation. Interestingly, even when changes in substrate supply do alter the synthesis of defenses, this may be a result of complex regulatory controls rather than simple mass action considerations. For example, the genes for both tomato proteinase inhibitor II and petunia chalcone synthase, a key regulatory enzyme in flavonoid biogenesis, have sucrose-responsive promoter elements[268,285,514] and so react to increased substrate supply with enhanced rates of transcription. In summary, while the carbon–nutrient balance hypothesis correctly predicts the response of many plant defensive metabolites to environmental changes, its predictions may only be valid when substrate supply actually limits the manufacture of carbon-based defenses, and when the costs of storage, transport, and maintenance are not prohibitively high.

VIII. CONCLUSIONS

The concept of cost has played a major role in the development of theories on the evolution and distribution of plant antiherbivore substances. In this review, the metabolic costs or "direct" costs of synthesizing, storing, transporting, and maintaining chemical defenses are surveyed in light of recent advances in the biochemistry and physiology of plant secondary metabolism. The major conclusions can be summarized as follows:

1. Terpenoids and alkaloids are generally more expensive to produce per gram than phenolics and other nitrogen-containing compounds, based on raw materials costs measured in terms of the amount of glucose required. However, under conditions of nitrogen scarcity, the costs of making all nitrogen-containing defenses will be greatly elevated. Variations in the actual concentrations of defenses in plant tissue may cause large differences in cost as expressed per unit of plant weight.

2. In the last few years, substantial progress has been made in understanding the enzymology and regulation of plant secondary metabolism, permitting a preliminary assessment of the cost of the biosynthetic machinery needed for making plant defenses. The mRNA and enzymes necessary for biosynthesis appear to be present in plants for only brief periods in development, thus minimizing the potential costs of replacement due to turnover. However, the use of complex mechanisms for regulating biosynthesis and the fact that there is little apparent sharing of enzymes between pathways of defense and those of primary metabolism may raise biosynthetic machinery costs.

3. Storage of defenses requires significant investments of plant resources, both for construction of the storage structures themselves and for other adaptations necessary to avoid autotoxicity (e.g., production of detoxification enzymes, modification of potential physiological targets). Storage costs may be especially high for lipophilic defenses that are sequestered in complex secretory structures.

4. The transport of antiherbivore defenses seems less expensive than most other aspects of defense cost, although recent results indicate that both intracellular and intercellular transport often occur via active processes mediated by selective membrane carriers.

5. The cost of maintenance also appears to be low, since neither volatilization, leaching, exudation, nor metabolic turnover has been shown unequivocally to cause significant losses of defense compounds.

6. Plants may employ various strategies to lower their defense costs, including the metabolic degradation of defense compounds that are no longer needed. Current evidence suggests that various types of defenses, including many that contain nitrogen, are commonly degraded during development, thereby recirculating the stored energy and nutrients into primary metabolic pathways. The recruitment of antiherbivore substances for additional ecological or physiological functions could also lead to a substantial savings in defense costs. Unfortunately, little is known about whether defensive metabolites commonly serve multiple functions in plants, which is a serious impediment to evaluating their cost.

7. The carbon–nutrient balance hypothesis successfully predicts the response of defense compounds to changes in light and nutrient availability in many plant species, although a number of exceptions have recently been reported. The failures of this hypothesis may be due to 1) its assumption that the supply of raw materials is always a limiting factor in the biosynthesis of defense compounds and 2) the fact that it only takes into account the cost of raw materials for biosynthesis and not the costs of biosynthetic machinery, storage, transport, or maintenance.

Clearly, the metabolic cost of plant chemical defense has a number of separate components, not all of which are well understood. It is therefore premature to make generalizations about the relative expenses incurred by particular classes of defensive substances based on only one component of cost. Before we are able to calculate defense costs accurately, more information is needed on all aspects of this subject. Happily, new insights on the cost of synthesis should continue to emerge from ongoing investigations concerning the formation of certain economically-valuable plant secondary metabolites. However, to determine the resources required for storage, transport, and maintenance, research must be encouraged that is specifically directed at understanding the energetics of these processes. Such studies might be carried out very profitably with plants that have been genetically transformed to alter their defensive phenotype. Some of these are in fact already available; for example, the gene encoding tomato proteinase inhibitor I has been inserted and expressed in tobacco, nightshade, and alfalfa.[271,272,375] Direct comparisons of transformed and untransformed plants should provide exciting opportunities for evaluating the various components of metabolic defense costs without the complications of most other genetic differences. Transformed plants could also be extremely useful in assessing the costs of defense measured in terms of reproductive fitness. A sound knowledge of both fitness costs and metabolic costs should be invaluable in helping to understand the evolution and distribution of antiherbivore defenses in plants.

ACKNOWLEDGMENTS

I thank Ian Baldwin, John Bryant, Rod Croteau, Marcel Dicke, Murray Isman, Jean Langenheim, Manuel Lerdau, Simon Mole, Paul Reichardt, and Tom Savage for their invaluable comments and Liz Bernays for inviting this contribution and then waiting patiently for it to finally materialize. Thanks are also due to Joyce Tamura-Brown for typing the final draft and to Rebecca Steever, Colleen Fowles, and Ruth Younce for preparing the figures. My research described here was supported by a grant from the Department of Energy (DE-FG06-88ER13869) to R. Croteau and by Project 0268 of the Agricultural Research Center, Washington State University.

APPENDIX: A SUMMARY OF THE METHODS USED FOR CALCULATING BIOSYNTHETIC COSTS IN TABLES 1 AND 2

The raw materials costs for the biosynthesis of a variety of primary and secondary metabolites are listed in Tables 1 and 2. Following the methods developed by Penning de Vries et al.[391] and Mooney and coworkers,[357,541] these costs are expressed in terms of the total grams of glucose needed to make each gram of product. Each of the starting materials, reactants, and cofactors consumed in biosynthesis (but not the enzymes involved) was constructed separately from glucose via established pathways, and the individual requirements for glucose were then added together.

All calculations were based on the most up-to-date biosynthetic information available. General pathways were taken from standard texts,[197,498,523] while the specific pathways for each compound were obtained from the references given in the list below. Although an attempt was made to choose substances whose biosynthesis had been fully worked out, this was not possible for all of the major classes of defenses. Thus, where intermediates or cofactor requirements were unknown, inferences were made based on analogous transformations in better studied pathways. These are noted under the individual compounds in the list below. The results in Tables 1 and 2 are sometimes at variance with the biosynthetic costs computed by other authors for the same compounds.[74,304,357,391,541] These differences are due to the use of newer biosynthetic information in the present treatment and to slight variations in some of the assumptions made regarding the pathways of cofactor biosynthesis and nutrient assimilation.

Of the major cofactors, ATP was assumed to be generated exclusively from glucose via glycolysis and the citric acid cycle coupled to electron transport in the mitochondrion. Since a total of 36 moles of ATP is derived from 1 mole of glucose, the cost of an ATP molecule was considered to be 1/36 (= 0.0278) glucose units. NADPH was postulated to be produced solely in the pentose phosphate pathway, in which the consumption of glucose-6-phosphate (the product of glucose + 1 ATP) generates 12 NADPH. Thus, the cost of one molecule of NADPH was set at $(1 + 1/36) \div 12 = 0.0856$ glucose units. The cost of adding a methyl group from the cofactor *S*-adenosylmethionine was taken as 12 ATP (= 0.334 glucose units).[12] For any molecule of ATP, reduced pyridine nucleotide, or other organic substance produced during biosynthesis, its cost in glucose units was subtracted from the total quantity of glucose required by the pathway. This includes the NADH formed during glycolysis, which generates 2 ATP after transport into the mitochondrion and so has a value of 0.0556 glucose units.

Turning to nutrients, nitrogen was assumed to be supplied as nitrate rather than as ammonium. The assimilation of nitrate into organic compounds necessitates 1) reduction to nitrite (1 NADPH needed), 2) the reduction of nitrite to ammonium, which was assumed to take place in nonphotosynthetic tissues (3 NADPH needed), and 3) the incorporation of ammonium into glutamate by the action of glutamine synthase and glutamate-oxoglutarate amidotransferase (1 ATP + 1 NADPH needed).[498] Hence, the net cost per nitrogen atom is 0.456 glucose units. For sulfur assimilation, sulfate was used as the starting material. The formation of sulfide requires the activation of sulfate to form 3′-phosphoadensoine-5′-phosphosulfate (3 ATP needed) followed by an eight electron reduction (which in nonphotosynthetic cells would need 4 NADPH), giving a net cost of 0.426 glucose units per sulfur atom.

REFERENCES FOR THE BIOSYNTHESIS OF SPECIFIC COMPOUNDS

Terpenoids

- Camphor[95,97,113]
- Menthone[99]
- Aucubin:[265] The cofactor requirements for several steps of the pathway, including the closure of the heterocyclic ring, have not been firmly established.
- Caryophyllene[114]
- Germacrone:[69] The last two steps of the pathway were inferred based on analogies to hydroxylation[278] and dehydrogenation[113] reactions in monoterpene biosynthesis.
- Pachydictyol A: The cyclization of geranylgeranyl pyrophosphate and the capture of the resulting carbocation by water were modeled after reactions occurring in the biosynthesis of other diterpenes.[539]
- Abietic acid:[458] Cyclization of geranylgeranyl pyrophosphate was assumed to proceed via copalyl pyrophosphate[539] and abietadiene, which is then oxidized to the final product by reactions like those involved in gibberellin biosynthesis.[198]
- Cucurbitacin B and Papyriferic acid: For these and nearly all other triterpenes, the biosynthetic steps following squalene oxide are completely unknown. A hypothetical sequence of cyclizations, oxidations, and esterifications was proposed for each of these compounds based on similar reactions occurring in the formation of sterols and other terpenoids.[197] For cucurbitacin B, some evidence has been obtained for the nature of the initial cyclization steps.[19]

Phenolics

The basic sequence of the shikimate pathway, including the role of arogenate as an intermediate is described in References 34, 263, and 264.

- Salicin:[207] Recent evidence suggests that the degradation of the side chain that occurs in the conversion of phenylpropanoids (C_6-C_3) to benzoic acid derivatives (C_6-C_1) does not involve CoA esters, but proceeds by nonoxidative cleavage reactions.[452]
- Chlorogenic acid[207,522,567]
- Daphnetin[36] and Psoralen:[217] Formation of the lactone ring was postulated to occur from an intermediate glucose ester.[61]
- Tellimagrandin II:[208,213,232] Gallic acid was assumed to arise from the dehydrogenation of 3-dehydroshikimate rather than from the β-oxidation of caffeic acid. The reaction that links the two galloyl moieties in the final product has not been studied.
- Procyanidin polymer:[232,242,479] The mechanism by which the monomers are coupled and the cofactor requirements (if any) of this step are completely unknown.
- Apigenin and 5,6,7,8-tetramethoxyflavone[242,293]
- Daidzein[294,537]

Alkaloids

- Thebaine:[477,566,568] The final step of thebaine biogenesis is thought to entail the formation of the oxygen bridge between the two aromatic rings. This reaction was assumed to be catalyzed by an NADPH-requiring cytochrome P-450 oxygenase in analogy with a similar transformation described at the intermolecular level.[568]
- Ajmalicine:[205,244,245] Most of the steps in the conversion of strictosidine to cathenamine are only poorly known. A hypothetical reaction sequence was formulated based on the biogenesis of other alkaloids. For references to the construction of the iridoid moiety, see Aucubin above.
- Lupanine:[552,553] The final steps of the pathway following oxosparteine have not been well studied but were conjectured to involve simple redox processes.
- Nicotine:[175,206,314,529,530] The pyridine ring was assumed to be derived from glyceraldehyde-3-phosphate and aspartate, rather than from tryptophan.[197,206] Quinolinic acid was postulated to be converted directly to nicotinic acid via nicotinic acid mononucleotide, rather than passing through all the steps of the pyridine nucleotide cycle.[529,531] In the biosynthesis of the pyrrolidine ring, the conversion of glutamate to ornithine was hypothesized to proceed through N-acetylglutamate and its derivatives, rather than via glutamate-5-semialdehyde.
- Caffeine[376,449,494]

Other Nitrogenous Defenses

- Prunasin and Linamarin[91,124,219]
- Methyl glucosinolate and 3,4-Dihydroxylbenzyl glucosinolate:[124,369] The previously hypothesized sequence of intermediates between the N-hydroxylation and S-alkylation steps was modified to reflect the latest findings on cyanogenic glycoside biosynthesis.[219] The reduced sulfur atom was assumed to arise from conjugation with a cysteine residue,[369] which is eventually released as pyruvate, in analogy to what occurs during methionine formation.[523]
- Mimosine[420,426]

- Canavanine:[427] The formation of this nonprotein amino acid is believed to involve the reduction of aspartate to homoserine, the attachment of an amino group to generate canaline, and the conversion of canaline to canavanine in the urea cycle, in analogy with the conversion of ornithine to arginine. However, recent evidence concerning the extreme toxicity of canaline to plants casts doubt on the role of this intermediate in canavanine biosynthesis.[425]

- Proteinase inhibitors I and II from tomato:[200,201] Costs were calculated for the synthesis of the respective preproteins rather than the post-translationally modified final products. However, for the portion of the protein that is post-translationally cleaved, costs were computed only for peptide bond formation and not for the biosynthesis of the individual amino acids, based on the presumption that these would be recycled readily into other proteins after hydrolysis. The cost of constructing a peptide bond was taken to be 4 ATP, 2 ATP for attaching the amino acid to its corresponding tRNA and 2 ATP for elongating the existing polypeptide chain on the ribosome (equivalent to the 2 GTP molecules actually required for this process).[523]

REFERENCES

1. **Adamczeski, M., Ni, J. X., Jaber, H., Huang, J., Kang, R., and Nakatsu, T.,** A novel hydrolyzable tannin and related compounds isolated from the leaf surface of *Chrysolepis sempervirens, J. Nat. Prod.*, 55, 521, 1992.
2. **Adewusi, S. R. A.,** Turnover of dhurrin in green sorghum seedlings, *Plant Physiol.*, 94, 1219, 1990.
3. **Aerts, R. J., van der Leer, T., van der Heijden, R., and Verpoorte, R.,** Developmental regulation of alkaloid production in *Cinchona* seedlings, *J. Plant Physiol.*, 136, 86, 1990.
4. **Akhila, A. and Gupta, M. M.,** Biosynthesis and translocation of diosgenin in *Costus speciosus, J. Plant Physiol.*, 130, 285, 1987.
5. **Al-Naib, F, A.-G. and Rice, E. L.,** Allelopathic effects of *Platanus occidentalis. Bull. Torr. Bot. Club,* 98, 75, 1971.
6. **Alhach, R. F., Juarez, A. T., and Lime, B. J.,** Time of naringin production in grapefruit, *J. Am. Soc. Hort. Sci.*, 94, 605, 1969.
7. **Alonso, W. R. and Croteau, R.,** Purification and characterization of the monoterpene cyclase γ-terpinene synthase from *Thymus vulgaris, Arch. Biochem. Biophys.*, 286, 511, 1991.
8. **Amman, M., Wanner, G., and Zenk, M. H.,** Intracellular compartmentation of two enzymes of berberine biosynthesis in plant cell cultures, *Planta*, 167, 310, 1986.
9. **Amrhein, N. and Diederich, E.,** Turnover of isoflavones in *Cicer arietinum* L., *Naturwiss.*, 67, 40, 1980.
10. **Anhalt, S. and Weissenbock, G.,** Subcellular localization of luteolin glucuronides and related enzymes in rye mesophyll, *Planta*, 187, 83, 1992.
11. **ap Rees, T.,** Compartmentation of plant metabolism, in *The Biochemistry of Plants: A Comprehensive Treatise*, Vol. 12, *Physiology of Metabolism*, Davies, D. D., Ed., Academic Press, San Diego, 1987, 87.
12. **Atkinson, D. E.,** *Cellular Energy Metabolism and its Regulation*, Academic Press, New York, 1977, ch. 3.
13. **Baas, W. J. and Figdor, C. G.,** Triterpene composition of *Hoya australis* cuticular wax in relation to leaf age, *Z. Pflanzenphysiol.*, 87, 243, 1978.
14. **Bach, T. J.,** Hydroxymethylglutaryl-CoA reductase, a key enzyme in phytosterol synthesis?, *Lipids*, 21, 82, 1986.
15. **Bach, T. J., Boronat, A., Caelles, C., Ferrer, A., Weber, T., and Wettstein, A.,** Aspects related to mevalonate biosynthesis in plants, *Lipids*, 26, 637, 1991.
16. **Baldwin, I. T.,** Mechanism of damage-induced alkaloid production in wild tobacco, *J. Chem. Ecol.*, 15, 1661, 1989.
17. **Baldwin, I. T., Schultz, J. C., and Ward, D.,** Patterns and sources of leaf tannin variation in yellow birch (*Betula allegheniensis*) and sugar maple (*Acer saccharum*), *J. Chem. Ecol.*, 13, 1069, 1987.
18. **Baldwin, I. T., Sims, C. L., and Kean, S. E.,** The reproductive consequences associated with inducible alkaloidal responses in wild tobacco, *Ecology*, 71, 252, 1990.

19. **Balliano, G., Caputo, O., Viola, F., Delprino, L., and Cattel, L.,** The transformation of 10α-cucurbita-5,24-dien-3β-ol into cucurbitacin C by seedlings of *Cucumis sativus, Phytochemistry,* 22, 909, 1983.

20. **Banthorpe, D. V. and Ekundayo, O.,** Biosynthesis of (+)-car-3-ene in *Pinus* species, *Phytochemistry,* 15, 109, 1976.

21. **Banthorpe, D. V., Doonan, H. J., and Wirz-Justice, A.,** Terpene biosynthesis. Part V. Interconversions of some monoterpenes in higher plants and their possible role as precursors of carotenoids, *J. Chem. Soc. Perkin Trans. I,* 1764, 1972.

22. **Barron, D., Varin, L., Ibrahim, R. K., Harborne, J. B., and Williams, C. A.,** Sulphated flavonoids — an update, *Phytochemistry,* 27, 2375, 1988.

23. **Barz, W.,** Degradation of polyphenols in plants and plant cell suspension cultures, *Physiol. Veg.,* 15, 261, 1977.

24. **Barz, W. and Hoesel, W.,** Metabolism and degradation of phenolic compounds in plants, in *Biochemistry of Plant Phenolics,* Vol. 12, *Recent Advances in Phytochemistry,* Swain, T., Harborne, J. B., and Van Sumere, C. F., Eds., Plenum Press, New York, 1979, 339.

25. **Barz, W. and Hosel, W.,** Uber den Umsatz von Flavonolen und Isoflavonen in *Cicer arietinum, Phytochemistry,* 10, 335, 1971.

26. **Barz, W. and Koster, J.,** Turnover and degradation of secondary (natural) products, in *The Biochemistry of Plants: A Comprehensive Treatise,* Vol. 7, *Secondary Plant Products,* Conn, E. E., Ed., Academic Press, New York, 1981, 35.

27. **Barz, W., Beimen, A., Drager, B., Jaques, U., Otto, C., Super, E., and Upmeier, B.,** Turnover and storage of secondary products in cell cultures, in *Secondary Products from Plant Tissue Culture,* Vol. 30, Annu. Proc. Phytochem. Soc. Eur., Charlwood, B. V. and Rhodes, M. J. C., Eds., Clarendon Press, Oxford, 1990, 79.

28. **Barz, W., Koster, J., Weltring, K.-M., and Strack, D.,** Recent advances in the metabolism and degradation of phenolic compounds in plants and animals, in *The Biochemistry of Plant Phenolics,* Vol. 25, Annu. Proc. Phytochem. Soc. Eur., Van Sumere, C. F. and Lea, P. J., Eds., Clarendon Press, Oxford, 1985, 307.

29. **Bazzaz, F. A., Chiariello, N. R., Coley, P. D., and Pitelka, L. F.,** Allocating resources to reproduction and defense, *Bioscience,* 37, 58, 1987.

30. **Bedbrook, J. R., Smith, S. M., and Ellis, R. J.,** Molecular cloning and sequencing of cDNA encoding the precursor to the small subunit of chloroplast ribulose-1,5,-bisphosphate carboxylase, *Nature,* 287, 692, 1980.

31. **Belingheri, L., Pauly, G., Gleizes, M., and Marpeau, A.,** Isolation by aqueous two-polymer phase system and identification of endomembranes from *Citrofortunella mitis* fruits for sesquiterpene hydrocarbon synthesis, *J. Plant Physiol.,* 132, 80, 1988.

32. **Benayoun, J. and Ikan, R.,** The formation of terpenoids and their role in the metabolism of *Pinus halepensis* Mill., *Ann. Bot.,* 45, 645, 1980.

33. **Bennett, B. C., Bell, C. R., and Boulware, R. T.,** Geographic variation in alkaloid content of *Sanguinaria canadensis* (Papaveraceae), *Rhodora,* 92, 57, 1990.

34. **Bentley, R.,** The shikimate pathway — a metabolic tree with many branches, *Crit. Rev. Biochem. Mol. Biol.,* 25, 307, 1990.

35. **Berenbaum, M.,** Brementown revisited: interactions among allelochemicals in plants, in *Chemically Mediated Interactions between Plants and Other Organisms,* Vol. 19, *Recent Advances in Phytochemistry,* Cooper-Driver, G. A., Swain, T., and Conn, E. E., Eds., Plenum Press, New York, 1985, 139.

36. **Berenbaum, M. R.,** Coumarins, in *Herbivores: Their Interactions with Secondary Plant Metabolites,* Vol. 2, 2nd ed., Rosenthal, G. A. and Berenbaum, M. R., Eds., Academic Press, San Diego, 1991, 221.

37. **Berenbaum, M. R.,** Charge of the light brigade: phototoxicity as a defense against insects, in *Light-Activated Pesticides,* ACS Symposium Series, No. 339, Heitz, J. R. and Downum, K. R., Eds., American Chemical Society, Washington, D.C., 1987, 206.

38. **Berenbaum, M. R. and Larson, R. A.,** Flux of singlet oxygen from leaves of phototoxic plants, *Experientia,* 44, 1030, 1988.

39. **Berenbaum, M. R. and Zangerl, A. R.,** Genetics of secondary metabolism and herbivore resistance in plants, in *Herbivores: Their Interactions with Secondary Plant Metabolites,* Vol. 2, 2nd ed., Rosenthal, G. A. and Berenbaum, M. R., Eds., Academic Press, San Diego, 1992, 415.

40. **Berenbaum, M. R., Zangerl, A. R., and Nitao, J. K.,** Constraints on chemical coevolution: wild parsnips and the parsnip webworm, *Evolution,* 40, 1215, 1986.

41. **Bergstrom, G.,** Chemical ecology of terpenoid and other fragrances of angiosperm flowers, in *Ecological Chemistry and Biochemistry of Plant Terpenoids,* Vol. 31, Annu. Proc. Phytochem. Soc. Eur., Harborne, J. B. and Tomas-Barberan, F. A., Eds., Clarendon Press, Oxford, 1991, 287.

42. **Berhow, M. A. and Vandercook, C. E.,** Sites of naringin biosynthesis in grapefruit seedlings, *J. Plant Physiol.,* 138, 176, 1991.

43. **Bicchi, C., D'Amato, A., Frattini, C., Cappelletti, E. M., Caniato, R., and Filippini, R.,** Chemical diversity of the contents from the secretory structures of *Heracleum sphondylium* subsp. *sphondylium, Phytochemistry,* 29, 1883, 1990.

44. **Biggs, D. R., Welle, R., and Grisebach, H.,** Intracellular localization of prenyltransferases of isoflavonoid phytoalexin biosynthesis in bean and soybean, *Planta*, 181, 244, 1990.
45. **Bjorkman, C., Larsson, S., and Gref, R.,** Effects of nitrogen fertilization on pine needle chemistry and sawfly performance, *Oecologia*, 86, 202, 1991.
46. **Black, T. H.,** The preparation and reactions of diazomethane, *Aldrichimica Acta*, 16, 3, 1983.
47. **Blom, T. J. M., Sierra, M., van Vliet, T. B., Franke-van Dijk, M. E. I., de Koning, P., van Iren, F., Verpoorte, R., and Libbenga, K. R.,** Uptake and accumulation of ajmalicine into isolated vacuoles of cultured cells of *Catharanthus roseus* (L.) G. Don. and its conversion into serpentine, *Planta*, 183, 170, 1991.
48. **Bones, A. M., Thangstad, O. P., Haugen, O. A., and Espevik, T.,** Fate of myrosin cells: characterization of monoclonal antibodies against myrosinase, *J. Exp. Bot.*, 42, 1541, 1991.
49. **Bonora, A., Botta, B., Menziani-Andreoli, E., and Bruni, A.,** Organ-specific distribution and accumulation of protoanemonin in *Ranunculus ficaria* L., *Biochem. Physiol. Pflanzen*, 183, 443, 1988.
50. **Booth, E. J., Walker, K. C., and Griffiths, D. W.,** A time-course study of the effect of sulphur on glucosinolates in oilseed rape (*Brassica napus*) from the vegetative stage to maturity, *J. Sci. Food Agric.*, 56, 479, 1991.
51. **Boufford, D. E., Kjaer, A., Madsen, J. O., and Skrydstrup, T.,** Glucosinolates in Bretschneideraceae, *Biochem. Syst. Ecol.*, 17, 375, 1989.
52. **Bough, W. A. and Gander, J. E.,** Exogenous L-tyrosine metabolism and dhurrin turnover in sorghum seedlings, *Phytochemistry*, 10, 67, 1971.
53. **Bowers, M. D. and Stamp, N. E.,** Chemical variation within and between individuals of *Plantago lanceolata* (Plantaginaceae), *J. Chem. Ecol.*, 18, 985, 1992.
54. **Bracher, D. and Kutchan, T. M.,** Strictosidine synthase from *Rauvolfia serpentina*: analysis of a gene involved in indole alkaloid biosynthesis, *Arch. Biochem. Biophys.*, 294, 717, 1992.
55. **Breccia, A. and Badiello, R.,** The role of general metabolites in the biosynthesis of natural products. I. The terpene marrubiin, *Z. Naturforsch.*, 22, 44, 1967.
56. **Briggs, M. A. and Schultz, J. C.,** Chemical defense production in *Lotus corniculatus* L. II. Trade-offs among growth, reproduction and defense, *Oecologia*, 83, 32, 1990.
57. **Brindle, P. A., Kuhn, P. J., and Threlfall, D. R.,** Biosynthesis and metabolism of sesquiterpenoid phytoalexins and triterpenoids in potato cell suspension cultures, *Phytochemistry*, 27, 133, 1988.
58. **Brooks, C. J. W. and Watson, D. G.,** Terpenoid phytoalexins, *Nat. Prod. Rep.*, 8, 367, 1991.
59. **Brouillard, R.,** Flavonoids and flower colour, in *The Flavonoids*, Harborne, J. B., Ed., Chapman and Hall, London, 1988, 525.
60. **Brown, D. G.,** The cost of plant defense: an experimental analysis with inducible proteinase inhibitors in tomato, *Oecologia*, 76, 467, 1988.
61. **Brown, S. A.,** Coumarins, in *The Biochemistry of Plants: A Comprehensive Treatise*, Vol. 7, *Secondary Plant Products*, Conn, E. E., Ed., Academic Press, New York, 1981, 269.
62. **Bryant, J. P., Chapin, F. S., III, and Klein, D. R.,** Carbon/nutrient balance of boreal plants in relation to vertebrate herbivory, *Oikos*, 40, 357, 1983.
63. **Bryant, J. P., Reichardt, P. B., Clausen, T. P., Provenza, F. D., and Kuropat, P. J.,** Woody plant–mammal interactions, in *Herbivores: Their Interactions with Secondary Plant Metabolites*, Vol. 2, 2nd ed., Rosenthal, G. A. and Berenbaum, M. R., Eds., Academic Press, San Diego, 1992, 343.
64. **Bu'lock, J. D.,** *The Biosynthesis of Natural Products*, McGraw-Hill, London, 1965, 80.
65. **Burbott, A. J. and Loomis, W. D.,** Evidence for metabolic turnover of monoterpenes in peppermint, *Plant Physiol.*, 44, 173, 1969.
66. **Caelles, C., Ferrer, A., Balcells, L., Hegardt, F. G., and Boronat, A.,** Isolation and structural characterization of a cDNA encoding *Arabidopsis thaliana* 3-hydroxy-3-methylglutaryl coenzyme A reductase, *Plant Mol. Biol.*, 13, 627, 1989.
67. **Caldwell, M. M., Robberecht, R., and Flint, S. D.,** Internal filters: prospects for UV-acclimation in higher plants, *Physiol. Plant.*, 58, 445, 1983.
68. **Caldwell, M. M., Teramura, A. H., and Tevini, M.,** The changing solar ultraviolet climate and the ecological consequences for higher plants, *Trends Ecol. Evol.*, 4, 363, 1989.
69. **Cane, D. E.,** Enzymatic formation of sesquiterpenes, *Chem. Rev.*, 90, 1089, 1990.
70. **Carlton, R. R., Waterman, P. G., and Gray, A. I.,** Variation of leaf gland volatile oil within a population of sweet gale (*Myrica gale*) (Myricaceae), *Chemoecology*, 3, 45, 1992.
71. **Castillo, J., Benavente, O., and del Rio, J. A.,** Naringin and neohesperidin levels during development of leaves, flower buds, and fruits of *Citrus aurantium*, *Plant Physiol.*, 99, 67, 1992.
72. **Castric, P. A., Farnden, K. J. F., and Conn, E. E.,** Cyanide metabolism in higher plants. V. The formation of asparagine from β-cyanoalanine, *Arch. Biochem. Biophys.*, 152, 62, 1972.
73. **Cates, R. G.,** The interface between slugs and wild ginger: some evolutionary aspects, *Ecology*, 56, 391, 1975.
74. **Chapin, F. S., III,** The cost of tundra plant structures: evaluation of concepts and currencies, *Am. Nat.*, 133, 1, 1989.

75. **Chapin, F. S., III,** The mineral nutrition of wild plants, *Annu. Rev. Ecol. Syst.,* 11, 233, 1980.

76. **Chappell, J., VonLanken, C., and Vogeli, U.,** Elicitor-inducible 3-hydroxy-3-methylglutaryl coenzyme A reductase activity is required for sesquiterpene accumulation in tobacco cell suspension cultures, *Plant Physiol.,* 97, 693, 1991.

77. **Chew, F. S.,** Biological effects of glucosinolates, in *Biologically Active Natural Products: Potential Use in Agriculture,* ACS Symposium Series, No. 380, Cutler, H. G., Ed., American Chemical Society, Washington, D.C., 1988, 155.

78. **Chew, F. S. and Rodman, J. E.,** Plant resources for chemical defense, in *Herbivores: Their Interaction with Secondary Plant Metabolites,* 1st ed., Rosenthal, G. A. and Janzen, D. H., Eds., Academic Press, New York, 1979, 271.

79. **Chiariello, N. R., Mooney, H. A., and Williams, K.,** Growth, carbon allocation and cost of plant tissues, in *Plant Physiological Ecology: Field Methods and Instrumentation,* Pearcy, R. W., Ehleringer, J. R., Mooney, H. A., and Rundel, P. W., Eds., Chapman and Hall, London, 1989, 327.

80. **Choi, D., Ward, B. L., and Bostock, R. M.,** Differential induction and suppression of potato 3-hydroxy-3-methylglutaryl coenzyme A reductase genes in response to *Phytophthora infestans* and to its elicitor arachidonic acid, *Plant Cell,* 4, 1333, 1992.

81. **Chye, M.-L., Tan, C.-T., and Chua, N.-H.,** Three genes encode 3-hydroxy-3-methylglutaryl-coenzyme A reductase in *Hevea brasiliensis: hmg1* and *hmg3* are differentially expressed, *Plant Mol. Biol.,* 19, 473, 1992.

82. **Clark, L. E. and Clark, W. D.,** Seasonal variation in leaf exudate flavonoids of *Isocoma acradenia* (Asteraceae), *Biochem. Syst. Ecol.,* 18, 145, 1990.

83. **Clegg, D. O., Conn, E. E., and Janzen, D. H.,** Developmental fate of the cyanogenic glucoside linamarin in Costa Rican wild lima bean seeds, *Nature,* 278, 343, 1979.

84. **Cleveland, T. E., Thornburg, R. W., and Ryan, C. A.,** Molecular characterization of a wound-inducible inhibitor I gene from potato and the processing of its mRNA and protein, *Plant Mol. Biol.,* 8, 199, 1987.

85. **Clossais-Besnard, N. and Larher, F.,** Physiological role of glucosinolates in *Brassica napus.* Concentration and distribution pattern of glucosinolates among plant organs during a complete life cycle, *J. Sci. Food Agric.,* 56, 25, 1991.

86. **Coley, P. D.,** Costs and benefits of defense by tannins in a neotropical tree, *Oecologia,* 70, 238, 1986.

87. **Coley, P. D., Bryant, J. P., and Chapin, F. S., III,** Resource availability and plant antiherbivore defense, *Science,* 230, 895, 1985.

88. **Compton, S. G., Newsome, D., and Jones, D. A.,** Selection for cyanogenesis in the leaves and petals of *Lotus corniculatus* L. at high latitudes, *Oecologia,* 60, 353, 1983.

89. **Conn, E. E.,** Cyanogenic compounds, *Annu. Rev. Plant Physiol.,* 31, 433, 1980.

90. **Conn, E. E.,** Compartmentation of secondary compounds, in *Membranes and Compartmentation in the Regulation of Plant Functions,* Vol. 24, Annu. Proc. Phytochem. Soc. Eur., Boudet, A. M., Alibert, G., Marigo, G., and Lea, P. J., Eds., Clarendon Press, Oxford, 1984, 1.

91. **Conn, E. E.,** The metabolism of a natural product: lessons learned from cyanogenic glycosides, *Planta Med.,* 57, S1, 1991.

92. **Cronshaw, J.,** Phloem structure and function, *Annu. Rev. Plant Physiol.,* 32, 465, 1981.

93. **Croteau, R.,** Biosynthesis and catabolism of monoterpenoids, *Chem. Rev.,* 87, 929, 1987.

94. **Croteau, R.,** Catabolism of monoterpenes in essential oil plants, in *Flavors and Fragrances: A World Perspective,* Mookherjee, B. D. and Willis, B. J., Eds., Elsevier, Amsterdam, 1988, 65.

95. **Croteau, R. and Karp, F.,** Biosynthesis of monoterpenes: preliminary characterization of bornyl pyrophosphate synthetase from sage (*Salvia officinalis*) and demonstration that geranyl pyrophosphate is the preferred substrate for cyclization, *Arch. Biochem. Biophys.,* 198, 512, 1979.

96. **Croteau, R. and Loomis, W. D.,** Biosynthesis of mono- and sesquiterpenes in peppermint from mevalonate-2-^{14}C, *Phytochemistry,* 11, 1055, 1972.

97. **Croteau, R. and Shaskus, J.,** Biosynthesis of monoterpenes: demonstration of a geranyl pyrophosphate:(–)-bornyl pyrophosphate cyclase in soluble enzyme preparations from tansy (*Tanacetum vulgare*), *Arch. Biochem. Biophys.,* 236, 535, 1985.

98. **Croteau, R. and Sood, V. K.,** Metabolism of monoterpenes: evidence for the function of monoterpene catabolism in peppermint (*Mentha piperita*) rhizomes, *Plant Physiol.,* 77, 801, 1985.

99. **Croteau, R. and Venkatachalam, K. V.,** Metabolism of monoterpenes: demonstration that (+)-*cis*-isopulegone, not piperitenone, is the key intermediate in the conversion of (–)-isopiperitenone to (+)-pulegone in peppermint (*Mentha piperita*), *Arch. Biochem. Biophys.,* 249, 306, 1986.

100. **Croteau, R. and Winters, J. N.,** Demonstration of the intercellular compartmentation of *l*-menthone metabolism in peppermint (*Mentha piperita*) leaves, *Plant Physiol.,* 69, 975, 1982.

101. **Croteau, R. B., Wheeler, C. J., Cane, D. E., Ebert, R., and Ha, H.-J.,** Isotopically sensitive branching in the formation of cyclic monoterpenes: proof that (–)-α-pinene and (–)-β-pinene are synthesized by the same monoterpene cyclase via deprotonation of a common intermediate, *Biochemistry,* 26, 5383, 1987.

102. **Croteau, R., Burbott, A. J., and Loomis, W. D.,** Biosynthesis of mono- and sesqui-terpenes in peppermint from glucose-^{14}C and ^{14}CO$_2$, *Phytochemistry,* 11, 2459, 1972.

103. **Croteau, R., El-Bialy, H., and Dehal, S. S.,** Metabolism of monoterpenes: metabolic fate of (+)-camphor in sage (*Salvia officinalis*), *Plant Physiol.*, 84, 649, 1987.

104. **Croteau, R., El-Bialy, H., and El-Hindawi, S.,** Metabolism of monoterpenes: lactonization of (+)-camphor and conversion of the corresponding hydroxy acid to the glucoside-glucose ester in sage (*Salvia officinalis*), *Arch. Biochem. Biophys.*, 228, 667, 1984.

105. **Croteau, R., Gurkewitz, S., Johnson, M. A., and Fisk, H. J.,** Biochemistry of oleoresinosis: monoterpene and diterpene biosynthesis in lodgepole pine saplings infected with *Ceratocystis clavigera* or treated with carbohydrate elicitors, *Plant Physiol.*, 85, 1123, 1987.

106. **Croteau, R., Karp, F., Wagschal, K. C., Satterwhite, D. M., Hyatt, D. C., and Skotland, C. B.,** Biochemical characterization of a spearmint mutant that resembles peppermint in monoterpene content, *Plant Physiol.*, 96, 744, 1991.

107. **Croteau, R., Sood, V. K., Renstrom, B., and Bhushan, R.,** Metabolism of monoterpenes: early steps in the metabolism of *d*-neomenthyl-β-D-glucoside in peppermint (*Mentha piperita*) rhizomes, *Plant Physiol.*, 76, 647, 1984.

108. **Cutler, A. J., Sternberg, M., and Conn, E. E.,** Properties of a microsomal enzyme system from *Linum usitatissimum* (linen flax) which oxidizes valine to acetone cyanohydrin and isoleucine to 2-methylbutanone cyanohydrin, *Arch. Biochem. Biophys.*, 238, 272, 1985.

109. **Daddona, P. E., Wright, J. L., and Hutchinson, C. R.,** Alkaloid catabolism and mobilization in *Catharanthus roseus*, *Phytochemistry*, 15, 941, 1976.

110. **Davies, D. D.,** Factors affecting protein turnover in plants, in *Nitrogen Assimilation of Plants*, Hewitt, E. J. and Cutting, C. V., Eds., Academic Press, London, 1979, 369.

111. **De Luca, V., Fernandez, J. A., Campbell, D., and Kurz, W. G. W.,** Developmental regulation of enzymes of indole alkaloid biosynthesis in *Catharanthus roseus*, *Plant Physiol.*, 86, 447, 1988.

112. **De-Eknamkul, W. and Zenk, M. H.,** Purification and properties of 1,2-dehydroreticuline reductase from *Papaver somniferum* seedlings, *Phytochemistry*, 31, 813, 1992.

113. **Dehal, S. S. and Croteau, R.,** Metabolism of monoterpenes: specificity of the dehydrogenases responsible for the biosynthesis of camphor, 3-thujone, and 3-isothujone, *Arch. Biochem. Biophys.*, 258, 287, 1987.

114. **Dehal, S. S. and Croteau, R.,** Partial purification and characterization of two sesquiterpene cyclases from sage (*Salvia officinalis*) which catalyze the respective conversion of farnesyl pyrophosphate to humulene and caryophyllene, *Arch. Biochem. Biophys.*, 261, 346, 1988.

115. **del Moral, R. and Muller, C. H.,** Fog drip: a mechanism of toxin transport from *Eucalyptus globulus*, *Bull. Torr. Bot. Club*, 96, 467, 1969.

116. **del Moral, R., Willis, R. J., and Ashton, D. H.,** Suppression of coastal heath vegetation by *Eucalyptus baxteri*, *Aust. J. Bot.*, 26, 203, 1978.

117. **Dell, B. and McComb, A. J.,** Plant resins — their formation, secretion and possible functions, in *Advances in Botanical Research*, Vol. 6, Woolhouse, H. W., Ed., Academic Press, London, 1978, 277.

118. **Dell, B. and McComb, A. J.,** Glandular hairs, resin production, and habitat of *Newcastelia viscida* E. Pritzel (Dicrastylidaceae), *Aust. J. Bot.*, 23, 373, 1975.

119. **Dement, W. A., Tyson, B. J., and Mooney, H. A.,** Mechanism of monoterpene volatilization in *Salvia mellifera*, *Phytochemistry*, 14, 2555, 1975.

120. **Desailly, I., Fliniaux, M.-A., and Jacquin-Dubreuil, A.,** Etude de la distribution des alcaloides derives de l'acide tropique chez *Datura stramonium* L., par dosage immunoenzymatique: localisation tissulaire et subcellulaire, *C. R. Acad. Sci. Paris*, 306, 591, 1988.

121. **Deus-Neumann, B. and Zenk, M. H.,** A highly selective alkaloid uptake system in vacuoles of higher plants, *Planta*, 162, 250, 1984.

122. **Deus-Neumann, B. and Zenk, M. H.,** Accumulation of alkaloids in plant vacuoles does not involve an ion-trap mechanism, *Planta*, 167, 44, 1986.

123. **Dewick, P. M.,** Pterocarpan biosynthesis: chalcone and isoflavone precursors of demethylhomopterocarpin and maackiain in *Trifolium pratense*, *Phytochemistry*, 14, 979, 1975.

124. **Dewick, P. M.,** The biosynthesis of cyanogenic glycosides and glucosinolates, *Nat. Prod. Rep.*, 1, 545, 1984.

125. **Dirzo, R. and Harper, J. L.,** Experimental studies on slug–plant interactions. IV. The performance of cyanogenic and acyanogenic morphs of *Trifolium repens* in the field, *J. Ecol.*, 70, 119, 1982.

126. **Dirzo, R. and Harper, J. L.,** Experimental studies on slug–plant interactions. III. Differences in the acceptability of individual plants of *Trifolium repens* to slugs and snails, *J. Ecol.*, 70, 101, 1982.

127. **Dixon, A. F. G.,** Aphids and translocation, in *Transport in Plants I, Phloem Transport,* Vol. 1, *Encyclopedia of Plant Physiology*, New Series, Zimmermann, M. H. and Milburn, J. A., Eds., Springer-Verlag, Berlin, 1975, 154.

128. **Dixon, R. A. and Lamb, C. J.,** Molecular communication in interactions between plants and microbial pathogens, *Annu. Rev. Plant Physiol. Plant Mol. Biol.*, 41, 339, 1990.

129. **Dixon, R. A., Bailey, J. A., Bell, J. N., Bolwell, G. P., Cramer, C. L., Edwards, K., Hamdan, M. A. M. S., Lamb, C. J., Robbins, M. P., Ryder, T. B., and Schuch, W.,** Rapid changes in gene expression in response to microbial elicitation, *Philos. Trans. R. Soc. Lond. B*, 314, 411, 1986.

130. **Dolman, D. M., Knight, D. W., Salan, U., and Toplis, D.,** A quantitative method for the estimation of parthenolide and other sesquiterpene lactones containing α-methylenebutyrolactone functions present in feverfew, *Tanacetum parthenium, Phytochem. Anal.,* 3, 26, 1992.

131. **Dooner, H. K., Robbins, T. P., and Jorgensen, R. A.,** Genetic and developmental control of anthocyanin biosynthesis, *Annu. Rev. Genet.,* 25, 173, 1991.

132. **Dovrat, A. and Goldschmidt, J.,** Cultivation aspects of *Catharanthus roseus* for roots, *Acta Hort.,* 73, 263, 1978.

133. **Downum, K. R.,** Light-activated plant defence, *New Phytol.,* 122, 401, 1992.

134. **Downum, K. R. and Rodriguez, E.,** Toxicological action and ecological importance of plant photosensitizers, *J. Chem. Ecol.,* 12, 823, 1986.

135. **Dreyer, D. L., Jones, K. C., and Molyneux, R. J.,** Feeding deterrency of some pyrrolizidine, indolizidine, and quinolizidine alkaloids towards pea aphid (*Acyrthosiphon pisum*) and evidence for phloem transport of indolizidine alkaloid swainsonine, *J. Chem. Ecol.,* 11, 1045, 1985.

136. **Dudai, N., Putievsky, E., Ravid, U., Palevitch, D., and Halevy, A. H.,** Monoterpene content in *Origanum syriacum* as affected by environmental conditions and flowering, *Physiol. Plant.,* 84, 453, 1992.

137. **Dudley, M. W., Dueber, M. T., and West, C. A.,** Biosynthesis of the macrocyclic diterpene casbene in castor bean (*Ricinus communis* L.) seedlings: changes in enzyme levels induced by fungal infection and intracellular localization of the pathway, *Plant Physiol.,* 81, 335, 1986.

138. **Dugan, R. E.,** Regulation of HMG-CoA reductase, in *Biosynthesis of Isoprenoid Compounds,* Vol. 1, Porter, J. W. and Spurgeon, S. L., Eds., John Wiley & Sons, New York, 1981, 95.

139. **Duperon, R., Thiersault, M., and Duperon, P.,** High level of glycosylated sterols in species of *Solanum* and sterol changes during the development of the tomato, *Phytochemistry,* 23, 743, 1984.

140. **Dussourd, D. E. and Denno, R. F.,** Deactivation of plant defense: correspondence between insect behavior and secretory canal architecture, *Ecology,* 72, 1383, 1991.

141. **Dustin, C. D. and Cooper-Driver, G. A.,** Changes in phenolic production in the hay-scented fern (*Dennstaedtia punctilobula*) in relation to resource availability, *Biochem. Syst. Ecol.,* 20, 99, 1992.

142. **Edwards, R. and Dixon, R. A.,** Purification and characterization of *S*-adenosyl-*L*-methionine:caffeic acid 3-*O*-methyltransferase from suspension cultures of alfalfa (*Medicago sativa* L.), *Arch. Biochem. Biophys.,* 287, 372, 1991.

143. **Effertz, B. and Weissenbock, G.,** ^{14}C-Phenylalanine incorporation into C-glycosylflavones of developing primary oat leaves, *Z. Pflanzenphysiol.,* 92, 319, 1979.

144. **Ehmke, A., von Borstel, K., and Hartmann, T.,** Alkaloid N-oxides as transport and vacuolar storage compounds of pyrrolizidine alkaloids in *Senecio vulgaris* L., *Planta,* 176, 83, 1988.

145. **Ellis, W. M., Keymer, R. J., and Jones, D. A.,** On the polymorphism of cyanogenesis in *Lotus corniculatus* L., *Heredity,* 39, 45, 1977.

146. **Eltayeb, E. A. and Roddick, J. G.,** Biosynthesis and degradation of α-tomatine in developing tomato fruits, *Phytochemistry,* 24, 253, 1985.

147. **Ernst, W. H. O., Kuiters, A. T., Nelissen, H. J. M., and Tolsma, D. J.,** Seasonal variation in phenolics in several savanna tree species in Botswana, *Acta Bot. Neerl.,* 40, 63, 1991.

148. **Evans, R. C., Tingey, D. T., Gumpertz, M. L., and Burns, W. F.,** Estimates of isoprene and monoterpene emission rates in plants, *Bot. Gaz.,* 143, 304, 1982.

149. **Facchini, P. J. and Chappell, J.,** Gene family for an elicitor-induced sesquiterpene cyclase in tobacco, *Proc. Natl. Acad. Sci. U.S.A.,* 89, 11088, 1992.

150. **Fagerstrom, T.,** Anti-herbivory chemical defense in plants: a note on the concept of cost, *Am. Nat.,* 133, 281, 1989.

151. **Fahn, A.,** *Secretory Tissues in Plants,* Academic Press, London, 1979.

152. **Fahn, A.,** Secretory tissues in vascular plants, *New Phytol.,* 108, 229, 1988.

153. **Fairbairn, J. W. and El-Masry, S.,** The alkaloids of *Papaver somniferum* L. V. Fate of the "end-product" alkaloid morphine, *Phytochemistry,* 6, 499, 1967.

154. **Fairbairn, J. W. and Paterson, A.,** Alkaloids as possible intermediaries in plant metabolism, *Nature,* 210, 1163, 1966.

155. **Fairbairn, J. W. and Suwal, P. N.,** The alkaloids of hemlock (*Conium maculatum* L.). II. Evidence for a rapid turnover of the major alkaloids, *Phytochemistry,* 1, 38, 1961.

156. **Fairbairn, J. W. and Wassel, G.,** The alkaloids of *Papaver somniferum* L. I. Evidence for a rapid turnover of the major alkaloids, *Phytochemistry,* 3, 253, 1964.

157. **Fajer, E. D.,** The effects of enriched CO_2 atmospheres on plant–insect herbivore interactions: growth responses of larvae of the specialist butterfly, *Junonia coenia* (Lepidoptera: Nymphalidae), *Oecologia,* 81, 514, 1989.

158. **Fajer, E. D., Bowers, M. D., and Bazzaz, F. A.,** The effect of nutrients and enriched CO_2 environments on production of carbon-based allelochemicals in *Plantago*: a test of the carbon/nutrient balance hypothesis, *Am. Nat.,* 140, 707, 1992.

159. **Fajer, E. D., Bowers, M. D., and Bazzaz, F. A.,** The effects of enriched carbon dioxide atmospheres on plant–insect herbivore interactions, *Science*, 243, 1198, 1989.

160. **Farmer, E. E. and Ryan, C. A.,** Octadecanoid precursors of jasmonic acid activate the synthesis of wound-inducible proteinase inhibitors, *Plant Cell*, 4, 129, 1992.

161. **Farmer, E. E., Moloshok, T. D., Saxton, M. J., and Ryan, C. A.,** Oligosaccharide signaling in plants: specificity of oligouronide-enhanced plasma membrane protein phosphorylation, *J. Biol. Chem.*, 266, 3140, 1991.

162. **Feeny, P.,** Plant apparency and chemical defense, in *Biochemical Interactions Between Plants and Insects*, Vol. 10, *Recent Advances in Phytochemistry*, Wallace, J. W. and Mansell, R. L., Eds., Plenum Press, New York, 1976, 1.

163. **Ferrer, M. A., Pedreno, M. A., Munoz, R., and Barcelo, A. R.,** Constitutive isoflavones as modulators of indole-3-acetic acid oxidase activity of acidic cell wall isoperoxidases from lupin hypocotyls, *Phytochemistry*, 31, 3681, 1992.

164. **Fischer, N. H.,** Plant terpenoids as allelopathic agents, in *Ecological Chemistry and Biochemistry of Plant Terpenoids*, Vol. 31, Annu. Proc. Phytochem. Soc. Eur., Harborne, J. B. and Tomas-Barberan, F. A., Eds., Clarendon Press, Oxford, 1991, 377.

165. **Flesch, V., Jacques, M., Cosson, L., Teng, B. P., Petiard, V., and Balz, J. P.,** Relative importance of growth and light level on terpene content of *Ginkgo biloba*, *Phytochemistry*, 31, 1941, 1992.

166. **Fluck, H.,** Intrinsic and extrinsic factors affecting the production of secondary plant products, in *Chemical Plant Taxonomy*, Swain, T., Ed., Academic Press, London, 1963, 167.

167. **Foulds, W. and Grime, J. P.,** The response of cyanogenic and acyanogenic phenotypes of *Trifolium repens* to soil moisture supply, *Heredity*, 28, 181, 1972.

168. **Fowden, L. and Lea, P. J.,** Mechanism of plant avoidance of autotoxicity by secondary metabolites, especially by nonprotein amino acids, in *Herbivores: Their Interaction with Secondary Plant Metabolites*, 1st ed., Rosenthal, G. A. and Janzen, D. H., Eds., Academic Press, New York, 1979, 135.

169. **Fox, L. R.,** Defense and dynamics in plant–herbivore systems, *Am. Zool.*, 21, 853, 1981.

170. **Fox, M. G. and French, J. C.,** Systematic occurrence of sterols in latex of Araceae: subfamily Colocasioideae, *Am. J. Bot.*, 75, 132, 1988.

171. **Francis, M. J. O. and O'Connell, M.,** The incorporation of mevalonic acid into rose petal monoterpenes, *Phytochemistry*, 8, 1705, 1969.

172. **Frank, A. W. and Marion, L.,** The biogenesis of alkaloids. XVI. Hordenine metabolism in barley, *Can. J. Chem.*, 34, 1641, 1956.

173. **Frehner, M. and Conn, E. E.,** The linamarin β-glucosidase in Costa Rican wild lima beans (*Phaseolus lunatus* L.) is apoplastic, *Plant Physiol.*, 84, 1296, 1987.

174. **Frehner, M., Scalet, M., and Conn, E. E.,** Pattern of the cyanide-potential in developing fruits: implications for plants accumulating cyanogenic monoglucosides (*Phaseolus lunatus*) or cyanogenic diglucosides in their seeds (*Linum usitatissimum, Prunus amygdalus*), *Plant Physiol.*, 94, 28, 1990.

175. **Friesen, J. B. and Leete, E.,** Nicotine synthase — an enzyme from *Nicotiana* species which catalyzes the formation of (*S*)-nicotine from nicotinic acid and 1-methyl-Δ'-pyrrolinium chloride, *Tetr. Lett.*, 31, 6295, 1990.

176. **Fritzemeier, K.-H., Cretin, C., Kombrink, E., Rohwer, F., Taylor, J., Scheel, D., and Hahlbrock, K.,** Transient induction of phenylalanine ammonia-lyase and 4-coumarate:CoA ligase mRNAs in potato leaves infected with virulent or avirulent races of *Phytophthora infestans*, *Plant Physiol.*, 85, 34, 1987.

177. **Fujimori, N. and Ashihara, H.,** Adenine metabolism and the synthesis of purine alkaloids in flowers of *Camellia*, *Phytochemistry*, 29, 3513, 1990.

178. **Funk, C., Koepp, A. E., and Croteau, R.,** Catabolism of camphor in tissue cultures and leaf disks of common sage (*Salvia officinalis*), *Arch. Biochem. Biophys.*, 294, 306, 1992.

179. **Gambliel, H. and Croteau, R.,** Pinene cyclases I and II: two enzymes from sage (*Salvia officinalis*) which catalyze stereospecific cyclizations of geranyl pyrophosphate to monoterpene olefins of opposite configuration, *J. Biol. Chem.*, 259, 740, 1984.

180. **Gehlert, R., Schoppner, A., and Kindl, H.,** Stilbene synthase from seedlings of *Pinus sylvestris*: purification and induction in response to fungal infection, *Mol. Plant–Microbe Interact.*, 3, 444, 1990.

181. **Geissman, T. A. and Crout, D. H. G.,** *Organic Chemistry of Secondary Plant Metabolism*, Freeman, Cooper and Co., San Francisco, 1969.

182. **Gerats, A. G. M. and Martin, C.,** Flavonoid synthesis in *Petunia hybrida*: genetics and molecular biology of flower colour, in *Phenolic Metabolism in Plants*, Vol. 26, *Recent Advances in Phytochemistry*, Stafford, H. A. and Ibrahim, R. K., Eds., Plenum Press, New York, 1992, 165.

183. **Gershenzon, J.,** unpublished results.

184. **Gershenzon, J.,** Changes in the levels of plant secondary metabolite production under water and nutrient stress, in *Phytochemical Adaptations to Stress*, Vol. 18, *Recent Advances in Phytochemistry*, Timmermann, B. N., Steelink, C., and Loewus, F. A., Eds., Plenum Press, New York, 1984, 273.

185. **Gershenzon, J. and Croteau, R.,** Regulation of monoterpene biosynthesis in higher plants, in *Biochemistry of the Mevalonic Acid Pathway to Terpenoids,* Vol. 24, *Recent Advances in Phytochemistry,* Towers, G. H. N. and Stafford, H. A., Eds., Plenum Press, New York, 1990, 83.

186. **Gershenzon, J. and Croteau, R.,** Terpenoid biosynthesis: the basic pathway and formation of monoterpenes, sesquiterpenes, and diterpenes, in *Lipid Metabolism in Plants,* Moore, T. S., Jr., Ed., CRC Press, Boca Raton, FL, 1993, 333.

187. **Gershenzon, J. and Croteau, R.,** unpublished results.

188. **Gershenzon, J. and Croteau, R.,** Terpenoids, in *Herbivores: Their Interactions with Secondary Plant Metabolites,* Vol. 1, 2nd ed., Rosenthal, G. A. and Berenbaum, M. R., Eds., Academic Press, San Diego, 1991, 165.

189. **Ghani, A.,** The site of synthesis and secondary transformation of hyoscyamine in *Solandra grandiflora, Phytochemistry,* 25, 617, 1986.

190. **Ghini, A. A., Burton, G., and Gros, E. G.,** Biodegradation of the indolic system of gramine in *Hordeum vulgare, Phytochemistry,* 30, 779, 1991.

191. **Giaquinta, R. T.,** Phloem loading of sucrose, *Annu. Rev. Plant Physiol.,* 34, 347, 1983.

192. **Gijzen, M., Lewinsohn, E., and Croteau, R.,** Antigenic cross-reactivity among monoterpene cyclases from grand fir and induction of these enzymes upon stem wounding, *Arch. Biochem. Biophys.,* 294, 670, 1992.

193. **Gijzen, M., McGregor, I., and Seguin-Swartz, G.,** Glucosinolate uptake by developing rapeseed embryos, *Plant Physiol.,* 89, 260, 1989.

194. **Givovich, A., Morse, S., Cerda, H., Niemeyer, H. M., Wratten, S. D., and Edwards, P. J.,** Hydroxamic acid glucosides in honeydew of aphids feeding on wheat, *J. Chem. Ecol.,* 18, 841, 1992.

195. **Gleizes, M., Pauly, G., Carde, J.-P., Marpeau, A., and Bernard-Dagan, C.,** Monoterpene hydrocarbon biosynthesis by isolated leucoplasts of *Citrofortunella mitis, Planta,* 159, 373, 1983.

196. **Goodwin, T. W. and Mercer, E. I.,** *Introduction to Plant Biochemistry,* 1st ed., Pergamon Press, Oxford, 1972.

197. **Goodwin, T. W. and Mercer, E. I.,** *Introduction to Plant Biochemistry,* 2nd ed., Pergamon Press, Oxford, 1983.

198. **Graebe, J. E.,** Gibberellin biosynthesis and control, *Annu. Rev. Plant Physiol.,* 38, 419, 1987.

199. **Graham, J. S., Hall, G., Pearce, G., and Ryan, C. A.,** Regulation of synthesis of proteinase inhibitors I and II mRNAs in leaves of wounded tomato plants, *Planta,* 169, 399, 1986.

200. **Graham, J. S., Pearce, G., Merryweather, J., Titani, K., Ericsson, L., and Ryan, C. A.,** Wound-induced proteinase inhibitors from tomato leaves. I. The cDNA-deduced primary structure of pre-inhibitor I and its post-translational processing, *J. Biol. Chem.,* 260, 6555, 1985.

201. **Graham, J. S., Pearce, G., Merryweather, J., Titani, K., Ericsson, L., and Ryan, C. A.,** Wound-induced proteinase inhibitors from tomato leaves. II. The cDNA-deduced primary structure of pre-inhibitor II, *J. Biol. Chem.,* 260, 6561, 1985.

202. **Gray, J. C.,** Control of isoprenoid biosynthesis in higher plants, *Adv. Bot. Res.,* 14, 25, 1987.

203. **Griffith, G. D., Griffith, T., and Byerrum, R. U.,** Nicotinic acid as a metabolite of nicotine in *Nicotiana rustica, J. Biol. Chem.,* 235, 3536, 1960.

204. **Grob, K. and Matile, P.,** Vacuolar location of glucosinolates in horseradish root cells, *Plant Sci. Lett.,* 14, 327, 1979.

205. **Groger, D.,** Alkaloids derived from tryptophan, in *Biochemistry of Alkaloids,* Mothes, K., Schutte, H. R., and Luckner, M., Eds., VCH Publishers, Berlin, 1985, 272.

206. **Gross, D.,** Alkaloids derived from nicotinic acid, in *Biochemistry of Alkaloids,* Mothes, K., Schutte, H. R., and Luckner, M., Eds., VCH Publishers, Berlin, 1985, 163.

207. **Gross, G. G.,** Biosynthesis and metabolism of phenolic acids and monolignols, in *Biosynthesis and Biodegradation of Wood Components,* Higuchi, T., Ed., Academic Press, Orlando, 1985, 229.

208. **Gross, G. G.,** Enzymatic synthesis of gallotannins and related compounds, in *Phenolic Metabolism in Plants,* Vol. 26, *Recent Advances in Phytochemistry,* Stafford, H. A. and Ibrahim, R. K., Eds., Plenum Press, New York, 1992, 297.

209. **Gulmon, S. L. and Mooney, H. A.,** Costs of defense and their effects on plant productivity, in *On the Economy of Plant Form and Function,* Givnish, T. J., Ed., Cambridge University Press, Cambridge, 1986, 681.

210. **Gundlach, H., Muller, M. J., Kutchan, T. M., and Zenk, M. H.,** Jasmonic acid is a signal transducer in elicitor-induced plant cell cultures, *Proc. Natl. Acad. Sci. U.S.A.,* 89, 2389, 1992.

211. **Gupton, C. L.,** Phenotypic recurrent selection for increased leaf weight and decreased alkaloid content of burley tobacco, *Crop Sci.,* 21, 921, 1981.

212. **Gustafson, G. and Ryan, C. A.,** Specificity of protein turnover in tomato leaves: accumulation of proteinase inhibitors, induced with the wound hormone, PIIF, *J. Biol. Chem.,* 251, 7004, 1976.

213. **Haddock, E. A., Gupta, R. K., Al-Shafi, S. M. K., Layden, K., Haslam, E., and Magnolato, D.,** The metabolism of gallic acid and hexahydroxydiphenic acid in plants: biogenetic and molecular taxonomic considerations, *Phytochemistry,* 21, 1049, 1982.

214. **Hagerman, A. E. and Butler, L. G.**, Tannins and lignins, in *Herbivores: Their Interactions with Secondary Plant Metabolites*, Vol. 1, 2nd ed., Rosenthal, G. A. and Berenbaum, M. R., Eds., Academic Press, San Diego, 1991, 355.

215. **Hahlbrock, K. and Grisebach, H.**, Enzymic controls in the biosynthesis of lignin and flavonoids, *Annu. Rev. Plant Physiol.*, 30, 105, 1979.

216. **Hahlbrock, K. and Grisebach, H.**, Biosynthesis of flavonoids, in *The Flavonoids*, Part 2, Harborne, J. B., Mabry, T. J., and Mabry, H., Eds., Academic Press, New York, 1975, 866.

217. **Hahlbrock, K. and Scheel, D.**, Physiology and molecular biology of phenylpropanoid metabolism, *Annu. Rev. Plant Physiol. Plant Mol. Biol.*, 40, 347, 1989.

218. **Halkier, B. A. and Moller, B. L.**, Biosynthesis of the cyanogenic glucoside dhurrin in seedlings of *Sorghum bicolor* (L.) Moench and partial purification of the enzyme system involved, *Plant Physiol.*, 90, 1552, 1989.

219. **Halkier, B. A. and Moller, B. L.**, The biosynthesis of cyanogenic glucosides in higher plants: identification of three hydroxylation steps in the biosynthesis of dhurrin in *Sorghum bicolor* (L.) Moench and the involvement of 1-*aci*-nitro-2-(*p*-hydroxyphenyl)ethane as an intermediate, *J. Biol. Chem.*, 265, 21114, 1990.

220. **Hall, G. D. and Langenheim, J. H.**, Temporal changes in the leaf monoterpenes of *Sequoia sempervirens*, *Biochem. Syst. Ecol.*, 14, 61, 1986.

221. **Hall, R. D. and Yeoman, M. M.**, The influence of intracellular pools of phenylalanine derivatives upon the synthesis of capsaicin by immobilized cell cultures of the chilli pepper, *Capsicum frutescens*, *Planta*, 185, 72, 1991.

222. **Hallahan, T. W. and Croteau, R.**, Monoterpene biosynthesis: demonstration of a geranyl pyrophosphate:sabinene hydrate cyclase in soluble enzyme preparations from sweet marjoram (*Majorana hortensis*), *Arch. Biochem. Biophys.*, 264, 618, 1988.

223. **Hanover, J. W.**, Genetics of terpenes. I. Gene control of monoterpene levels in *Pinus monticola* Dougl., *Heredity*, 21, 73, 1966.

224. **Harborne, J. B.**, *Introduction to Ecological Biochemistry*, 3rd ed., Academic Press, New York, 1988.

225. **Harborne, J. B.**, Recent advances in chemical ecology, *Nat. Prod. Rep.*, 3, 323, 1986.

226. **Harborne, J. B.**, Natural fungitoxins, in *Biologically Active Natural Products*, Vol. 27, Annu. Proc. Phytochem. Soc. Eur., Hostettmann, K. and Lea, P. J., Eds., Clarendon Press, Oxford, 1987, 195.

227. **Hargreaves, J. A., Brown, G. A., and Holloway, P. J.**, The structural and chemical characteristics of the leaf surface of *Lupinus albus* L. in relation to the distribution of antifungal compounds, in *The Plant Cuticle*, Cutler, D. F., Alvin, K. L., and Price, C. E., Eds., Academic Press, London, 1982, 331.

228. **Harker, C. L., Ellis, T. H. N., and Coen, E. S.**, Identification and genetic regulation of the chalcone synthase multigene family in pea, *Plant Cell*, 2, 185, 1990.

229. **Hartmann, T.**, Alkaloids, in *Herbivores: Their Interactions with Secondary Plant Metabolites*, Vol. 1, 2nd ed., Rosenthal, G. A. and Berenbaum, M. R., Eds., Academic Press, San Diego, 1991, 79.

230. **Hartmann, T., Ehmke, A., Eilert, U., von Borstel, K., and Theuring, C.**, Sites of synthesis, translocation and accumulation of pyrrolizidine alkaloid N-oxides in *Senecio vulgaris* L., *Planta*, 177, 98, 1989.

231. **Hasegawa, S., Herman, Z., Orme, E., and Ou, P.**, Biosynthesis of limonoids in *Citrus*: sites and translocation, *Phytochemistry*, 25, 2783, 1986.

232. **Haslam, E.**, *Plant Polyphenols, Vegetable Tannins Revisited*, Cambridge University Press, Cambridge, 1989.

233. **Hatano, T., Kira, R., Yoshizaki, M., and Okuda, T.**, Seasonal changes in the tannins of *Liquidambar formosana* reflecting their biogenesis, *Phytochemistry*, 25, 2787, 1986.

234. **Hauffe, K. D., Hahlbrock, K., and Scheel, D.**, Elicitor-stimulated furanocoumarin biosynthesis in cultured parsley cells: *S*-adenosyl-*L*-methionine:bergaptol and *S*-adenosyl-*L*-methionine:xanthotoxol *O*-methyltransferases, *Z. Naturforsch.*, 41c, 228, 1986.

235. **Hauser, M.-T. and Wink, M.**, Uptake of alkaloids by latex vesicles and isolated mesophyll vacuoles of *Chelidonium majus* (Papaveraceae), *Z. Naturforsch.*, 45c, 949, 1990.

236. **Hawkins, A. J. S.**, Protein turnover: a functional appraisal, *Funct. Ecol.*, 5, 222, 1991.

237. **Hay, M. E. and Fenical, W.**, Marine plant–herbivore interactions: the ecology of chemical defense, *Annu. Rev. Ecol. Syst.*, 19, 111, 1988.

238. **Hay, M. E. and Steinberg, P. D.**, The chemical ecology of plant–herbivore interactions in marine versus terrestrial communities, in *Herbivores: Their Interactions with Secondary Plant Metabolites*, Vol. 2, 2nd ed., Rosenthal, G. A., and Berenbaum, M. R., Eds., Academic Press, San Diego, 1992, 371.

239. **Hay, M. E., Duffy, J. E., and Pfister, C. A.**, Chemical defense against different marine herbivores: are amphipods insect equivalents?, *Ecology*, 68, 1567, 1987.

240. **Hefendehl, F. W., Underhill, E. W., and von Rudloff, E.**, The biosynthesis of the oxygenated monoterpenes in mint, *Phytochemistry*, 6, 823, 1967.

241. **Heintze, A., Gorlach, J., Leuschner, C., Hoppe, P., Hagelstein, P., Schulze-Siebert, D., and Schultz, G.**, Plastidic isoprenoid synthesis during chloroplast development: change from metabolic autonomy to a division-of-labor stage, *Plant Physiol.*, 93, 1121, 1990.

242. **Heller, W. and Forkmann, G.**, Biosynthesis, in *The Flavonoids*, Harborne, J. B., Ed., Chapman and Hall, London, 1988, 399.

243. **Helmlinger, J., Rausch, T., and Hilgenberg, W.,** Localization of newly synthesized indole-3-methylglucosinolate (=glucobrassicin) in vacuoles from horseradish (*Armoracia rusticana*), *Physiol. Plant.*, 58, 302, 1983.

244. **Hemscheidt, T. and Zenk, M. H.,** Glucosidases involved in indole alkaloid biosynthesis of *Catharanthus* cell cultures, *FEBS Lett.*, 110, 187, 1980.

245. **Hemscheidt, T. and Zenk, M. H.,** Partial purification and characterization of a NADPH dependent tetrahydroalstonine synthase from *Catharanthus roseus* cell suspension cultures, *Plant Cell Rep.*, 4, 216, 1985.

246. **Henry, M., Rochd, M., and Bennini, B.,** Biosynthesis and accumulation of saponins in *Gypsophila paniculata*, *Phytochemistry*, 30, 1819, 1991.

247. **Herrmann, A., Schulz, W., and Hahlbrock, K.,** Two alleles of the single-copy chalcone synthase gene in parsley differ by a transposon-like element, *Mol. Gen. Genet.*, 212, 93, 1988.

248. **Hershko, A. and Ciechanover, A.,** The ubiquitin system for protein degradation, *Annu. Rev. Biochem.*, 61, 761, 1992.

249. **Hoelz, H., Kreis, W., Haug, B., and Reinhard, E.,** Storage of cardiac glycosides in vacuoles of *Digitalis lanata* mesophyll cells, *Phytochemistry*, 31, 1167, 1992.

250. **Hofman, P. J. and Menary, R. C.,** Variations in morphine, codeine and thebaine in the capsules of *Papaver somniferum* L. during maturation, *Aust. J. Agric. Res.*, 31, 313, 1979.

251. **Hoglund, A.-S., Lenman, M., and Rask, L.,** Myrosinase is localized to the interior of myrosin grains and is not associated to the surrounding tonoplast membrane, *Plant Sci.*, 85, 165, 1992.

252. **Hopfinger, J. A., Kumamoto, J., and Scora, R. W.,** Diurnal variation in the essential oils of Valencia orange leaves, *Am. J. Bot.*, 66, 111, 1979.

253. **Hori, K. and Atalay, R.,** Biochemical changes in the tissue of Chinese cabbage injured by the bug *Lygus disponsi*, *Appl. Entomol. Zool.*, 15, 234, 1980.

254. **Hosel, W.,** Glycosylation and glycosidases, in *The Biochemistry of Plants: A Comprehensive Treatise*, Vol. 7, *Secondary Plant Products*, Conn, E. E., Ed., Academic Press, New York, 1981, 725.

255. **Hrazdina, G.,** Compartmentation in aromatic metabolism, in *Phenolic Metabolism in Plants*, Vol. 26, *Recent Advances in Phytochemistry*, Stafford, H. A. and Ibrahim, R. K., Eds., Plenum Press, New York, 1992, 1.

256. **Hrazdina, G. and Jensen, R. A.,** Spatial organization of enzymes in plant metabolic pathways, *Annu. Rev. Plant Physiol. Plant Mol. Biol.*, 43, 241, 1992.

257. **Ibrahim, R. K.,** Immunolocalization of flavonoid conjugates and their enzymes, in *Phenolic Metabolism in Plants*, Vol. 26, *Recent Advances in Phytochemistry*, Stafford, H. A. and Ibrahim, R. K., Eds., Plenum Press, New York, 1992, 25.

258. **Ibrahim, R. K., De Luca, V., Khouri, H., Latchinian, L., Brisson, L., and Charest, P. M.,** Enzymology and compartmentation of polymethylated flavonol glucosides in *Chrysosplenium americanum*, *Phytochemistry*, 26, 1237, 1987.

259. **Jacobs, M. and Rubery, P. H.,** Naturally occurring auxin transport regulators, *Science*, 241, 346, 1988.

260. **Janiszowska, W. and Szakiel, A.,** The transport of [3-^{3}H]oleanolic acid and its monoglycosides to isolated vacuoles of protoplasts from *Calendula officinalis* leaves, *Phytochemistry*, 31, 2993, 1992.

261. **Janzen, D. H.,** Behavior of *Hymenaea courbaril* when its predispersal seed predator is absent, *Science*, 189, 145, 1975.

262. **Jaques, U., Koster, J., and Barz, W.,** Differential turnover of isoflavone 7-O-glucoside-6''-O-malonates in *Cicer arietinum* roots, *Phytochemistry*, 24, 949, 1985.

263. **Jensen, R. A.,** The shikimate/arogenate pathway: link between carbohydrate metabolism and secondary metabolism, *Physiol. Plant.*, 66, 164, 1986.

264. **Jensen, R. A.,** Tyrosine and phenylalanine biosynthesis: relationship between alternative pathways, regulation and subcellular location, in *The Shikimic Acid Pathway*, Vol. 20, *Recent Advances in Phytochemistry*, Conn, E. E., Ed., Plenum Press, New York, 1986, 57.

265. **Jensen, S. R.,** Plant iridoids, their biosynthesis and distribution in angiosperms, in *Ecological Chemistry and Biochemistry of Plant Terpenoids*, Vol. 31, Annu. Proc. Phytochem. Soc. Eur., Harborne, J. B. and Tomas-Barberan, F. A., Eds., Clarendon Press, Oxford, 1991, 133.

266. **Johnson, E. L. and Elsohly, M. A.,** Content and *de novo* synthesis of cocaine in embryos and endosperms from fruit of *Erythroxylum coca* Lam., *Ann. Bot.*, 68, 451, 1991.

267. **Johnson, N. D. and Bentley, B. L.,** Symbiotic N_2-fixation and the elements of plant resistance to herbivores: lupine alkaloids and tolerance to defoliation, in *Microbial Mediation of Plant–Herbivore Interactions*, Barbosa, P., Krischik, V. A., and Jones, C. G., Eds., John Wiley & Sons, New York, 1991, 45.

268. **Johnson, R. and Ryan, C. A.,** Wound-inducible potato inhibitor II genes: enhancement of expression by sucrose, *Plant Mol. Biol.*, 14, 527, 1990.

269. **Johnson, R. H. and Lincoln, D. E.,** Sagebrush and grasshopper responses to atmospheric carbon dioxide concentration, *Oecologia*, 84, 103, 1990.

270. **Johnson, R. H. and Lincoln, D. E.**, Sagebrush carbon allocation patterns and grasshopper nutrition: the influence of CO_2 enrichment and soil mineral limitation, *Oecologia*, 87, 127, 1991.

271. **Johnson, R., Lee, J. S., and Ryan, C. A.**, Regulation of expression of a wound-inducible tomato inhibitor I gene in transgenic nightshade plants, *Plant Mol. Biol.*, 14, 349, 1990.

272. **Johnson, R., Narvaez, J., An, G., and Ryan, C.**, Expression of proteinase inhibitors I and II in transgenic tobacco plants: effects on natural defense against *Manduca sexta* larvae, *Proc. Natl. Acad. Sci. U.S.A.*, 86, 9871, 1989.

273. **Kakes, P.**, An analysis of the costs and benefits of the cyanogenic system in *Trifolium repens* L., *Theor. Appl. Genet.*, 77, 111, 1989.

274. **Kakes, P. and Hakvoort, H.**, Is there rhodanese activity in plants?, *Phytochemistry*, 31, 1501, 1992.

275. **Kalberer, P.**, Breakdown of caffeine in the leaves of *Coffea arabica* L., *Nature*, 205, 597, 1965.

276. **Karabourniotis, G., Papadopoulos, K., Papamarkou, M., and Manetas, Y.**, Ultraviolet-B radiation absorbing capacity of leaf hairs, *Physiol. Plant.*, 86, 414, 1992.

277. **Karp, R., Harris, J. L., and Croteau, R.**, Metabolism of monoterpenes: demonstration of the hydroxylation of (+)-sabinene to (+)-*cis*-sabinol by an enzyme preparation from sage (*Salvia officinalis*) leaves, *Arch. Biochem. Biophys.*, 256, 179, 1987.

278. **Karp, R., Mihaliak, C. A., Harris, J. L., and Croteau, R.**, Monoterpene biosynthesis: specificity of the hydroxylations of (−)-limonene by enzyme preparations from peppermint (*Mentha piperita*), spearmint (*Mentha spicata*) and perilla (*Perilla frutescens*) leaves, *Arch. Biochem. Biophys.*, 276, 219, 1990.

279. **Kavanaugh, D., Berge, M. A., and Rosenthal, G. A.**, A higher plant enzyme exhibiting broad acceptance of stereoisomers, *Plant Physiol.*, 94, 67, 1990.

280. **Keene, C. K. and Wagner, G. J.**, Direct demonstration of duvatrienediol biosynthesis in glandular heads of tobacco trichomes, *Plant Physiol.*, 79, 1026, 1985.

281. **Kefeli, V. I. and Dashek, W. V.**, Non-hormonal stimulators and inhibitors of plant growth and development, *Biol. Rev.*, 59, 273, 1984.

282. **Kelsey, R. G., Reynolds, G. W., and Rodriguez, E.**, The chemistry of biologically active constituents secreted and stored in plant glandular trichomes, in *Biology and Chemistry of Plant Trichomes*, Rodriguez, E., Healey, P. L., and Mehta, I., Eds., Plenum Press, New York, 1984, 187.

283. **Kemp, M. S. and Burden, R. S.**, Phytoalexins and stress metabolites in the sapwood of trees, *Phytochemistry*, 25, 1261, 1986.

284. **Khouri, H. E., De Luca, V., and Ibrahim, R. K.**, Enzymatic synthesis of polymethylated flavonols in *Chrysosplenium americanum*. III. Purification and kinetic analysis of *S*-adenosyl-*L*-methionine:3-methylquercetin 7-*O*-methyltransferase, *Arch. Biochem. Biophys.*, 265, 1, 1988.

285. **Kim, S.-R., Costa, M. A., and An, G.**, Sugar response element enhances wound response of potato proteinase inhibitor II promoter in transgenic tobacco, *Plant Mol. Biol.*, 17, 973, 1991.

286. **King, R. R., Calhoun, L. A., Singh, R. P., and Boucher, A.**, Sucrose esters associated with glandular trichomes of wild *Lycopersicon* species, *Phytochemistry*, 29, 2115, 1990.

287. **Kitamura, Y., Miura, H., and Sugii, M.**, Change of atropine esterase activity in the regenerated plants of *Duboisia myoporoides* during development, and its relation to alkaloid accumulation, *J. Plant Physiol.*, 133, 316, 1988.

288. **Kjonaas, R. B., Venkatachalam, K. V., and Croteau, R.**, Metabolism of monoterpenes: oxidation of isopiperitenol to isopiperitenone, and subsequent isomerization to piperitenone by soluble enzyme preparations from peppermint (*Mentha piperita*) leaves, *Arch. Biochem. Biophys.*, 238, 49, 1985.

289. **Kjonaas, R., Martinkus-Taylor, C., and Croteau, R.**, Metabolism of monoterpenes: conversion of *l*-menthone to *l*-menthol and *d*-neomenthol by stereospecific dehydrogenases from peppermint (*Mentha piperita*) leaves, *Plant Physiol.*, 69, 1013, 1982.

291. **Kleinig, H.**, The role of plastids in isoprenoid biosynthesis, *Annu. Rev. Plant Physiol. Plant Mol. Biol.*, 40, 39, 1989.

292. **Koch, B., Nielsen, V. S., Halkier, B. A., Olsen, C. E., and Moller, B. L.**, The biosynthesis of cyanogenic glucosides in seedlings of cassava (*Manihot esculenta* Crantz), *Arch. Biochem. Biophys.*, 292, 141, 1992.

293. **Kochs, G. and Grisebach, H.**, Induction and characterization of a NADPH-dependent flavone synthase from cell cultures of soybean, *Z. Naturforsch.*, 42c, 343, 1987.

294. **Kochs, G. and Grisebach, H.**, Phytoalexin synthesis in soybean: purification and reconstitution of cytochrome P450 3,9-dihydroxypterocarpan 6a-hydroxylase and separation from cytochrome P450 cinnamate 4-hydroxylase, *Arch. Biochem. Biophys.*, 273, 543, 1989.

295. **Koehn, R. K.**, The cost of enzyme synthesis in the genetics of energy balance and physiological performance, *Biol. J. Linn. Soc.*, 44, 231, 1991.

296. **Koes, R. E., Spelt, C. E., and Mol, J. N. M.**, The chalcone synthase multigene family of *Petunia hybrida* (V30): differential, light-regulated expression during flower development and UV light induction, *Plant Mol. Biol.*, 12, 213, 1989.

297. **Koes, R. E., Spelt, C. E., Mol, J. N. M., and Gerats, A. G. M.,** The chalcone synthase multigene family of *Petunia hybrida* (V30): sequence homology, chromosomal localization and evolutionary aspects, *Plant Mol. Biol.,* 10, 159, 1987.

298. **Korcz, A., Markiewicz, M., Pulikowska, J., and Twardowski, T.,** Species-specific inhibitory effect of lupine alkaloids on translation in plants, *J. Plant Physiol.,* 128, 433, 1987.

299. **Kreis, W. and Reinhard, E.,** Selective uptake and vacuolar storage of primary cardiac glycosides by suspension-cultured *Digitalis lanata* cells, *J. Plant Physiol.,* 128, 311, 1987.

300. **Krischik, V. A. and Denno, R. F.,** Individual, population, and geographic patterns in plant defense, in *Variable Plants and Herbivores in Natural and Managed Systems,* Denno, R. F. and McClure, M. S., Eds., Academic Press, New York, 1983, 463.

301. **Kuhlemeier, C.,** Transcriptional and post-transcriptional regulation of gene expression in plants, *Plant Mol. Biol.,* 19, 1, 1992.

302. **Kuntz, M., Romer, S., Suire, C., Hugueney, P., Weil, J. H., Schantz, R., and Camara, B.,** Identification of a cDNA for the plastid-located geranylgeranyl pyrophosphate synthase from *Capsicum annuum*: correlative increase in enzyme activity and transcript level during fruit ripening, *Plant J.,* 2, 25, 1992.

303. **Lamb, B., Guenther, A., Gay, D., and Westberg, H.,** A national inventory of biogenic hydrocarbon emissions, *Atmos. Environ.,* 21, 1695, 1987.

304. **Lambers, H. and Rychter, A. M.,** The biochemical background of variation in respiration rate: respiratory pathways and chemical composition, in *Causes and Consequences of Variation in Growth Rate and Productivity of Higher Plants,* Lambers, H., Ed., SPB Academic Publ., The Hague, 1989, 199.

305. **Langenheim, J. H., Lincoln, D. E., Stubblebine, W. H., and Gabrielli, A. C.,** Evolutionary implications of leaf resin pocket patterns in the tropical tree *Hymenaea* (Caesalpinioideae: Leguminosae), *Am. J. Bot.,* 69, 595, 1982.

306. **Langenheim, J. H., Stubblebine, W. H., and Foster, C. E.,** Effect of moisture stress on composition and yield in leaf resin of *Hymenaea courbaril, Biochem. Syst. Ecol.,* 7, 21, 1979.

307. **Langenheim, J. H., Stubblebine, W. H., Lincoln, D. E., and Foster, C. E.,** Implications of variation in resin composition among organs, tissues and populations in the tropical legume *Hymenaea, Biochem. Syst. Ecol.,* 6, 299, 1978.

308. **Lanz, T., Schroder, G., and Schroder, J.,** Differential regulation of genes for reservatrol synthase in cell cultures of *Arachis hypogaea* L., *Planta,* 181, 169, 1990.

309. **Lapinjoki, S. P., Elo, H. A., and Taipale, H. T.,** Development and structure of resin glands on tissues of *Betula pendula* Roth. during growth, *New Phytol.,* 117, 219, 1991.

310. **Larson, R. A.,** The antioxidants of higher plants, *Phytochemistry,* 27, 969, 1988.

311. **Lee, C. I. and Tukey, H. B., Jr.,** Effect of intermittent mist on development of fall color in foliage of *Euonymus alatus* Sieb. 'Compactus', *J. Am. Soc. Hort. Sci.,* 97, 97, 1971.

312. **Lee, J. S., Brown, W. E., Graham, J. S., Pearce, G., Fox, E. A., Dreher, T. W., Ahern, K. G., Pearson, G. D., and Ryan, C. A.,** Molecular characterization and phylogenetic studies of a wound-inducible proteinase inhibitor I gene in *Lycopersicon* species, *Proc. Natl. Acad. Sci. U.S.A.,* 83, 7277, 1986.

313. **Leete, E.,** Biosynthesis of alkaloids, in *Biosynthesis,* Vol. 4, A Specialist Periodical Report, Bu'lock, J. D., Ed., The Chemical Society, London, 1976, 97.

314. **Leete, E.,** The biosynthesis of nicotine and related alkaloids in intact plants, isolated plant parts, tissue cultures, and cell-free systems, in *Secondary Metabolite Biosynthesis and Metabolism,* Petroski, R. J. and McCormick, S. P. Eds., Plenum Press, New York, 1992, 121.

315. **Leete, E. and Bell, V. M.,** The biogenesis of the *Nicotiana* alkaloids. VIII. The metabolism of nicotine in *N. tabacum, J. Am. Chem. Soc.,* 81, 4358, 1959.

316. **Leete, E. and Chedekel, M. R.,** Metabolism of nicotine in *Nicotiana glauca, Phytochemistry,* 13, 1853, 1974.

317. **Leete, E. and Chedekel, M. R.,** The aberrant formation of (–)-N-methylanabasine from N-methyl-Δ-piperideinium chloride in *Nicotiana tabacum* and *N. glauca, Phytochemistry,* 11, 2751, 1972.

318. **Lerdau, M.,** Future discounts and resource allocation in plants, *Funct. Ecol.,* 6, 371, 1992.

319. **Lersten, N. R. and Curtis, J. D.,** Polyacetylene reservoir (duct) development in *Ambrosia trifida* (Asteraceae) staminate flowers, *Am. J. Bot.,* 76, 1000, 1989.

320. **Les, D. H. and Sheridan, D. J.,** Biochemical heterophylly and flavonoid evolution in North American *Potamogeton* (Potamogetonaceae), *Am. J. Bot.,* 77, 453, 1990.

321. **Levitt, M. J.,** Rapid methylation of micro amounts of nonvolatile acids, *Anal. Chem.,* 45, 618, 1973.

322. **Lewinsohn, E., Gijzen, M., and Croteau, R.,** Regulation of monoterpene biosynthesis in conifer defense, in *Regulation of Isopentenoid Metabolism,* ACS Symposium Series, No. 497, Nes, W. D., Parish, E. J., and Trzaskos, J. M., Eds., American Chemical Society, Washington, D.C., 1992, 8.

323. **Lewinsohn, E., Gijzen, M., and Croteau, R.,** Wound-inducible pinene cyclase from grand fir: purification, characterization and renaturation after SDS-PAGE, *Arch. Biochem. Biophys.,* 293, 167, 1992.

324. **Lewis, N. G. and Yamamoto, E.,** Tannins — their place in plant metabolism, in *Chemistry and Significance of Condensed Tannins,* Hemingway, R. W. and Karchesy, J. J., Eds., Plenum Press, New York, 1989, 23.

325. **Lieberei, R., Nahrstedt, A., Selmar, D., and Gasparotto, L.,** Occurrence of lotaustralin in the genus *Hevea* and changes of HCN-potential in developing organs of *Hevea brasiliensis, Phytochemistry*, 25, 1573, 1986.

326. **Lieberei, R., Selmar, D., and Biehl, B.,** Metabolization of cyanogenic glucosides in *Hevea brasiliensis, Plant Syst. Evol.*, 150, 49, 1985.

327. **Liedvogel, B.,** Acetyl coenzyme A and isopentenylpyrophosphate as lipid precursors in plant cells — biosynthesis and compartmentation, *J. Plant Physiol.*, 124, 211, 1986.

328. **Lincoln, D. E. and Couvet, D.,** The effect of carbon supply on allocation to allelochemicals and caterpillar consumption of peppermint, *Oecologia*, 78, 112, 1989.

329. **Lois, A. F. and West, C. A.,** Regulation of expression of the casbene synthetase gene during elicitation of castor bean seedlings with pectic fragments, *Arch. Biochem. Biophys.*, 276, 270, 1990.

330. **Lois, R., Dietrich, A., Hahlbrock, K., and Schulz, W.,** A phenylalanine ammonia-lyase gene from parsley: structure, regulation and identification of elicitor and light responsive *cis*-acting elements, *EMBO J.*, 8, 1641, 1989.

331. **Louda, S. and Mole, S.,** Glucosinolates: chemistry and ecology, in *Herbivores: Their Interactions with Secondary Plant Metabolites*, Vol. 1, 2nd ed., Rosenthal, G. A. and Berenbaum, M. R., Eds., Academic Press, San Diego, 1991, 123.

332. **Louda, S. M., Farris, M. A., and Blua, M. J.,** Variation in methylglucosinolate and insect damage to *Cleome serrulata* (Capparaceae) along a natural soil moisture gradient, *J. Chem. Ecol.*, 13, 569, 1987.

333. **Luckner, M.,** *Secondary Metabolism in Microorganisms, Plants, and Animals,* 2nd ed., Springer-Verlag, Berlin, 1984.

334. **Luckner, M., Diettrich, B., and Lerbs, W.,** Cellular compartmentation and chanelling of secondary metabolism in microorganisms and higher plants, in *Progress in Phytochemistry*, Vol. 6, Reinhold, L., Harborne, J. B., and Swain, T., Eds., Pergamon Press, Oxford, 1980, 103.

335. **Luttge, U. and Schnepf, E.,** Elimination processes by glands: organic substances, in *Transport in Plants II, Part B, Tissues and Organs,* Vol. 2B, *Encyclopedia of Plant Physiology, New Series,* Luttge, U. and Pitman, M. G., Eds., Springer-Verlag, Berlin, 1976, 244.

336. **Majak, W., Quinton, D. A., and Broersma, K.,** Cyanogenic glycoside levels in Saskatoon serviceberry, *J. Range Manage.*, 33, 197, 1980.

337. **Malcolm, S. B.,** Cardenolide-mediated interactions between plants and herbivores, in *Herbivores: Their Interactions with Secondary Plant Metabolites*, Vol. 1, 2nd ed., Rosenthal, G. A. and Berenbaum, M. R., Eds., Academic Press, San Diego, 1991, 251.

338. **Marcinowski, S. and Grisebach, H.,** Turnover of coniferin in pine seedlings, *Phytochemistry*, 16, 1665, 1977.

339. **Margna, U. and Vainjarv, T.,** Buckwheat seedling flavonoids do not undergo rapid turnover, *Biochem. Physiol. Pflanzen*, 176, 44, 1981.

340. **Margna, U., Margna, E., and Paluteder, A.,** Localization and distribution of flavonoids in buckwheat seedling cotyledons, *J. Plant Physiol.*, 136, 166, 1990.

341. **Marner, F.-J. and Kerp, B.,** Composition of iridals, unusual triterpenoids from sword-lilies, and the seasonal dependence of their content in various parts of different *Iris* species, *Z. Naturforsch.*, 47c, 21, 1992.

342. **Matern, U., Heller, W., and Himmelspach, K.,** Conformational changes of apigenin 7-O-(6-O-malonylglucoside), a vacuolar pigment from parsley, with solvent composition and proton concentration, *Eur. J. Biochem.*, 133, 439, 1983.

343. **Matern, U., Reichenbach, C., and Heller, W.,** Efficient uptake of flavonoids into parsley (*Petroselinum hortense*) vacuoles requires acylated glycosides, *Planta*, 167, 183, 1986.

344. **Mathews, A. and Rai, P. V.,** Mimosine content of *Leucaena leucocephala* and the sensitivity of *Rhizobium* to mimosine, *J. Plant Physiol.*, 117, 377, 1985.

345. **Mathur, A. K., Ahuja, P. S., Pandey, B., Kukreja, A. K., and Mandal, S.,** Screening and evaluation of somaclonal variations for quantitative and qualitative traits in an aromatic grass, *Cymbopogon winterianus* Jowitt, *Plant Breeding*, 101, 321, 1988.

346. **Matile, P.,** The sap of plant cells, *New Phytol.*, 105, 1, 1987.

347. **Matsuda, K. and Senbo, S.,** Chlorogenic acid as a feeding deterrent for the Salicaceae-feeding leaf beetle, *Lochmaeae capreae cribrata* (Coleoptera: Chrysomelidae) and other species of leaf beetles, *Appl. Entomol. Zool.*, 21, 411, 1986.

348. **Mazzafera, P., Crozier, A., and Magalhaes, A. C.,** Caffeine metabolism in *Coffea arabica* and other species of coffee, *Phytochemistry*, 30, 3913, 1991.

349. **McCaskill, D. G., Martin, D. L., and Scott, A. I.,** Characterization of alkaloid uptake by *Catharanthus roseus* (L.) G. Don. protoplasts, *Plant Physiol.*, 87, 402, 1988.

350. **McCullough, D. G. and Kulman, H. M.,** Differences in foliage quality of young jack pine (*Pinus banksiana* Lamb.) on burned and clearcut sites: effects on jack pine budworm (*Choristoneura pinus pinus* Freeman), *Oecologia*, 87, 135, 1991.

351. **McDermitt, D. K. and Loomis, R. S.,** Elemental composition of biomass and its relation to energy content, growth efficiency, and growth yield, *Ann. Bot.*, 48, 275, 1981.

352. **McGrain, A. K., Chen, J. C., Wilson, K. A., and Tan-Wilson, A. L.,** Degradation of trypsin inhibitors during soybean germination, *Phytochemistry*, 28, 1013, 1989.

353. **McIntosh, L., Poulsen, C., and Bogorad, L.,** Chloroplast gene sequence for the large subunit of ribulose bisphosphatecarboxylase of maize, *Nature*, 288, 536, 1980.

354. **McKey, D.,** The distribution of secondary compounds within plants, in *Herbivores: Their Interaction with Secondary Plant Metabolites*, 1st ed., Rosenthal, G. A. and Janzen, D. H., Eds., Academic Press, New York, 1979, 55.

355. **Meinzer, F. C., Wisdom, C. S., Gonzalez-Coloma, A, Rundel, P. W., and Shultz, L. M.,** Effects of leaf resin on stomatal behaviour and gas exchange of *Larrea tridentata* (DC.) Cov., *Funct. Ecol.*, 4, 579, 1990.

356. **Mende, P. and Wink, M.,** Uptake of the quinolizidine alkaloid lupanine by protoplasts and isolated vacuoles of suspension-cultured *Lupinus polyphyllus* cells. Diffusion or carrier-mediated transport?, *J. Plant Physiol.*, 129, 229, 1987.

357. **Merino, J., Field, C., and Mooney, H. A.,** Construction and maintenance costs of mediterranean-climate evergreen and deciduous leaves. II. Biochemical pathway analysis, *Oecol. Plant.*, 5, 211, 1984.

358. **Metcalf, R. L., Rhodes, A. M., Metcalf, R. A., Ferguson, J., Metcalf, E. R., and Lu, P.-Y.,** Cucurbitacin contents and Diabroticite (Coleoptera: Chrysomelidae) feeding upon *Cucurbita* spp., *Environ. Entomol.*, 11, 931, 1982.

359. **Meyer, J. J. M., Grobbelaar, N., Vleggaar, R., and Louw, A. I.,** Fluoroacetyl-coenzyme A hydrolase-like activity in *Dichapetalum cymosum*, *J. Plant Physiol.*, 139, 369, 1992.

360. **Miersch, J., Juhlke, C., Sternkopf, G., and Krauss, G.-J.,** Metabolism and exudation of canavanine during development of alfalfa (*Medicago sativa* L. cv. Verko), *J. Chem. Ecol.*, 18, 2117, 1992.

361. **Mihaliak, C. A. and Lincoln, D. E.,** Changes in leaf mono- and sesquiterpene metabolism with nitrate availability and leaf age in *Heterotheca subaxillaris*, *J. Chem. Ecol.*, 15, 1579, 1989.

362. **Mihaliak, C. A., Gershenzon, J., and Croteau, R.,** Lack of rapid monoterpene turnover in rooted plants: implications for theories of plant chemical defense, *Oecologia*, 87, 373, 1991.

363. **Miller, J. M. and Conn, E. E.,** Metabolism of hydrogen cyanide by higher plants, *Plant Physiol.*, 65, 1199, 1980.

364. **Miller, R. J., Jolles, C., and Rapoport, H.,** Morphine metabolism and normorphine in *Papaver somniferum*, *Phytochemistry*, 12, 597, 1973.

365. **Mkpong, O. E., Yan, H., Chism, G., and Sayre, R. T.,** Purification, characterization, and localization of linamarase in cassava, *Plant Physiol.*, 93, 176, 1990.

366. **Mo, Y., Nagel, C., and Taylor, L. P.,** Biochemical complementation of chalcone synthase mutants defines a role for flavonols in functional pollen, *Proc. Natl. Acad. Sci. U.S.A.*, 89, 7213, 1992.

367. **Molderez, M., Nagels, L., and Parmentier, F.,** Time-course tracer studies on the metabolism of cinnamic acid in *Cestrum poeppigii*, *Phytochemistry*, 17, 1747, 1978.

368. **Moller, B. L. and Conn, E. E.,** The biosynthesis of cyanogenic glucosides in higher plants: channeling of intermediates in dhurrin biosynthesis by a microsomal system from *Sorghum bicolor* (Linn) Moench, *J. Biol. Chem.*, 255, 3049, 1980.

369. **Moller. B. L.,** The involvement of *N*-hydroxyamino acids as intermediates in metabolic transformations, in *Cyanide in Biology*, Vennesland, B., Conn, E. E., Knowles, C. J., Westley, J., and Wissing, F., Eds., Academic Press, London, 1981, 197.

370. **Monfar, M., Caelles, C., Balcells, L., Ferrer, A., Hegardt, F. G., and Boronat, A.,** Molecular cloning and characterization of plant 3-hydroxy-3-methylglutaryl coenzyme A reductase, in *Biochemistry of the Mevalonic Acid Pathway to Terpenoids*, Vol. 24, *Recent Advances in Phytochemistry*, Towers, G. H. N. and Stafford, H. A., Eds., Plenum Press, New York, 1990, 83.

371. **Muday, G. K. and Herrmann, K. M.,** Wounding induces one of two isoenzymes of 3-deoxy-D-*arabino*-heptulosonate 7-phosphate synthase in *Solanum tuberosum* L., *Plant Physiol.*, 98, 496, 1992.

372. **Munck, S. L. and Croteau, R.,** Purification and characterization of the sesquiterpene cyclase patchoulol synthase from *Pogostemon cablin*, *Arch. Biochem. Biophys.*, 282, 58, 1990.

373. **Muzika, R. M., Pregitzer, K. S., and Hanover, J. W.,** Changes in terpene production following nitrogen fertilization of grand fir (*Abies grandis* (Dougl.) Lindl.) seedlings, *Oecologia*, 80, 485, 1989.

374. **Nachman, R. J. and Olsen, J. D.,** Ranunculin: a toxic constituent of the poisonous range plant bur buttercup (*Ceratocephalus testiculatus*), *J. Agric. Food Chem.*, 31, 1358, 1983.

375. **Narvaez-Vasquez, J., Orozco-Cardenas, M. L., and Ryan, C. A.,** Differential expression of a chimeric CaMV-tomato proteinase inhibitor I gene in leaves of transformed nightshade, tobacco and alfalfa plants, *Plant Mol. Biol.*, 20, 1149, 1992.

376. **Negishi, O., Ozawa, T., and Imagawa, H.,** Conversion of xanthosine into caffeine in tea plants, *Agric. Biol. Chem.*, 49, 251, 1985.

377. **Neumann, D.,** Storage of alkaloids, in *Biochemistry of Alkaloids*, Mothes, K., Schutte, H. R., and Luckner, M., Eds., VCH Publishers, Berlin, 1985, 49.

378. **Niemeyer, H. M.**, Hydroxamic acids (4-hydroxy-1,4-benzoxazin-3-ones), defence chemicals in the Gramineae, *Phytochemistry*, 27, 3349, 1988.
379. **Njar, V. C. O., Arnold, L. M., Banthorpe, D. V., Branch, S. A., Christie, A. C., and Marsh, D. C.**, Metabolism of exogenous monoterpenes and their epoxides in seedlings of *Pinus pinaster* Ait., *J. Plant Physiol.*, 135, 628, 1989.
380. **Noe, W. and Berlin, J.**, Induction of de-novo synthesis of tryptophan decarboxylase in cell suspensions of *Catharanthus roseus*, *Planta*, 166, 500, 1985.
381. **Ohigashi, H., Wagner, M. R., Matsumura, F., and Benjamin, D. M.**, Chemical basis of differential feeding behavior of the larch sawfly, *Pristiphora erichsonii* (Hartig), *J. Chem. Ecol.*, 7, 599, 1981.
382. **Owuor, P. O. and Chavanji, A. M.**, Caffeine contents of clonal tea; seasonal variations and effects of plucking standards under Kenyan conditions, *Food Chem.*, 20, 225, 1986.
383. **Paczkowski, C. and Wojciechowski, Z. A.**, The occurrence of UDPG-dependent glucosyltransferase specific for sarsasapogenin in *Asparagus officinalis*, *Phytochemistry*, 27, 2743, 1988.
384. **Park, H., Denbow, C. J., and Cramer, C. L.**, Structure and nucleotide sequence of tomato *HMG2* encoding 3-hydroxy-3-methyl-glutaryl coenzyme A reductase, *Plant Mol. Biol.*, 20, 327, 1992.
385. **Pate, J. S.**, Distribution of metabolites, in *Plant Physiology: A Treatise*, Vol. VIII, *Nitrogen Metabolism*, Steward, F. C., Ed., Academic Press, New York, 1983, 335.
386. **Pate, J. S.**, Transport and partitioning of nitrogenous solutes, *Annu. Rev. Plant Physiol.*, 31, 313, 1980.
387. **Pearce, G., Ryan, C. A., and Liljegren, D.**, Proteinase inhibitors I and II in fruit of wild tomato species: transient components of a mechanism for defense and seed dispersal, *Planta*, 175, 527, 1988.
388. **Pearce, G., Strydom, D., Johnson, S., and Ryan, C. A.**, A polypeptide from tomato leaves induces wound-inducible proteinase inhibitor proteins, *Science*, 253, 895, 1991.
389. **Pena-Cortes, H., Sanchez-Serrano, J. J., Mertens, R., Willmitzer, L., and Prat, S.**, Abscisic acid is involved in the wound-induced expression of the proteinase inhibitor II gene in potato and tomato, *Proc. Natl. Acad. Sci. U.S.A.*, 86, 9851, 1989.
390. **Penning de Vries, F. W. T.**, The cost of maintenance processes in plant cells, *Ann. Bot.*, 39, 77, 1975.
391. **Penning de Vries, F. W. T., Brunsting, A. H. M., and van Laar, H. H.**, Products, requirements and efficiency of biosynthesis: a quantitative approach, *J. Theor. Biol.*, 45, 339, 1974.
392. **Peters, N. K., Frost, J. W., and Long, S. R.**, A plant flavone, luteolin, induces expression of *Rhizobium meliloti* nodulation genes, *Science*, 233, 977, 1986.
393. **Phillips, D. A.**, Flavonoids: plant signals to soil microbes, in *Phenolic Metabolism in Plants*, Vol. 26, *Recent Advances in Phytochemistry*, Stafford, H. A. and Ibrahim, R. K., Eds., Plenum Press, New York, 1992, 201.
394. **Pimentel, D. and Bellotti, A. C.**, Parasite–host population systems and genetic stability, *Am. Nat.*, 110, 877, 1976.
395. **Poulsen, C. and Verpoorte, R.**, Roles of chorismate mutase, isochorismate synthase and anthranilate synthase in plants, *Phytochemistry*, 30, 377, 1991.
396. **Poulton, J. E.**, Transmethylation and demethylation reactions in the metabolism of secondary plant products, in *The Biochemistry of Plants: A Comprehensive Treatise*, Vol. 7, *Secondary Plant Products*, Conn, E. E., Ed., Academic Press, New York, 1981, 667.
397. **Poulton, J. E.**, Cyanogenesis in plants, *Plant Physiol.*, 94, 401, 1990.
398. **Preisig, C. L., VanEtten, H. D., and Moreau, R. A.**, Induction of 6a-hydroxymaackiain 3-O-methyltransferase and phenylalanine ammonia-lyase mRNA translational activities during the biosynthesis of pisatin, *Arch. Biochem. Biophys.*, 290, 468, 1991.
399. **Proksch, M., Strack, D., and Weissenbock, G.**, Incorporation of [^{14}C]phenylalanine and [^{14}C]cinnamic acid into leaf pieces and mesophyll protoplasts from oat primary leaves for studies on flavonoid metabolism at the tissue and cell level, *Z. Naturforsch.*, 36c, 222, 1981.
400. **Purohit, S., Laloraya, M. M., and Bharti, S.**, Effect of phenolic compounds on abscisic acid-induced stomatal movement: structure–activity relationship, *Physiol. Plant.*, 81, 79, 1991.
401. **Putnam, A. R. and Tang, C.-S.**, Eds., *The Science of Allelopathy*, John Wiley & Sons, New York, 1986.
402. **Ramachandra Reddy, A. and Das, V. S. R.**, Partial purification and characterization of 3-hydroxy-3-methylglutaryl coenzyme A reductase from the leaves of guayule (*Parthenium argentatum*), *Phytochemistry*, 25, 2471, 1986.
403. **Ramachandra Reddy, A. and Das, V. S. R.**, Chloroplast autonomy for the biosynthesis of isopentenyl diphosphate in guayule (*Parthenium argentatum* Gray), *New Phytol.*, 106, 457, 1987.
404. **Raven, J. A.**, The role of vacuoles, *New Phytol.*, 106, 357, 1987.
405. **Redmond, J. W., Batley, M., Djordjevic, M. A., Innes, R. W., Kuempel, P. L., and Rolfe, B. G.**, Flavones induce expression of nodulation genes in *Rhizobium*, *Nature*, 323, 632, 1986.
406. **Rees, S. B. and Harborne, J. B.**, The role of sesquiterpene lactones and phenolics in the chemical defence of the chicory plant, *Phytochemistry*, 24, 2225, 1985.

407. **Regnier, F. E., Waller, G. R., Eisenbraun, E. J., and Auda, H.,** The biosynthesis of methylcyclopentane monoterpenoids. II. Nepetalactone, *Phytochemistry*, 7, 221, 1968.
408. **Reichardt, P. B., Bryant, J. P., Anderson, B. J., Phillips, D., Clausen, T. P., Meyer, M., and Frisby, K.,** Germacrone defends Labrador tea from browsing by snowshoe hares, *J. Chem. Ecol.*, 16, 1961, 1990.
409. **Reichardt, P. B., Bryant, J. P., Clausen, T. P., and Wieland, G. D.,** Defense of winter-dormant Alaska paper birch against snowshoe hares, *Oecologia*, 65, 58, 1984.
410. **Reichardt, P. B., Chapin, F. S., III, Bryant, J. P., Mattes, B. R., and Clausen, T. P.,** Carbon/nutrient balance as a predictor of plant defense in Alaskan balsam poplar: potential importance of metabolite turnover, *Oecologia*, 88, 401, 1991.
411. **Renaudin, J. P. and Guern, J.,** Transport and vacuolar storage of secondary metabolites in plant cell cultures, in *Secondary Products from Plant Tissue Culture*, Vol. 30, Annu. Proc. Phytochem. Soc. Eur., Charlwood, B. V. and Rhodes, M. J. C., Eds., Clarendon Press, Oxford, 1990, 59.
412. **Rhoades, D. F.,** Evolution of plant chemical defense against herbivores, in *Herbivores: Their Interaction with Secondary Plant Metabolites*, 1st ed., Rosenthal, G. A. and Janzen, D. H., Eds., Academic Press, New York, 1979, 3.
413. **Rhoades, D. F.,** Integrated antiherbivore, antidesiccant and ultraviolet screening properties of creosotebush resin, *Biochem. Syst. Ecol.*, 5, 281, 1977.
414. **Rhoades, D. F. and Cates, R. G.,** A general theory of plant antiherbivore chemistry, in *Biochemical Interactions Between Plants and Insects*, Vol. 10, *Recent Advances in Phytochemistry*, Wallace, J. W. and Mansell, R. L., Eds., Plenum Press, New York, 1976, 168.
415. **Rice, E. L.,** *Allelopathy*, 2nd ed., Academic Press, Orlando, 1984.
416. **Ripa, P. V., Martin, E. A., Cocciolone, S. M., and Adler, J. H.,** Fluctuation of phytoecdysteroids in developing shoots of *Taxus cuspidata*, *Phytochemistry*, 29, 425, 1990.
417. **Roberts, M. R., Homeyer, B. C., and Pham, T. D. T.,** Further studies of sequestration of alkaloids in *Papaver somniferum* L. latex vacuoles, *Z. Naturforsch.*, 46c, 377, 1991.
418. **Robinson, T.,** *The Organic Constituents of Higher Plants*, 2nd ed., Burgess, Minneapolis, 1967, 72.
419. **Robinson, T.,** Metabolism and function of alkaloids in plants, *Science*, 184, 434, 1974.
420. **Robinson, T.,** *The Biochemistry of Alkaloids*, 2nd ed., Springer-Verlag, Berlin, 1981, 38.
421. **Roddick, J. G.,** The steroidal glycoalkaloid α-tomatine, *Phytochemistry*, 13, 9, 1974.
422. **Roitman, J. N., Wollenweber, E., and Arriaga-Giner, F. J.,** Xanthones and triterpene acids as leaf exudate constituents in *Orphium frutescens*, *J. Plant Physiol.*, 139, 632, 1992.
423. **Rosenthal, G. A.,** Nitrogen allocation for L-canavanine synthesis and its relationship to chemical defense of the seed, *Biochem. Syst. Ecol.*, 5, 219, 1977.
424. **Rosenthal, G. A.,** Investigations of canavanine biochemistry in the jack bean plant, *Canavalia ensiformis* (L.) DC. I. Canavanine utilization in the developing plant, *Plant Physiol.*, 46, 273, 1970.
425. **Rosenthal, G. A.,** Metabolism of L-canavanine and L-canaline in leguminous plants, *Plant Physiol.*, 94, 1, 1990.
426. **Rosenthal, G. A.,** *Plant Nonprotein Amino and Imino Acids: Biological, Biochemical and Toxicological Properties*, Academic Press, New York, 1982, 183.
427. **Rosenthal, G. A.,** L-Canavanine metabolism in jack bean, *Canavalia ensiformis* (L.) DC. (Leguminosae), *Plant Physiol.*, 69, 1066, 1982.
428. **Rosenthal, G. A. and Berenbaum, M. R.,** Eds., *Herbivores: Their Interactions with Secondary Plant Metabolites*, Vol. 1, 2nd ed., Academic Press, San Diego, 1991.
429. **Rosenthal, G. A. and Berge, M. A.,** Catabolism of L-canavanine and L-canaline in the jack bean, *Canavalia ensiformis* (L.) DC. (Leguminosae), *J. Agric. Food Chem.*, 37, 591, 1989.
430. **Rosenthal, G. A. and Rhodes, D.,** L-Canavanine transport and utilization in developing jack bean, *Canavalia ensiformis* (L.) DC. [Leguminosae], *Plant Physiol.*, 76, 541, 1984.
431. **Rosenthal, G. A., Berge, M. A., and Bleiler, J. A.,** A novel mechanism for detoxification of L-canaline, *Biochem. Syst. Ecol.*, 17, 203, 1989.
432. **Rosenthal, G. A., Berge, M. A., Ozinskas, A. J., and Hughes, C. G.,** Ability of L-canavanine to support nitrogen metabolism in the jack bean, *Canavalia ensiformis* (L.) DC., *J. Agric. Food Chem.*, 36, 1159, 1988.
433. **Ross, J. D. and Sombrero, C.,** Environmental control of essential oil production in Mediterranean plants, in *Ecological Chemistry and Biochemistry of Plant Terpenoids*, Vol. 31, Annu. Proc. Phytochem. Soc. Eur., Harborne, J. B. and Tomas-Barberan, F. A., Eds., Clarendon Press, Oxford, 1991, 83.
434. **Rousi, M., Tahvanainen, J., and Uotila, I.,** A mechanism of resistance to hare browsing in winter-dormant European white birch (*Betula pendula*), *Am. Nat.*, 137, 64, 1991.
435. **Rowell-Rahier, M. and Pasteels, J. M.,** Phenolglucosides and interactions at three trophic levels: Salicaceae — herbivores — predators, in *Insect–Plant Interactions*, Vol. II, Bernays, E. A., Ed., CRC Press, Boca Raton, FL, 1990, 75.
436. **Russin, W. A., Uchytil, T. F., and Durbin, R. D.,** Isolation of structurally intact secretory cavities from leaves of African marigold, *Tagetes erecta* L. (Asteraceae), *Plant Sci.*, 85, 115, 1992.

437. **Russin, W. A., Uchytil, T. F., Feistner, G., and Durbin, R. D.,** Developmental changes in content of foliar secretory cavities of *Tagetes erecta* (Asteraceae), *Am. J. Bot.,* 75, 1787, 1988.
438. **Ryan, C. A.,** Protease inhibitors in plants: genes for improving defenses against insects and pathogens, *Annu. Rev. Phytopathol.,* 28, 425, 1990.
439. **Ryan, C. A.,** The search for the proteinase inhibitor-inducing factor, PIIF, *Plant Mol. Biol.,* 19, 123, 1992.
440. **Ryan, C. A. and An, G.,** Molecular biology of wound-inducible proteinase inhibitors in plants, *Plant Cell Environ.,* 11, 345, 1988.
441. **Sabine, J. R.,** *3-Hydroxy-3-methylglutaryl Coenzyme A Reductase,* CRC Press, Boca Raton, FL, 1983.
442. **Salatino, A., Monteiro, W. R., and Bomtempi, N., Jr.,** Histochemical localization of phenolic deposits in shoot apices of common species of Asteraceae, *Ann. Bot.,* 61, 557, 1988.
443. **Sander, H.,** Studien uber Bildung und Abbau von Tomatin in der Tomatenpflanze, *Planta,* 47, 374, 1956.
444. **Sander, H. and Hartmann, T.,** Site of synthesis, metabolism and translocation of senecionine N-oxide in cultured roots of *Senecio erucifolius, Plant Cell Tiss. Org. Cult.,* 18, 19, 1989.
445. **Sato, H., Taguchi, G., Fukui, H., and Tabata, M.,** Role of malic acid in solubilizing excess berberine accumulating in vacuoles of *Coptis japonica, Phytochemistry,* 31, 3451, 1992.
446. **Savelkoul, F. H. M. G., Van Der Poel, A. F. B., and Tamminga, S.,** The presence and inactivation of trypsin inhibitors, tannins, lectins and amylase inhibitors in legume seeds during germination. A review, *Plant Foods Hum. Nutr.,* 42, 71, 1992.
447. **Scheel, D. and Parker, J. E.,** Elicitor recognition and signal transduction in plant defense gene activation, *Z. Naturforsch.,* 45c, 569, 1990.
448. **Schindler, T. and Kotzias, D.,** Comparison of monoterpene volatilization and leaf-oil composition of conifers, *Naturwiss.,* 76, 475, 1989.
449. **Schlee, D.,** Alkaloids derived from purines, in *Biochemistry of Alkaloids,* Mothes, K., Schutte, H. R., and Luckner, M., Eds., VCH Publishers, Berlin, 1985, 338.
450. **Schloman, W. W., Jr., Garrot, D. J., Jr., Ray, D. T., and Bennett, D. J.,** Seasonal effects on guayule resin composition, *J. Agric. Food Chem.,* 34, 177, 1986.
451. **Schnepf, E.,** Gland cells, in *Dynamic Aspects of Plant Ultrastructure,* Robards, A. W., Ed., McGraw-Hill, New York, 1974, 331.
452. **Schnitzler, J.-P., Madlung, J., Rose, A., and Seitz, H. U.,** Biosynthesis of *p*-hydroxybenzoic acid in elicitor-treated carrot cell cultures, *Planta,* 188, 594, 1992.
453. **Schulz, M. and Weissenbock, G.,** Dynamics of the tissue-specific metabolism of luteolin glucuronides in the mesophyll of rye primary leaves (*Secale cereale*), *Z. Naturforsch.,* 43c, 187, 1988.
454. **Schulz, M. and Weissenbock, G.,** Three specific UDP glucuronate:flavone-glucuronosyl-transferases from primary leaves of *Secale cereale, Phytochemistry,* 27, 1261, 1988.
455. **Schulze-Siebert, D. and Schultz, G.,** Full autonomy in isoprenoid synthesis in spinach chloroplasts, *Plant Physiol. Biochem.,* 25, 145, 1987.
456. **Schulze-Siebert, D., Heintze, A., and Schultz, G.,** Substrate flow from photosynthetic carbon metabolism to chloroplast isoprenoid synthesis in spinach: evidence for a plastidic phosphoglycerate mutase, *Z. Naturforsch.,* 42c, 570, 1987.
457. **Scora, R. W. and Mann, J. D.,** Essential oil synthesis in *Monarda punctata, Lloydia,* 30, 236, 1967.
458. **Seaman, F., Bohlmann, F., Zdero, C., and Mabry, T. J.,** *Diterpenes of Flowering Plants, Compositae (Asteraceae),* Springer-Verlag, New York, 1990, 427.
459. **Seigler, D. A.,** Cyanide and cyanogenic glycosides, in *Herbivores: Their Interactions with Secondary Plant Metabolites,* Vol. 1, 2nd ed., Rosenthal, G. A. and Berenbaum, M. R., Eds., Academic Press, San Diego, 1991, 35.
460. **Seigler, D. and Price, P. W.,** Secondary compounds in plants: primary functions, *Am. Nat.,* 110, 101, 1976.
461. **Seigler, D. S.,** Primary roles for secondary compounds, *Biochem. Syst. Ecol.,* 5, 195, 1977.
462. **Selmar, D., Frehner, M., and Conn, E. E.,** Purification and properties of endosperm protoplasts of *Hevea brasiliensis* L., *J. Plant Physiol.,* 135, 105, 1989.
463. **Selmar, D., Grocholewski, S., and Seigler, D. S.,** Cyanogenic lipids: utilization during seedling development of *Ungnadia speciosa, Plant Physiol.,* 93, 631, 1990.
464. **Selmar, D., Lieberei, R., and Biehl, B.,** Mobilization and utilization of cyanogenic glycosides: the linustatin pathway, *Plant Physiol.,* 86, 711, 1988.
465. **Selmar, D., Lieberei, R., Biehl, B., Nahrstedt, A., Schmidtmann, V., and Wray, V.,** Occurrence of the cyanogen linustatin in *Hevea brasiliensis, Phytochemistry,* 26, 2400, 1987.
466. **Selmar, D., Lieberei, R., Junqueira, N., and Biehl, B.,** Changes in cyanogenic glucoside content in seeds and seedlings of *Hevea* species, *Phytochemistry,* 30, 2135, 1991.
467. **Sharma, R. D.,** Isoflavone content of Bengalgram (*Cicer arietinum*) at various stages of germination, *J. Plant Foods,* 3, 259, 1981.
468. **Shirai, R. and Kurihara, T.,** Distribution of rhodanese in plants, *Bot. Mag.,* 104, 341, 1991.
469. **Simms, E. L. and Fritz, R. S.,** The ecology and evolution of host-plant resistance to insects, *Trends Ecol. Evol.,* 5, 356, 1990.

470. **Simms, E. L. and Rausher, M. D.,** The evolution of resistance to herbivory in *Ipomoea purpurea*. II. Natural selection by insects and costs of resistance, *Evolution*, 43, 573, 1989.
471. **Simms, E. L. and Rausher, M. D.,** Costs and benefits of plant resistance to herbivory, *Am. Nat.*, 130, 570, 1987.
472. **Siqueira, J. O., Safir, G. R., and Nair, M. G.,** Stimulation of vesicular-arbuscular mycorrhiza formation and growth of white clover by flavonoid compounds, *New Phytol.*, 118, 87, 1991.
473. **Skogsmyr, I. and Fagerstrom, T.,** The cost of anti-herbivory defence: an evaluation of some ecological and physiological factors, *Oikos*, 64, 451, 1992.
474. **Smith, D. A. and Banks, S. W.,** Biosynthesis, elicitation and biological activity of isoflavonoid phytoalexins, *Phytochemistry*, 25, 979, 1986.
475. **Solomonson, L. P.,** Cyanide as a metabolic inhibitor, in *Cyanide in Biology*, Vennesland, B., Conn, E. E., Knowles, C. J., Westley, J., and Wissing, F., Eds., Academic Press, London, 1981, 11.
476. **Srere, P. A.,** Complexes of sequential metabolic enzymes, *Annu. Rev. Biochem.*, 56, 89, 1987.
477. **Stadler, R. and Zenk, M. H.,** A revision of the generally accepted pathway for the biosynthesis of the benzyltetrahydroisoquinoline alkaloid reticuline, *Liebigs Ann. Chem.*, 555, 1990.
478. **Stafford, H. A.,** Compartmentation in natural product biosynthesis by multienzyme complexes, in *The Biochemistry of Plants: A Comprehensive Treatise*, Vol. 7, *Secondary Plant Products*, Conn, E. E., Ed., Academic Press, New York, 1981, 117.
479. **Stafford, H. A.,** The enzymology of proanthocyanidin biosynthesis, in *Chemistry and Significance of Condensed Tannins*, Hemingway, R. W. and Karchesy, J. J., Eds., Plenum Press, New York, 1989, 47.
480. **Stahl-Biskup, E.,** Monoterpene glycosides, state-of-the-art, *Flav. Fragr. J.*, 2, 75, 1987.
481. **Steel, C. C. and Drysdale, R. B.,** Electrolyte leakage from plant and fungal tissues and disruption of liposome membranes by α-tomatine, *Phytochemistry*, 27, 1025, 1988.
482. **Steiner, A. M.,** Zum Umsatz von Anthocyan-3-monoglucosiden in Petalen von *Petunia hybrida*, *Z. Pflanzenphysiol.*, 65, 210, 1971.
483. **Steiner, A. M.,** Der Umsatz von Anthocyanen bei *in vivo* und *in vitro* kultivierten Bluten von *Petunia hybrida*, *Z. Pflanzenphysiol.*, 69, 55, 1973.
484. **Stermer, B. A. and Bostock, R. M.,** Involvement of 3-hydroxy-3-methylglutaryl coenzyme A reductase in the regulation of sesquiterpenoid phytoalexin synthesis in potato, *Plant Physiol.*, 84, 404, 1987.
485. **Stermitz, F. R., Arslanian, R. L., and Castro, O.,** Flavonoids from the leaf surface of *Godmania aesculifolia* (Bignoniaceae), *Biochem. Syst. Ecol.*, 20, 481, 1992.
486. **Stich, K., Eidenberger, T., Wurst, F., and Forkmann, G.,** Flavonol synthase activity and the regulation of flavonol and anthocyanin biosynthesis during flower development in *Dianthus caryophyllus* L. (Carnation), *Z. Naturforsch.*, 47c, 553, 1992.
487. **Strack, D., Heilemann, J., Momken, M., and Wray, V.,** Cell wall-conjugated phenolics from Coniferae leaves, *Phytochemistry*, 27, 3517, 1988.
488. **Strobel, H., Knobloch, K., and Ziegler, E.,** Uber die atherischen Ole von *Chrysanthemum balsamita* L., *Z. Naturforsch.*, 42c, 502, 1987.
489. **Sukhov, G. V.,** The use of radiocarbon in the study of biosynthesis of terpenes, in *Radioisotopes in Scientific Research*, Extermann, R. C., Ed., Pergamon Press, New York, 1958, 535.
490. **Sun, Y. and Hrazdina, G.,** Isolation and characterization of a UDPglucose:flavonol O^3-glucosyltransferase from illuminated red cabbage (*Brassica oleracea* cv Red Danish) seedlings, *Plant Physiol.*, 95, 570, 1991.
491. **Suzuki, T. and Waller, G. R.,** Biosynthesis and biodegradation of caffeine, theobromine, and theophylline in *Coffea arabica* L. fruits, *J. Agric. Food Chem.*, 32, 845, 1984.
492. **Suzuki, T. and Waller, G. R.,** Biodegradation of caffeine: formation of theophylline and theobromine from caffeine in mature *Coffea arabica* fruits, *J. Sci. Food Agric.*, 35, 66, 1984.
493. **Suzuki, T. and Waller, G. R.,** Total nitrogen and purine alkaloids in the tea plant throughout the year, *J. Sci. Food Agric.*, 37, 862, 1986.
494. **Suzuki, T., Ashihara, H., and Waller, G. R.,** Purine and purine alkaloid metabolism in *Camellia* and *Coffea* plants, *Phytochemistry*, 31, 2575, 1992.
495. **Svoboda, K. P., Hay, R. K. M., and Waterman, P. G.,** Growing summer savory (*Satureja hortensis*) in Scotland: quantitative and qualitative analysis of the volatile oil and factors influencing oil production, *J. Sci. Food Agric.*, 53, 193, 1990.
496. **Swiezewska, E., Dallner, G., Andersson, B., and Ernster, L.,** Biosynthesis of ubiquinone and plastoquinone in the endoplasmic reticulum-Golgi membranes of spinach leaves, *J. Biol. Chem.*, 268, 1494, 1993.
497. **Sze, H.,** H^+-translocating ATPases of the plasma membrane and tonoplast of plant cells, *Physiol. Plant.*, 61, 683, 1984.
498. **Taiz, L. and Zeiger, E.,** *Plant Physiology*, Benjamin Cummings, Redwood City, CA, 1991.
499. **Takahama, U.,** Hydrogen peroxide-dependent oxidation of flavonoids and hydroxycinnamic acid derivatives in epidermal and guard cells of *Tradescantia virginiana* L., *Plant Cell Physiol.*, 29, 475, 1988.
500. **Tallamy, D. W. and Raupp, M. J.,** Eds., *Phytochemical Induction by Herbivores*, John Wiley & Sons, New York, 1991.

501. **Tang, C.-S.,** Continuous trapping techniques for the study of allelochemicals from higher plants, in *The Science of Allelopathy,* Putnam, A. R. and Tang, C.-S., Eds., John Wiley & Sons, New York, 1986, 113.

502. **Tang, C.-S. and Young, C.-C.,** Collection and identification of allelopathic compounds from the undisturbed root system of Bigalta limpograss (*Hemarthria altissima*), *Plant Physiol.,* 69, 155, 1982.

503. **Taylor, A. O. and Zucker, M.,** Turnover and metabolism of chlorogenic acid in *Xanthium* leaves and potato tubers, *Plant Physiol.,* 41, 1350, 1966.

504. **Thieme, H.,** Die phenolglykoside der Salicaceen, *Planta Med.,* 431, 1965.

505. **Thompson, A. C.,** Ed., *The Chemistry of Allelopathy: Biochemical Interactions Among Plants,* ACS Symposium Series, No. 268, American Chemical Society, Washington, D.C., 1985.

506. **Threlfall, D. R. and Whitehead, I. M.,** Terpenoid phytoalexins: aspects of biosynthesis, catabolism, and regulation, in *Ecological Chemistry and Biochemistry of Plant Terpenoids,* Vol. 31, Annu. Proc. Phytochem. Soc. Eur., Harborne, J. B. and Tomas-Barberan, F. A., Eds., Clarendon Press, Oxford, 1991, 159.

507. **Tingey, D. T., Manning, M., Grothaus, L. C., and Burns, W. F.,** Influence of light and temperature on monoterpene emission rates from slash pine, *Plant Physiol.,* 65, 797, 1980.

508. **Tingey, D. T., Turner, D. P., and Weber, J. A.,** Factors controlling the emissions of monoterpenes and other volatile organic compounds, in *Trace Gas Emissions by Plants,* Sharkey, T. D., Holland, E. A., and Mooney, H. A., Eds., Academic Press, San Diego, 1991, 93.

509. **Tomas-Barberan, F. A., Msonthi, J. D., and Hostettmann, K.,** Antifungal epicuticular methylated flavonoids from *Helichrysum nitens, Phytochemistry,* 27, 753, 1988.

510. **Toppel, G., Witte, L., and Hartmann, T.,** *N*-Oxidation and degradation of pyrrolizidine alkaloids during germination of *Crotalaria scassellatii, Phytochemistry,* 27, 3757, 1988.

511. **Towers, G. H. N.,** Significance of phototoxic phytochemicals in insect herbivory, *J. Chem. Ecol.,* 12, 813, 1986.

512. **Tso, T. C. and Jeffrey, R. N.,** Biochemical studies on tobacco alkaloids. I. The fate of labeled tobacco alkaloids supplied to *Nicotiana* plants, *Arch. Biochem. Biophys.,* 80, 46, 1959.

513. **Tso, T. C. and Jeffrey, R. N.,** Biochemical studies on tobacco alkaloids. IV. The dynamic state of nicotine supplied to *N. rustica, Arch. Biochem. Biophys.,* 92, 253, 1961.

514. **Tsukaya, H., Ohshima, T., Naito, S., Chino, M., and Komeda, Y.,** Sugar-dependent expression of the *CHS*-A gene for chalcone synthase from petunia in transgenic *Arabidopsis, Plant Physiol.,* 97, 1414, 1991.

515. **Tukey, H. B., Jr.,** Leaching of substances from plants, in *Biochemical Interactions Among Plants,* National Academy of Sciences, Washington, D.C., 1971, 25.

516. **Tuomi, J., Niemela, P., Chapin, F. S., III, Bryant, J. P., and Siren, S.,** Defensive responses of trees in relation to their carbon/nutrient balance, in *Mechanisms of Woody Plant Defenses Against Insects: Search for Pattern,* Mattson, W. J., Levieux, J., Bernard-Dagan, C., Eds., Springer-Verlag, New York, 1988, 57.

517. **Tyson, B. J., Dement, W. A., and Mooney, H. A.,** Volatilisation of terpenes from *Salvia mellifera, Nature,* 252, 119, 1974.

518. **Upmeier, B., Gross, W., Koster, S., and Barz, W.,** Purification and properties of *S*-adenosyl-*L*-methionine:nicotinic acid-*N*-methyltransferase from cell suspension cultures of *Glycine max* L., *Arch. Biochem. Biophys.,* 262, 445, 1988.

519. **Van Damme, E. J. M. and Peumans, W. J.,** Developmental changes and tissue distribution of lectin in *Galanthus nivalis* L. and *Narcissus* cv. Carlton, *Planta,* 182, 605, 1990.

520. **van der Werf, A., Hirose, T., and Lambers, H.,** Variation in root respiration; causes and consequences for growth, in *Causes and Consequences of Variation in Growth Rate and Productivity of Higher Plants,* Lambers, H., Ed., SPB Academic Publ., The Hague, 1989, 227.

521. **van Tunen, A. J., Koes, R. E., Spelt, C. E., van der Krol, R., Stuitje, A. R., and Mol, J. N. M.,** Cloning of the two chalcone flavanone isomerase genes from *Petunia hybrida*: coordinate, light-regulated and differential expression of flavonoid genes, *EMBO J.,* 7, 1257, 1988.

522. **Villegas, R. J. A. and Kojima, M.,** Purification and characterization of hydroxycinnamoyl D-glucose quinate hydroxycinnamoyl transferase in the root of sweet potato, *Ipomoea batatas* Lam., *J. Biol. Chem.,* 261, 8729, 1986.

523. **Voet, D. and Voet, J. G.,** *Biochemistry,* John Wiley & Sons, New York, 1990.

524. **Vogeli, U. and Chappell, J.,** Regulation of a sesquiterpene cyclase in cellulase-treated tobacco cell suspension cultures, *Plant Physiol.,* 94, 1860, 1990.

525. **Vogt, T., Gulz, P.-G., and Reznik, H.,** UV Radiation dependent flavonoid accumulation of *Cistus laurifolius* L., *Z. Naturforsch.,* 46c, 37, 1991.

526. **Von der Gathen, H. and Horster, H.,** Monoterpenglykoside, Experimente zur Klarung ihrer biologischen Bedeutung, in *Atherische Ole,* Kubeczka, K. H., Ed., Georg Thieme Verlag, Stuttgart, 1982, 206.

527. **von Borstel, K. and Hartmann, T.,** Selective uptake of pyrrolizidine *N*-oxides by cell suspension cultures from pyrrolizidine alkaloid producing plants, *Plant Cell Rep.,* 5, 39, 1986.

528. **Wagner, G. J.,** Compartmentation in plant cells: the role of the vacuole, in *Cellular and Subcellular Localization in Plant Metabolism,* Vol. 16, *Recent Advances in Phytochemistry,* Creasy, L. L. and Hrazdina, G., Eds., Plenum Press, New York, 1982, 1.

529. **Wagner, R., Feth, F., and Wagner, K. G.,** Regulation in tobacco callus of enzyme activities of the nicotine pathway. II. The pyridine-nucleotide cycle, *Planta*, 168, 408, 1986.

530. **Wagner, R., Feth, F., and Wagner, K. G.,** The regulation of enzyme activities of the nicotine pathway in tobacco, *Physiol. Plant.*, 68, 667, 1986.

531. **Wagner, R., Feth, F., and Wagner, K. G.,** The pyridine-nucleotide cycle in tobacco: enzyme activities for the recycling of NAD, *Planta*, 167, 226, 1986.

532. **Wagschal, K., Savage, T. J., and Croteau, R.,** Isotopically sensitive branching as a tool for evaluating multiple product formation by monoterpene cyclases, *Tetrahedron*, 47, 5933, 1991.

533. **Waller, G. R. and Lee, J. L.-C.,** Metabolism of the α-pyridone ring of ricinine in *Ricinus communis* L., *Plant Physiol.*, 44, 522, 1969.

534. **Waller, G. R. and Nowacki, E. K.,** *Alkaloid Biology and Metabolism in Plants*, Plenum Press, New York, 1978.

535. **Waterman, P. G. and Mole, S.,** Extrinsic factors influencing production of secondary metabolites in plants, in *Insect–Plant Interactions*, Vol. 1, Bernays, E. A., Ed., CRC Press, Boca Raton, FL, 1989, 107.

536. **Weeks, W. W. and Bush, L. P.,** Alkaloid changes in tobacco seeds during germination, *Plant Physiol.*, 53, 73, 1974.

537. **Welle, R. and Grisebach, H.,** Phytoalexin synthesis in soybean cells: elicitor induction of reductase involved in biosynthesis of 6'-deoxychalcone, *Arch. Biochem. Biophys.*, 272, 97, 1989.

538. **Werner, C. and Matile, P.,** Accumulation of coumarylglucosides in vacuoles of barley mesophyll protoplasts, *J. Plant Physiol.*, 118, 237, 1985.

539. **West, C. A.,** Biosynthesis of diterpenes, in *Biosynthesis of Isoprenoid Compounds*, Vol. 1, Porter, J. W. and Spurgeon, S. L., Eds., John Wiley & Sons, New York, 1981, 375.

540. **Whitehead, I. M., Threlfall, D. R., and Ewing, D. F.,** *Cis*-9,10-dihydrocapsenone: a possible catabolite of capsidiol from cell suspension cultures of *Capsicum annuum*, *Phytochemistry*, 26, 1367, 1987.

541. **Williams, K., Percival, F., Merino, J., and Mooney, H. A.,** Estimation of tissue construction cost from heat of combustion and organic nitrogen content, *Plant Cell Environ.*, 10, 725, 1987.

542. **Wilson, K. A., Papastoitsis, G., Hartl, P., and Tan-Wilson, A. L.,** Survey of the proteolytic activities degrading the Kunitz trypsin inhibitor and glycinin in germinating soybeans (*Glycine max*), *Plant Physiol.*, 88, 355, 1988.

543. **Wilson, T. M. and Russell, D. W.,** The localization, partial purification and regulation of pea plastid HMG-CoA reductase, *Biochem. Biophys. Res. Commun.*, 184, 530, 1992.

544. **Wink, M.,** Physiology of secondary product formation in plants, in *Secondary Products from Plant Tissue Culture*, Vol. 30, Annu. Proc. Phytochem. Soc. Eur., Charlwood, B. V. and Rhodes, M. J. C., Eds., Clarendon Press, Oxford, 1990, 23.

545. **Wink, M.,** Site of lupanine and sparteine biosynthesis in intact plants and *in vitro* organ cultures, *Z. Naturforsch.*, 42c, 868, 1987.

546. **Wink, M.,** Metabolism of quinolizidine alkaloids in plants and cell suspension cultures: induction and degradation, in *Primary and Secondary Metabolism of Plant Cell Cultures*, Neumann, K.-H., Barz, W., and Reinhard, E., Eds., Springer-Verlag, Berlin, 1985, 107.

547. **Wink, M. and Hartmann, T.,** Diurnal fluctuation of quinolizidine alkaloid accumulation in legume plants and photomixotrophic cell suspension cultures, *Z. Naturforsch.*, 37c, 369, 1982.

548. **Wink, M. and Mende, P.,** Uptake of lupanine by alkaloid-storing epidermal cells of *Lupinus polyphyllus*, *Planta Med.*, 53, 465, 1987.

549. **Wink, M. and Witte, L.,** Quinolizidine alkaloids as nitrogen source for lupin seedlings and cell cultures, *Z. Naturforsch.*, 40c, 767, 1985.

550. **Wink, M. and Witte, L.,** Turnover and transport of quinolizidine alkaloids. Diurnal fluctuations of lupanine in the phloem sap, leaves and fruits of *Lupinus albus* L., *Planta*, 161, 519, 1984.

551. **Wink, M. and Witte, L.,** Quinolizidine alkaloids in *Petteria ramentacea* and the infesting aphids, *Aphis cytisorum*, *Phytochemistry*, 24, 2567, 1985.

552. **Wink, M., Hartmann, T., and Schiebel, H.-M.,** A model mechanism for the enzymatic synthesis of lupin alkaloids, *Z. Naturforsch.*, 34c, 704, 1979.

553. **Wink, M., Hartmann, T., and Witte, L.,** Enzymatic synthesis of quinolizidine alkaloids in lupin chloroplasts, *Z. Naturforsch.*, 35c, 93, 1980.

554. **Wink, M., Hartmann, T., Witte, L., and Rheinheimer, J.,** Interrelationship between quinolizidine alkaloid producing legumes and infesting insects: exploitation of the alkaloid-containing phloem sap of *Cytisus scoparius* by the broom aphid *Aphis cytisorum*, *Z. Naturforsch.*, 37c, 1081, 1982.

555. **Wink, M.,** Quinolizidine alkaloids: biochemistry, metabolism, and function in plants and cell suspension cultures, *Planta Med.*, 53, 509, 1987.

556. **Wollenweber, E. and Jay, M.,** Flavones and flavonols, in *The Flavonoids*, Harborne, J. B., Ed., Chapman and Hall, London, 1988, 233.

557. **Wurtele, E. S., Nikolau, B. J., and Conn, E. E.,** Subcellular and developmental distribution of β-cyanoalanine synthase in barley leaves, *Plant Physiol.*, 78, 285, 1985.

558. **Yamamoto, H., Suzuki, M., Kitamura, T., Fukui, H., and Tabata, M.,** Energy-requiring uptake of protoberberine alkaloids by cultured cells of *Thalictrum flavum, Plant Cell Rep.*, 8, 361, 1989.

559. **Yang, Z., Park, H., Lacy, G. H., and Cramer, C. L.,** Differential activation of potato 3-hydroxy-3-methylglutaryl coenzyme A reductase genes by wounding and pathogen challenge, *Plant Cell*, 3, 397, 1991.

560. **Yip, W.-K. and Yang, S. F.,** Cyanide metabolism in relation to ethylene production in plant tissues, *Plant Physiol.*, 88, 473, 1988.

561. **Ylstra, B., Touraev, A., Moreno, R. M. B., Stoger, E., van Tunen, A. J., Vicente, O., Mol, J. N. M., and Heberle-Bors, E.,** Flavonols stimulate development, germination, and tube growth of tobacco pollen, *Plant Physiol.*, 100, 902, 1992.

562. **Yokouchi, Y. and Ambe, Y.,** Factors affecting the emission of monoterpenes from red pine (*Pinus densiflora*), *Plant Physiol.*, 75, 1009, 1984.

563. **Yoshida, D.,** Degradation and translocation of ^{15}N-labeled nicotine injected into intact tobacco leaves, *Plant Cell Physiol.*, 3, 391, 1952.

564. **Zador, E. and Jones, D.,** The biosynthesis of a novel nicotine alkaloid in the trichomes of *Nicotiana stocktonii, Plant Physiol.*, 82, 479, 1986.

565. **Zangerl, A. R. and Berenbaum, M. R.,** Furanocoumarin induction in wild parsnip: genetics and populational variation, *Ecology*, 71, 1933, 1990.

566. **Zenk, M. H.,** Biosynthesis of alkaloids using plant cell cultures, in *Plant Nitrogen Metabolism, Vol. 23, Recent Advances in Phytochemistry*, Poulton, J. E., Romeo, J. T., and Conn, E. E., Eds., Plenum Press, New York, 1989, 429.

567. **Zenk, M. H.,** Recent work on cinnamoyl CoA derivatives, in *Biochemistry of Plant Phenolics*, Vol. 12, *Recent Advances in Phytochemistry*, Swain, T., Harborne, J. B., and Van Sumere, C. F., Eds., Plenum Press, New York, 1979, 139.

568. **Zenk, M. H., Gerardy, R., and Stadler, R.,** Phenol oxidative coupling of benzylisoquinoline alkaloids is catalysed by regio- and stereo-selective cytochrome P-450 linked plant enzymes: salutaridine and berbamunine, *J. Chem. Soc. Chem. Commun.*, 1725, 1989.

569. **Zenk, M. H., Rueffer, M., Amann, M., Deus-Neumann, B., and Nagakura, N.,** Benzylisoquinoline biosynthesis by cultivated plant cells and isolated enzymes, *J. Nat. Prod.*, 48, 725, 1985.

570. **Ziegler, H.,** Nature of transported substances, in *Transport in Plants I, Phloem Transport*, Vol. 1, *Encyclopedia of Plant Physiology*, New Series, Zimmermann, M. H. and Milburn, J. A., Eds., Springer-Verlag, Berlin, 1975, 59.

571. **Zimowski, J.,** Occurrence of a glucosyltransferase specific for solanidine in potato plants, *Phytochemistry*, 30, 1827, 1991.

572. **Zobel, A. M. and Brown, S. A.,** Localization of daphnetin and umbelliferone in different tissues of *Daphne mezereum* shoots, *Can. J. Bot.*, 67, 1456, 1989.

573. **Zobel, A. M. and Brown, S. A.,** Dermatitis-inducing furanocoumarins on leaf surfaces of eight species of rutaceous and umbelliferous plants, *J. Chem. Ecol.*, 16, 693, 1990.

Chapter 6

LIFE HISTORY TRAITS OF INSECT HERBIVORES IN RELATION TO HOST QUALITY

Simon R. Leather

TABLE OF CONTENTS

I. INTRODUCTION

Many papers have been written concerning host quality and its effect on the reproductive parameters of insect herbivores. A wide variety of insect–plant interactions have been described, and an equally wide range of approaches have been used; for example, the effects of temperature and host quality on the fecundity of the grain aphid, *Sitobion avenae*;[1] oviposition preferences and larval performance measures in Lepidoptera and Hymenoptera;[35,201] effects of seasonal changes in host quality on aphid reproduction;[121,222,226] relationships between migration, flight, and host quality in Lepidoptera and aphids and many other herbivores.[82,129]

All these approaches have two things in common: first, they assume that the host plant or plants exhibit variability in quality, and second, they assume that the insect herbivore is able to distinguish between hosts of differing quality, and, most importantly, is able, when there is a choice, to respond to these differences.

In this paper, the concept of host quality is defined, reviewed, and discussed in terms of its effects on the life history traits of insect herbivores, with particular reference to aphids and Lepidoptera.

II. HOST QUALITY

What is host quality? The term host quality, as used by entomologists, is a somewhat nebulous concept. It is a term used frequently in the insect–plant relationships literature (e.g., References 83, 91, 169, 185, 214, 237) but rarely if ever defined. Host quality appears to be a term of relatively recent usage. Prior to the use of the term, the concept was described somewhat more cumbersomely as the role of "nutrition",[7,61] although the criteria used to describe nutrition were far from precise (senescing leaves, mature leaves, young leaves, etc.). It is as if there were a group awareness of what nutrition was and that differences in host quality existed. The term "nutrition" was largely replaced by the term "food quality",[38,40,56,89,143,153] and this term is still used frequently.[9,19,24,41,102,104] The terms "food quality" and "host quality" are almost but not exactly analogous. The use of these two terms appears to depend on how the writer in question views the insect–plant relationship. If the relationship is viewed primarily as a predatory one, then the term "food quality" is more likely to be used. If, on the other hand, the relationship is viewed as a parasitic one, then "host quality" is the preferred term. Thus, entomologists working on sucking insects such as aphids and other bugs tend to use the term "host quality".[1,129] Workers using defoliating lepidoptera as their study animals tend to use the term "food quality".[102,179,181]

Definitions of host or food quality are hard to find. In the majority of textbooks where one would expect to find a definition under sections headed by those terms, no specific definitions are provided, despite whole chapters being dedicated to the subject.[170,187,189,195] Instead, much is made of the differing chemical compositions of host plants, and the reader is left to assume that nitrogen, for example, is the most important positive factor but that it can be modified adversely by plant secondary compounds, such as tannins. There have, however, been some attempts to define the concept. Scriber discussed the concept in relation to insect herbivores as host plant suitability, and defined it as the adequacy of the selected food to sustain growth, survival, and reproduction.[181] This is a somewhat parsimonious definition but has been and still is used widely. Singer also equated host quality with host suitability but failed to define either term.[185] Schwartz and Hobbs, discussing mammalian herbivores, although not defining host or food quality, state that, for herbivores, the value of a plant is related to its digestible energy content, and later, although not stating it as a definition, produced an equation for digestibility:[180]

$$\text{True digestion coefficient} = I_A - (F_A - E_A/I_A)$$

TABLE 1
Effect of Fertilizer Application on Nitrogen Concentration of *Tsuga* Needles and the Survival and Fecundity of *Fiorinia externa*

	Average nitrogen content (% dry wt)	Nymphal survival (%)	Fecundity
Fertilized	5.6 ± 0.4	81.5 ± 4.6	13.3 ± 2.1
Control	4.3 ± 0.4	68.5 ± 7.9	9.3 ± 1.9

Note: All fitness parameters between treatments were significant at $P < 0.005$.

From McClure, M. S., *Environ. Entomol.*, 12, 1811, 1983. With permission.

where I = intake, F = feces, E = endogenous, and A = amount. The best definition available is that of Crawley, who states that food quality can be defined strictly only in terms of herbivore fitness, and he defines fitness for a given genotype as its rate of increase [r] when fed a certain food.[37] He further states that a plant tissue is of high quality for a particular species when [r] is large and of low quality when [r] is small. He goes on to discuss the complexity of measuring food quality and of identifying one particular compound as a general index of food quality. He concludes that, as the nutritional properties of plant chemicals are not rigidly determined and may in one case be a nutrient and in another a toxin, a general index of food quality is an impossibility.[37] It is of interest to note that he obviously regards host quality and food quality as separate although closely interlinked attributes.

Host quality can thus be defined as those plant attributes, chemical or physical, that contribute either negatively or positively to the fitness of the insect population or individual that feeds upon the plant's tissues.

This view of host quality is considerably broader than that of food quality and includes a wide variety of factors other than the quality of the nutrients alone. For example, insects feeding on plants such as milkweed, which contain large amounts of toxic cardenolides, would appear to an observer concerned only with food quality to be at a disadvantage, as their growth on tissues high in cardenolides is lower than that on tissues containing low concentrations of them.[208] However, with respect to the concept of host quality defined above, there is no conflict, as the protection offered from predators by the sequestration of those compounds, compensates for the apparent lack of fitness as measured solely by larval growth but regained by larval survival.[27] There is, for example, a significant negative relationship between survival of young larvae of *Danaus plexippus* and the cardiac glycoside content of their host plants,[240] but this is presumably at least partly compensated for by the enhanced larval survival afforded by being toxic to vertebrate predators.[27] Other insect species demonstrate similar trade-offs (see Section VI).

It is obvious that plant chemistry is the major determinant of host quality, although the physical aspects of plant form and structure are also of great importance. The different classes of plant chemicals that influence insect performance, e.g., nitrogen (Table 1), are interlinked or interact. For example, leaf water content is very closely correlated with plant nitrogen content (Figure 1), and tannin content is often negatively correlated with nitrogen content.[181] Furthermore, the quality of a host plant can be affected by the insects feeding on it. Such induced changes have been recently reviewed.[85] To deal with the topic of host plant chemistry and its relationship to host quality is an enormous task that would require a separate review article for each element, compound or class of compounds to be discussed. This has already been done for several groups of compounds, such as the tannins.[84]

It will be assumed that the reader is generally aware of the potential effects of the various

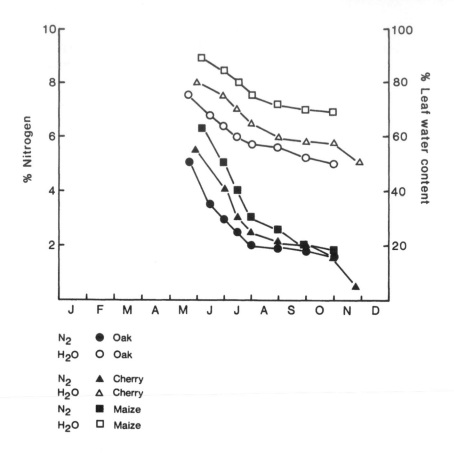

FIGURE 1. Seasonal changes in leaf water and nitrogen for several plant species. (From Scriber, J. M. and Slansky, F., *Annu. Rev. Entomol.*, 26, 183, 1981. With permission.)

chemical constituents of plants on insects. Instead, the quality of hosts will be discussed relative to the ways that good or poor plant quality affects the life history traits of insects that feed upon them.

III. LIFE HISTORY TRAITS

Adult insects can be divided into two main categories, those that feed after becoming adult and those that do not feed, or if they do, feed only minimally. The former are usually long-lived and obtain the nutrients required for reproduction as adults. The latter are usually short-lived and have laid down their reserves for reproduction during their larval life, as is the case with many Lepidoptera. There are, of course, exceptions. Aphids, for example, feed during adult life but are generally short lived, and larval (nymphal) experience plays a large part in how they function as adults.[59] Some Lepidoptera do feed as adults and not solely on carbohydrates.[65,111] It is possible to visualize an insect as a simplistic compartmentalized organism, with certain areas allocated to certain functions: flight and/or dispersal, metabolism (longevity), and most importantly, reproduction (Figure 2). How the resources of these components are allocated depends on a number of factors, which are discussed below.

A. COSTS OF REPRODUCTION
Female insects have two ovaries, each of which contains a number of ovarioles; it is the number and size of these that determine the potential and achieved fecundity of a particular

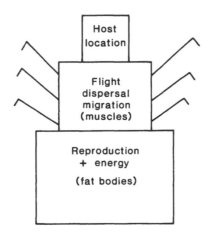

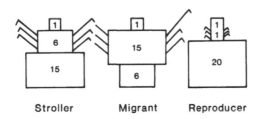

FIGURE 2. Stylized representation of resource allocation in an insect.

insect.[95] The more ovarioles an insect contains, the potentially more fecund it is.[227] Ovariole number can vary within and between species, for example, in aphids[227] and in tephritid flies of the genus *Dacus*.[72,73] Among Lepidoptera there is consistency throughout the whole order.[31]

If an animal species opts to produce many offspring, then this reproductive effort must be paid for in terms of a reduced life span (notwithstanding the generally positive correlation between offspring number and longevity),[203] reduced offspring size,[120] or reduced migratory activity.[129] It is in the balancing of these trade-offs that the concept of host quality becomes of paramount importance.[216]

1. Life Span and Pre-reproductive Delay

Insects and other animals must not only locate suitable breeding/oviposition sites but must allocate their energy reserves in such a way as to maximize their fitness.[71] The relationship between life span and reproductive effort is of particular importance, as there can be a trade-off between them.[126,203,217] However, the value of offspring is not a constant over the life span of an individual; the first offspring is more valuable than the second and so on, depending on the quality of the host plant and the dispersal ability of the mother.[15]

Reproducing insects can delay oviposition in expectation of a better oviposition site (host) but run the risk of not reproducing at all. If they reproduce immediately on a suboptimal host, they maximize reproductive potential but perhaps do so at the expense of offspring survival.[129,219] In general, a long pre-reproductive delay in insects is correlated with reduced achieved fecundity, as in the case of the moth *Panolis flammea* on a poor quality host.[127] This has also been demonstrated for the butterfly, *Battus philenor*.[174] Although achieved fecundity is reduced, the size of the offspring produced may be increased, thus maintaining reproductive effort at an absolute level.[120,177] This is discussed in greater detail later (see section VI. E). The pre-reproductive period can also be increased by a delay in mating; this too reduces achieved and potential fecundity.[119,127,158] However, delayed mating is rarely, if ever, linked with host quality per se and will not be discussed further.

2. Size and Fecundity

Until recently it was considered axiomatic for most arthropods that increased weight and/ or size resulted in increased fecundity.[78,221] Certainly, this has been reported for many insects, especially aphids. However, this relationship is not as straightforward as it appears.[117] Many studies have used potential fecundity (i.e., the number of oocytes, eggs, or offspring within an insect in the adult) as their index of fecundity for their estimate of lifetime fecundity. This is an adequate prediction in those insects that have produced all their eggs or embryos by the time adulthood is reached, such as in the pine looper moth, *Bupalus piniarius*,[18] and the sycamore aphid, *Drepanosiphum platanoidis*.[110] However, in species where ovulation does not cease after the adult molt and the number of steps between the dependent (offspring number) variable and independent variable (weight) is increased, this assumption becomes less reliable. For example, many researchers working on aphids use embryo or large embryo counts as indicators of fecundity.[20,49,66] In the case of aphids such as *Rhopalosiphum padi*, in which ovulation continues after the adult molt, and for which the initial fecundity estimate can be increased by 135%,[110] this method would produce erroneous results.

The size–fecundity relationship is affected by a number of factors, many of which can be linked either directly or indirectly with host quality. Moreover, the effect of host quality on this relationship can act through the larval stage as well as on the reproductive adult stage. In addition, even if large insects of the same species are more fecund than smaller ones of the same species, confounding factors can occur. For example, large individuals of the sawfly, *Neodiprion sertifer,* produce more eggs than small *N. sertifer*, but fewer of their eggs are fertile.[86] In some aphids (e.g., *Aphis fabae*), the relationship between size and fecundity is indirect and acts mainly through reproductive rate and mortality.[199] Small aphids have a faster rate of reproduction than large aphids but do not live as long and are less fecund. In Lepidoptera the situation can be even more confusing. For example, adult weight in the pine beauty moth, *P. flammea*, is a good indicator of the number of eggs contained within an adult female, the potential fecundity.[111] However, this has little bearing on the number of eggs that are actually laid, the achieved fecundity sometimes being as little as 1% of the potential fecundity.[127] In fact, it has been calculated that fewer than 15% of laboratory reared *P. flammea* lay even 70% of their maximum potential number of eggs.[117] In field conditions, this is likely to be even lower,[33] and the maximum fecundity estimate for field *P. flammea* is only 28% of its maximum reproductive potential.[127]

3. Longevity and Fecundity

Longevity is a complicating factor in the size–fecundity relationship for a number of insect species. For example, in the planthopper, *Prokelisia marginata*, a significant relationship between eggs produced and adult female weight has been demonstrated.[44] However, this relationship was arrived at by dividing total fecundity by total longevity and using the figure obtained as eggs per female per day. This is not strictly true, as the pre- and postreproductive periods were ignored and both are of great importance in determining insect fecundity (and see above).[101,127,199]

Longevity and the factors affecting longevity appear to be the most important factors influencing achieved fecundity in the Lepidoptera. In *P. flammea* there is no relationship between weight and achieved fecundity, but a marked positive correlation exists between longevity and achieved fecundity.[111] In addition, there is no relationship between longevity and weight, but a significant relationship between availability of adult food source and longevity.[111] In *Euxoa mesoria* there is a significant relationship between pupal weight and both achieved fecundity and longevity. However, if the moths are fed, the relationship between weight and longevity is lost, although longevity and achieved fecundity remain correlated.[29] Increased fecundity has also been found to be correlated with increased longevity in *Heliothis virescens* and *Sesamia nonagriodes*[2,136] and with the availability of an adult food source in the pierid butterfly, *Colias philodice eurytheme*.[193] Longevity is also affected by larval host plants, and this in turn affects the achieved fecundity of *S. nonagriodes* and *P. flammea* in both the laboratory and the field.[2,112,127] It is interesting to note that although virgin moths live longer than those that have mated in conditions when a suitable oviposition site is not available,[127] when mated moths delay mating because of poor host quality, life expectancy is sacrificed in favor of producing fewer but fitter eggs.[120]

B. POLYMORPHISMS

Polymorphisms are widespread in insects, and a number of insect characters can display variation. Some are very obvious, such as the striking color differences seen within species of ladybird beetles, such as *Adalia bipunctata*,[25] or the color morphs of a number of aphid species, which can range from green to brown, passing through a range of reds and pinks on the way, e.g., *Acyrthosiphon pisum*[144] and *Sitobion avenae*.[32] Another striking polymorphism apparent in insects relates to wing development; some insect species produce forms with or without wings or with nonfunctional wings. This is seen in the majority of aphids,[59] and is a common occurrence in other homopterans such as *P. marginata*,[45] in heteropteran species such as *Gerris* spp.,[212] and in orthopteran species such as *Gryllus firmus*.[178]

Less striking polymorphisms, but of at least equal importance, are the polymorphisms seen at the behavioral and physiological levels. For example, the moth *Malacosoma pluviale* occurs in an active and less active form;[229] tephritid flies and aphids have a varying number of ovarioles within species;[72,130] phenological polymorphism is seen in the spruce budworm, *Choristoneura occidentalis*,[213] and the burnet moth, *Zygaena trifolii*.[235] Polymorphism in enzymes has also been shown in many species, including aphids.[134,135] What is of interest to us, is how, if at all, these polymorphisms affect the fecundity or other life history traits of the insects in question. The role of host quality in determining some of these polymorphisms will be discussed here and in the following sections.

1. Color

Within herbivorous insects, the most commonly recognized color polymorphisms are found within the Lepidoptera. Many species mimic distasteful species and can show a wide range of wing pattern differences across a population, such as in *Heliconius melpomene* and *H. erato*.[205] This form of color polymorphism, however, has little to do with host quality per se, unless it is taken in the sense that the butterflies mimicking the distasteful species are implying that their host quality is low for the predators that feed upon them. Industrial melanism is linked more with escape from predation than with food or reproductive strategies, but is also common in Lepidoptera, the peppered moth (*Biston betularia*) being a well-known example.[94] Other color polymorphisms seen in adult wing patterns of butterflies have been linked with climate and are thought to be thermoregulatory adaptations[42] or linked with reduced migration and increased fecundity, as in the alba polymorphism of *Colias eurytheme*.[79] Other color polymorphisms appear at present to have no biological explanation.[141]

Color polymorphism also occurs in lepidopteran larvae. The larvae of the moth *Spodoptera exempta* occur in a light and a dark form. The dark form develops under crowded conditions, feeds more efficiently, and develops faster to an adult that is no less fecund than the normal color morph.[184] Dark forms of the larvae of the spruce budworm, *C. occidentalis*, also occur, and these are linked with the speed at which spring emergence occurs.[213]

A commonly overlooked color polymorphism occurs in several aphid species. These exist mainly in terms of brown and green, with the brown sometimes being pinkish or red. The reason for color polymorphism in aphids is poorly understood. Examples include the pea aphid, *Acyrthosiphon pisum*,[144] the sycamore aphid, *Drepanosiphum platanoidis*,[55] the nettle aphid, *Microlophium carnosum*,[21] the rose aphid *Macrosiphum rosae*,[140] the peach potato aphid, *Myzus persicae*,[145] and the grain aphid, *S. avenae*.[32] The color changes in *M. carnosum* and *D. platanoidis* have been attributed to temperature, *M. carnosum* becoming pink or purple in hot weather and *D. plantanoidis* going from green to red in warm weather.[55] The biological significance of the color forms is little understood, although the red form of *D. platanoidis* is more active than the green form.[55] Markkula, working with red and green forms of the pea aphid, *A. pisum*, was able to show that the green forms were more fecund than the red forms.[144] However, no other advantage of the red form (the unusual morph) over the green form was demonstrated, so what advantage the aphid gains by color polymorphism is unclear. Parasitism rates of *S. avenae* are higher in the normal green form than in the brown form, so this may explain the occurrence of the darker color morphs, to a certain extent.[5]

2. Wing Polymorphism

Wing polymorphism is a frequently observed trait in insects. The formation of different wing morphs is caused by a variety of stimuli. In aphids and other homopterans, winged forms can appear at certain stages of the life cycle as a response to changes in day length or temperature.[58,131-133] This occurs, for example, in the aphids *Megoura viciae* and *A. pisum*. They may appear in response to increases in density, as in the aphid *R. padi*[61] and the adelgid *Gilletteella cooleyii*[162] or in response to a deterioration in host quality.[59]

Wing polymorphism, unlike color polymorphism, has been strongly linked with reproductive success. The most common wing polymorphisms are either lack of wings (aptery) versus full wings or small, nonfunctional wings (brachyptery) versus functional wings (macroptery). The delphacids *P. dolus* and *P. marginata* have long- and short-winged forms, and the short-winged forms are more fecund than the long-winged forms.[47] Short-winged forms of the aphid *Drepanosiphum dixoni* are more fecund than the long-winged forms,[55a] and small-winged morphs of the crickets *Gryllus firmus* and *Allonemobius fasciatus* are more fecund than the large-winged forms.[178] In addition, the growth rate of the embryos within winged aphids is lower than that of embryos within apterous forms, and the offspring produced by alate mothers are initially smaller than those produced by apterous morphs, indicating that the possession of wings has a marked cost in terms of reproductive effort.[157] In some insects where neither winged form flies, no difference in fecundity is seen, although some of the developmental parameters are slower in the larger-winged morphs than in the short-winged morphs e.g., in the heteropteran bug, *Pyrrhocoris apterus*.[90] Wing size also varies within species of Lepidoptera, but no link with reproductive success has been sought. It is thought that the size of lepidopteran wings is linked with climate and aids thermoregulation.[42] A similar theory has been put forward to explain the presence of wing dimorphism in the waterstrider, *Gerris remigis*, where dispersal is almost absent. In this case, it has been thought that the winged forms are at a greater advantage in warmer habitats than the apterous forms, and no link with host quality has been postulated.[67]

There is thus a great deal of plasticity in resource allocation within a number of insect species in terms of flight and fecundity. The trade-offs between flight and reproduction are discussed below.

3. Ovariole Number

Ovariole number in insects is not always a constant for either family or species. In some groups, e.g., Lepidoptera, it is nearly always a constant, eight being the norm,[95] although in some rare individuals of the butterfly *Heliconius charitonius*, six, seven, and nine ovarioles have been reported.[64] Ovariole number also varies slightly in *Colias philodice eriphyle*, individuals with eight, nine, and twelve ovarioles having been observed.[190] No explanation for this phenomenon has been sought, and it is so rare as to have no bearing on reproductive

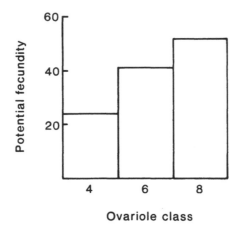

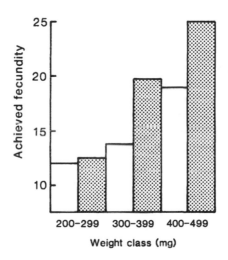

FIGURE 3. a. Relationship between potential fecundity and ovariole number in alate exules of *Rhopalosiphum padi*. b. Relationship between number of offspring produced in 7 d and adult weight for alate exules of *R. padi* with four ovarioles (open bars) and 6 ovarioles (hatched bars). (From Wellings, P. W., Leather, S. R., and Dixon, A. F. G., *J. Anim. Ecol.*, 49, 975, 1980. With permission.)

strategies. In other groups such as Aphididae and Diptera, there is considerable variation not just between species but also within species. In those insect species with a variable number of ovarioles, ovariole number is positively related to fecundity.[60,72,73,125,227]

In some insect groups, for example Diptera, the number of ovarioles is positively related to the weight of the insect,[16,73,125] but in others, such as aphids, no significant relationship between weight and ovariole number can be satisfactorily demonstrated.[125,216,227] Thus, in aphids, the more ovarioles an individual possesses, the more fecund that individual is likely to be.[125] Within the same ovariole class, large individuals are more fecund than smaller ones, but a female with six ovarioles and the same weight as one with ten ovarioles would be less fecund than the latter (Figure 3). Ovariole number in aphids is both a function of generational morph[130,227] and of host quality. Ovariole number in the majority of the aphid species studied is high at the beginning of the life cycle, namely in those generations emerging from eggs or, as in the case of *R. padi*, in the second generation from the egg. It then decreases as the season progresses and then shows an increase in the autumnal generations. The morphs associated with the highest host quality available, expanding buds and leaves in the spring and senescing leaf tissue in the autumn, are those with the greatest numbers of ovarioles.[50,54] By having a large reproductive potential, these aphids are able to fully exploit this high host quality and rapidly increase their numbers before the leaves mature and host quality falls. Aphids exploiting plants with a low host quality tend to have fewer ovarioles and to be less fecund (see Section V).[125,130]

4. Miscellaneous Polymorphisms

Lepidoptera can respond to the quality of the host plant by behavioral and physiological polymorphisms. For example, in the western tent caterpillar, *Malacosoma californicum pluviale*, and the gypsy moth, *Lymantria dispar*, larger eggs are produced in times of low host quality,[230] and the larvae that hatch from these eggs are more motile and have a greater capability for dispersal.[12,229] This has been demonstrated in a large number of insect species and has been fully reviewed elsewhere.[10]

Physiological polymorphisms, such as in the rate of larval development and duration of spring emergence of adult *Choristoneura* spp. have also been shown to exist,[213] but no link with host quality has been demonstrated.

C. DISPERSAL AND MIGRATION

In herbivorous insects, most dispersal is by flight, although of course there are exceptions. The gypsy moth, *L. dispar*, has wingless females, and the larvae are the dispersive form, using silk threads for ballooning to new hosts.[11] Dispersal in insects is also brought about by walking. There is, for example, considerable interplant movement by grain aphids in cereal fields on the ground.[188] Flight is, however, the most successful method of dispersal in insects. Aphids, once they have taken off, can be carried hundreds or even thousands of miles by air currents.[228,233]

Flight does impose a cost on an insect, and in female insects this is usually shown by a reduction in fecundity.[178] For example, unfed female *Spodoptera exempta* can suffer a reduction in fecundity of more than 50% if they fly for more than 10 h,[82] whereas if they are supplied with a food source such as sucrose, fecundity remains statistically unaffected by flight duration (Figure 4). How does host plant quality affect the migratory or dispersal strategies of insects? The following sections demonstrate some of the many ways in which insects respond to changes in host quality.

IV. SEASONAL PATTERNS IN FOLIAR NITROGEN LEVELS AND THEIR EFFECTS ON INSECT LIFE CYCLES

Seasonal changes in plant nitrogen levels have had a marked effect on insect life cycles. This is particularly true in the case of sap-feeding insects, which, feeding as they do on a dilute solution, have nitrogen acquisition as one of their major objectives. With nitrogen at a premium, it is in sap-feeding insects that the importance of host quality to migratory and reproductive strategies may be most easily demonstrated. The soluble nitrogen content of plant tissues varies throughout the year.[53,54,96,225] In the spring in temperate climates, when the leaves are growing rapidly, foliar nitrogen content is high. This falls away during the summer months as the leaves mature and rises again (at least in the phloem) in the autumn as the leaves begin to senesce and nitrogen reserves are mobilized within the plant in preparation for leaf fall.

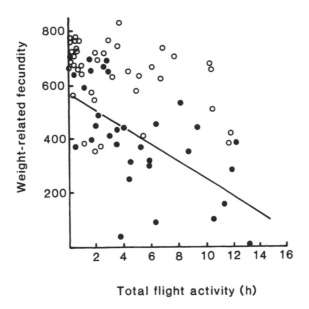

FIGURE 4. Relationship between weight-related fecundity (eggs/100 mg pharate adult weight) and total flight activity in unfed (closed circles) and fed (open circles) female Spodoptera exempta. (From Gunn, A., Gatehouse, A. G., and Woodrow, K. P., *Physiol. Entomol.*, 14, 419, 1989. With permission.)

This pattern of events has affected the life history strategies of a number of tree dwelling aphids. For example, the bird cherry-oat aphid, *R. padi*, a host-alternating aphid, feeds and reproduces on its primary host, *P. padus,* while soluble nitrogen levels are high in the spring.[54] As soluble nitrogen levels fall, winged forms are produced in response to the deterioration in host quality, and these migrate to their secondary graminaceous hosts, where host quality is higher.[123] In autumn, as host quality on these grass hosts declines, they return to their primary host, whose quality has once again improved.[54,119] The aphid *Uroleucon gravicorne*, although not strictly a host alternating aphid, usually moves from its overwintering host *Solidago* spp., to plants of several annual *Erigeron* species. Like *R. padi,* it responds to changes in soluble nitrogen levels of its host plant, and when nitrogen levels are low, reproductive performance is also low. Soluble nitrogen content declines throughout the season on *Erigeron* spp., and growth and development rates, as well as the fecundity of *U. gravicorne,* decrease at the same time.[152]

The sycamore aphid, *Drepanosiphum platanoidis*, has also to face the same deterioration in host quality as that faced by *R. padi*. However, instead of migrating to a plant of a higher host quality status, *D. platanoidis* reduces its metabolic rate and goes into a summer aestivation until the host quality of its food plant once again becomes suitable for growth and reproduction.[51,53] Yet another tree dwelling aphid, the birch aphid, *Euceraphis punctipennis*, has solved this same problem by exploiting the fact that birch trees (*Betula* spp.) have a certain number of shoot tips that produce new leaves throughout the season. These are high in soluble nitrogen content, and the adult birch aphids track these growing points throughout the season, moving within the tree and even between trees when necessary so as to deposit their offspring on suitable food material.[236] Aphid species on annual hosts are also greatly affected by the changes in soluble nitrogen concentration which occur as their hosts age. The cabbage aphid, *Brevicoryne brassicae*, shows a steady decline in growth rate as their Brussels sprout host plants age.[210,211]

Not all sap feeders show this same dependence on young flushing tissue or nitrogen-rich senescent tissue. However, in all cases, a nitrogen-rich feeding site is required. Even within the aphid species already discussed, some differences in reproductive strategy are shown. The aphid *R. padi* is frequently found on older cereal plants in the summer months but is habitually found on the basal parts of the stem where its growth and reproductive rates are highest.[121] This is also the area of the mature cereal plant where nitrogen levels are high.[92]

In the case of other sap suckers such as the delphacid, *P. marginata*, which feeds on the same plant species (*Spartina* spp.) for the whole year, a response is still seen to soluble nitrogen levels, the insects moving to streamside vegetation, which is of a higher host quality, in the spring and early summer and returning to overwinter on the nutritionally less suitable marshland *Spartina*, where greater protection is afforded.[44] Thus, like the birch aphid, these insects track their plant resources. The psyllids, *Psylla peregrina* and *P. subferruginea*, also show this ability to track nitrogen-rich resources. Early in the season, they are closely associated with the expanding buds of their host, *Crataegus monogyna*. After bud burst is completed, the only sources of high nitrogen flush are in the growing shoots and inflorescences, and it is on these points that the psyllids aggregate.[196] Insects that feed throughout the season on one portion of their host plants, such as the membracid, *Publilia reticulata* on *Veronia noveboracensis*, must either increase their feeding rate to maintain the same level of activity, go into summer estivation, or restrict their feeding to that portion of the leaf that affords the best nutrition. In the case of *P. reticulata*, feeding is confined to the major cross-veins, where sap flow is brisk; this enables the insect to maintain a high reproductive rate.[26] Other membracids with relatively long generation times have hatching times that permit nymphal development to take place at those times of the year when soluble nitrogen levels are high. The oak tree hopper, *Platycotis vitata*, has two generations a year, one in spring and the other in autumn, while adult females diapause in the winter and summer, depending on their generation, and the eggs are laid in spring and autumn, respectively, to take advantage of the flushes in soluble nitrogen in their host trees.[100]

Among chewing insects, a large number of species have adopted flush feeding strategies to achieve high growth rates, although of course they may then have to overcome high levels of secondary metabolites. The winter moth, *Operophtera brumata*, feeding on oak, is a bud feeder.[69] Others begin feeding on flush foliage and then move on to older, more abundant tissues once they have successfully survived the early, highly vulnerable establishment phase. For example, newly hatched larvae of the pine beauty moth, *P. flammea*, feed on young newly expanding needles but then move on to the previous year's growth to complete their development,[224] only returning to the current year's foliage in the late summer if no older needles are available.[224a]

With this type of feeding strategy, it is expected that the adult females would be able to distinguish between host plants of differing suitability. The pine beauty moth certainly selects its host plant in a way that maximizes larval growth and survival. The cues used involve the monoterpene profiles of the plants, more eggs being laid on plants with a high β- to α-pinene ratio than on plants with a low ratio.[113,116] There is no evidence available yet as to oviposition preferences relative to phenology in this case, as the eggs are laid when the plants are still dormant. Adults of the cinnabar moth, *Tyria jacobaeae*, which oviposit on growing plants of tansy ragwort, *Senecio jacobaea*, lay more eggs on individual plants with high nitrogen contents.[207] The noctuid moth, *Helicoverpa armigera*, uses cues such as flowering status of the various host plants to determine its oviposition choices. Flower structures are more suitable for larval growth and survival than vegetative structures, due to their high nitrogen content and sometimes, also, low levels of alkaloids and other secondary compounds.[70]

V. TRADE-OFFS BETWEEN MIGRATION AND REPRODUCTION IN RELATION TO HOST QUALITY

As discussed earlier, migration, whether it be by active flight, ballooning, or walking, must impose a cost on reproduction, if only because of the energy used for this purpose rather than for offspring production.

It is interesting to note that in many insects the decision to invest in a migratory form is usually caused by a deterioration in host quality. For example, the light brown apple moth, *Epiphyas postvittana*, produces smaller adults when the food available to the larvae is suboptimal. These smaller insects have a significantly lower wing-loading than large individuals, and are more adapted to dispersal and presumably to the location of additional suitable hosts for their offspring.[39,40] This is seen in a number of other lepidopteran species. The velvetbean caterpillar, *Anticarsia gemmatalis*, if deprived of food for any period during its larval development, produces adult moths with a low wing-loading and, consequently, greater powers of dispersal. The fall armyworm, *Spodoptera frugiperda*, the corn earworm, *Helicoverpa zea*, and the armyworm, *S. latifascia*, also show this response to larval starvation.[4]

One of the more obvious responses to changes in host quality evinced by certain species of insects is the switch from apterous nonwinged forms to winged forms. As well as being influenced by changes in photoperiod and temperature, wing formation in insects is also influenced by changes in food quality and by rearing density. Crowding during development in nongregarious species is usually an indicator that host quality is likely to decline in the near future, and the production of a winged dispersive form in these circumstances is adaptive. This change to the production of winged forms is common in many species of Homoptera. For example, many planthoppers, including *P. marginata*, produce winged forms in response to nymphal crowding, as do many species of aphids.[43,59] A reasonable hypothesis would be that insects capable of dispersal respond to a deterioration in host quality by producing either a specifically adapted morph or by utilizing a set of behavioral patterns to escape adverse conditions. This, then, would maximize reproductive potential while operating under the constraints of crowding or poor food. To illustrate this hypothesis, a few well documented examples from the literature will be used.

In aphids, where the possession of wings is the norm, migratory flight is taken when host quality deteriorates locally. For example, the sycamore aphid, *D. platanoidis*, responds to crowded conditions during development by leaving its host plant with little or no reproduction taking place, preferring to reproduce on an unexploited resource.[52] In those aphids that show alary polymorphism, the winged morphs also colonize habitats of a higher host quality than those on which they were reared. On arrival on the new host plant, they produce apterous offspring which are able to exploit the new habitat effectively.[227] These apterae are also able to detect and anticipate deterioration in their habitat through cues, such as crowding and lowered foliage nitrogen levels, and in turn produce alatae, able to fly to other hosts.[61]

Host location is, however, a risky process, and although the quality of their present host plant may be deteriorating, it will not always become totally unsuitable for reproduction and feeding. Thus, it is not to the advantage of the species for all of the alatae to disperse from the current host plant. A flexible response should be an advantage. This is certainly the case within some aphids. Winged *R. padi* and *S. avenae* are produced with a variable number of ovarioles; that is, the reproductive investment of the offspring of an apterous mother on a host plant of deteriorating quality is not a constant.[130] The reproductive investment of the aphids is linked with their migratory strategies. Alate aphids with a small number of ovarioles take off more readily, do so at a steeper angle, and delay wing muscle autolysis for longer than those aphids with a greater number of ovarioles. In addition, those aphids with a lower number of ovarioles possess more chemosensory organs and are thus better able to locate host plants. Thus, those individuals with a low reproductive investment are better suited to long distance dispersal than those with a greater reproductive investment.[215]

In addition to these differences, aphids with smaller reproductive investments are able to withstand starvation for longer periods of time than those with a large reproductive investment.[215] In addition, aphids with a large number of ovarioles (i.e., greater reproductive investment) are more likely to die at an early stage.[219] The number of individuals within each ovariole class produced by the mother appears to be adjusted so as to maximize the fitness of the offspring. Thus, on poor quality hosts where survival of apterous forms is likely to be low, a higher proportion of those that are born have a low number of ovarioles when compared to those born on high quality hosts.[216] Alate aphids landing on high quality hosts produce proportionately more offspring with a large number of ovarioles than do the wingless apterae. In addition, when aphids autolyze their wing muscles, their reproductive output become similar to that of apterous morphs.[156]

When apterous aphids come under the influence of nutritional stress, they are less able to avoid the effects than are winged forms. They can disperse to a certain extent by walking, but this is likely to take some time, and the aphid is likely to undergo a period of deprivation. The black bean aphid, *Aphis fabae*, can survive starvation as an adult for up to 3 d. This was shown to reduce lifetime fecundity and longevity. However, to compensate for this potential loss of fitness, the adult aphids increased their rate of reproduction above normal, when favorable host conditions became available.[126]

After being produced in response to host adversity, active winged insects migrate in search of new habitats. Not all the plants selected by migrants are of the same quality, and the postalighting response of a migrating insect is influenced by features of the host it has landed on. Characteristically, a complex suite of responses is found in herbivorous insects (Figure 5). For example, prealighting behavior often depends on different cues than oviposition behavior. The butterflies *Eurema brigitta* and *E. herla*, which are monophagous on *Cassia mimosoides*, alight on plants with similar leaf shapes and size and color. They then use chemical and textural cues to confirm their choice of host, moving on if the oviposition stimulants are not received.[139] Similar or more complex patterns occur in other plant-feeding insects. In aphids, plants are probed and sampled; the decision to remain is a result of the balance between food plant stimuli, resources remaining for migration, and those available for reproduction.[22,209] This is itself strongly influenced by the availability of wing muscles that can be autolyzed after settling, as is common in many species. In the aphid *R. padi*, alates that land on poor quality hosts are relatively less likely to autolyze their wing muscles, thus giving themselves the option to produce a few offspring and still make a further migratory flight.[129] It is noticeable that the emigrants of this species, which are produced in early summer when habitat quality is generally higher, are less variable in terms of ovariole number and do not show this adaptation.[129]

The seasonal adaptations of aphids to their expected environment are best demonstrated by looking at two studies involving the bird cherry aphid, *R. padi*. The dispersal and reproductive strategies of the three winged morphs can be compared. These are 1) the emigrants (arising from the primary host and migrating to the secondary host in spring), 2) the alate exules (migrating between secondary hosts during the summer), and 3) the gynoparae (arising from the secondary host and migrating to the primary host in autumn).[57,129] Several dispersal and reproductive characteristics were examined (Table 2). The emigrants, which migrate into a high quality habitat, that is, one with plenty of nitrogen-rich, rapidly growing graminaceous hosts, have a high reproductive investment with most of their embryos in an advanced stage of development and wing muscles that tend to autolyze sooner, rather than later. The alate exules, on the other hand, which are produced in times of adversity and are likely to be surrounded by poor quality hosts, have a low reproductive investment and wing muscles that are not programmed to autolyze quickly. They are thus able to delay wing muscle autolysis if the host on which they alight is of a poorer quality than that from which they emigrated. The environment to which they migrate is unpredictable, and they are adapted to this unpredictability. The gynoparae, on the other hand, which are migrating to a less common host (of good but temporally limited quality due to the imminence of leaf fall), have a low fecundity but produce their offspring very quickly once they have settled. Their fat reserves are high, which is appropriate, as they may take some time to locate a suitable host. Each morph appears to be well adapted to the expected environment.

Although aphids are so versatile, they are not the only insects that can respond to host quality in such a flexible manner. The pine beauty moth, *P. flammea*, is also able to assess host

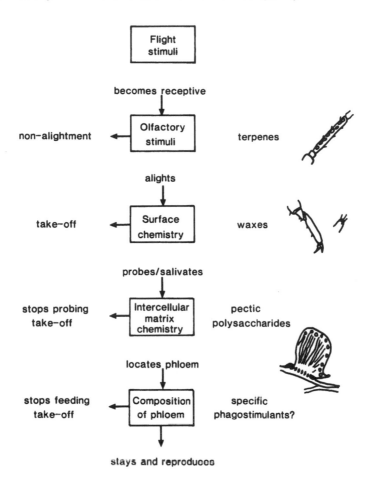

FIGURE 5. The sequence of potential activities of the alate aphid on possible host plants during the selection process. (From Blackman, R. L., *Aphids,* Ginn & Company, London, 1974, chap. 10. With permission.)

quality by monitoring the monoterpene composition of the plant foliage and will avoid poor quality hosts if possible.[113,116] If choice of host plant is restricted, then the female moths will eventually oviposit on the low quality hosts, but only after a prolonged pre-reproductive delay.[127]

Among butterflies, if the reproductive cost of dispersal is low, then many species will move between host patches so as to maximize offspring fitness. This has been shown in *Pieris rapae*, which, after initiating oviposition on a suitable host plant, will eventually leave for another host patch when its supply of chorionated eggs is exhausted.[34] Habitat quality, the relative proportion of suitable host plants, also affects dispersal. The checkerspot butterfly, *Euphydryas editha*, for example, will leave patches with few or none of its preferred host, *Pedicularis semibarbata*, but will remain in those patches where *P. semibarbata* is abundant.[200a]

Thus, herbivorous insects respond to detrimental changes in the quality of their host plants in different ways:

- They may produce forms especially adapted to dispersal;
- They may produce individuals able to survive adverse conditions;
- They may have genetic variation for movement patterns,
- They may, in extreme cases (such as the sycamore aphid, *D. platanoidis*), undertake a temporal migration for a summer estivation, where the insect becomes dormant and has no need to feed on the poor quality host.[51,198]

TABLE 2
Differences between the Three Winged Morphs of the Bird Cherry Aphid *Rhopalosiphum padi*

Characteristic	Emigrant	Alate exules	Gynoparae
Fecundity	71	51	14
Length of adult life	Medium	Long	Short
Length of reproductive life	Medium	Long	Short
Weight	Heavy	Light	Medium
Number of offspring born in first 2 d	Some	Few	Many
Number of offspring born in first 4 d	Few	Few	Few
No. well-developed embryos	Many	Few	Some
Total fecundity as well-developed embryos in teneral adults	Medium	Medium	Low
Range in number of ovarioles	×1.8	×2.5	×1.6
Average reproductive investment (mean ovariole number × mature embryos/ovariole)	13.3	5.3	7.1
Weight of offspring	Medium	Medium	Light
Ovulation after adult moult	Light	Some	None
Fat content	Low	Medium	Medium

From Dixon, A. F. G., *J. Anim. Ecol.*, 45, 1976. With permission.
From Leather, S. R., *Entomol. Exp Appl.*, 31, 386, 1982. With permission.
From Leather, S. R., Wellings, P. W., and Walters, K. F. A., *J. Nat. Hist.*, 22, 381, 1988. With permission.

After dispersal has occurred, the insect is faced with another problem, the exploitation of the host to which it has migrated.

VI. OVIPOSITION, LARVIPOSITION, AND FECUNDITY

Once a reproducing female insect has arrived at its new host, it must deposit its young, be they immobile eggs or mobile larvae. A number of decisions must be taken at this stage, and these are affected by the plant, not just in terms of its suitability as a host, but as to its attractiveness to the female insect. Although it is in the interest of the insect to produce as many offspring as possible in as short a time as possible, its responses are mediated by the quality of the host plant. The relationship between oviposition preference in insect herbivores and their offspring performance is recorded to range from good to poor, although most evidence points to a good correlation.[201,202] The reasons that this relationship is not as good as human observers would expect are probably linked to how preference and performance are related genetically; this has been fully reviewed elsewhere.[201] It must be added that some authors feel that one of the reasons why the fit is not as good as expected is that not all the attributes of offspring fitness are being measured.[185] The concept of perceived host quality has been discussed earlier but is exemplified by the oviposition behavior of the two chrysomelid beetles, *Phratora vitellinae* and *Galerucella lineola,* which both feed on *Salix* spp. *G. lineola* lays its eggs on species of *Salix,* which are low in salicylates and on which the larvae are able to grow and develop better than if they were feeding on salicylate-rich plants. *P. vitellinae,* on the other hand, lays its eggs on species high in salicylates, on which its larvae are at a disadvantage in terms of growth and development. However, the larvae of *P.vitellinae* have a better predator defense, as they produce a secretion when attacked that is high in salicylates, whereas larvae of *G. lineola* and *P. vitellinae,* feeding on hosts with low concentrations of salicylates, are virtually defenseless.[46] It is also felt by most authors that, although the correlation between oviposition preference and offspring performance is not perfect, the majority of insects requiring specialist hosts can depend on plant chemistry in order to select the most suitable host for the development and survival of their offspring.[30,176]

A. DETERRENTS AND ATTRACTANTS

It is likely that one of the major ways in which an insect assesses the host quality of a particular plant is by the volatile compounds present in the tissues of that plant. These compounds can act as either deterrents or attractants, or indeed as oviposition stimulants. It would be expected that those compounds acting as deterrents would be correlated with poor herbivore performance and that those acting as attractants would be postively correlated with herbivore performance. Evidence is mounting, however, that deterrents are not always correlated with negative postingestion effects, as would logically be expected.[17a] Instead, it has been suggested that avoidance responses to these chemicals have evolved under selection pressures that are not related to herbivore performance in respect to the deterrent.[17b] Insects with a wide host range appear to be markedly affected by some of these compounds, e.g., iridoid glycosides, which often have strong deterrent effects.[173]

The situation with those compounds that act as oviposition stimulants and/or attractants is somewhat more clear-cut. It is possible that certain volatile chemicals, although not directly acting as nutrition sources, are correlated with the presence or absence of true nutrients, such as protein and amino acids.

For example, the budworm, *C. fumiferana,* is stimulated to oviposit on host plants by the presence of the monoterpenes α- and β-pinene,[191] as is the pine beauty moth, *P. flammea,* which lays more eggs on pine plants with high β- to α-pinene ratios.[113,116] The pyralid moth, *Dioryctria amatella,* lays more eggs on pine trees containing high levels of monoterpenes.[68] The cinnabar moth, *Tyria jacobaeae,* on the other hand, lays its eggs on plants high in organic nitrogen and sugars, once it has become oriented on a host plant, using the secondary compounds as primary stimulants.[207] The butterfly *Pieris rapae* lays its eggs on tall plants high in nitrogen, although the primary stimulants are various glucosinolates,[137] and the monarch butterfly is primarily stimulated to oviposit by the presence of cardenolide glucosides.[159]

B. OFFSPRING FITNESS

Offspring fitness can be measured in a number of ways, but generally, high larval growth rates, rapid larval development, and high larval survival, leading to a large fecund adult, are indicative of a good quality host plant.

1. Mother Knows Best

There is some debate about whether egg distribution on a single host plant indicates sites where offspring fitness is enhanced. A recent review has appeared to suggest that ovipositing insects are relatively poor at discerning suitable host plants for their offspring,[34a] and this is supported to some extent by other authors, such as Thompson.[201] However, the real area of debate is about whether plant chemistry is the major determinant of host plant selection and offspring fitness. The suggestion that deterrents may have little to do with offspring fitness is indeed a tenable hypothesis,[17a] but chemical attractants do appear to be more positively correlated with offspring fitness than are the deterrents.[113,114]

It has been suggested that, rather than host plant chemistry (= quality) being the criterion for host plant selection, other abiotic factors may be the determining measure.[34a] For example, Moore, Myers, and Eng, working with the western tent caterpillar, *Malacosoma californicum pluviale,* suggested that microclimate was more important than host quality.[150] They questioned the idea that egg distribution reflected host quality variation within the host tree, as they found that eggs were laid mostly on the sunny side of the tree. From this they suggested that it was not host chemistry that determined oviposition site but instead temperature. However, the same piece of work indicated that leaves on the sunny side of the tree had higher nitrogen

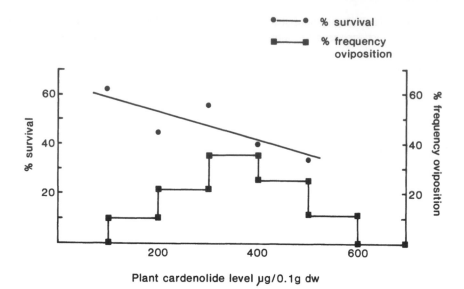

FIGURE 6. Relationship between mean oviposition frequency and early larval mortality of *Danaus plexippus* and the cardenolide content (μg/0.1 g dry weight) of *Asclepias* spp. plants. (From Oyeyele, S. O. and Zalucki, M. P., *Ecol. Entomol.*, 15, 177, 1990. With permission.)

levels and that the male pupae produced from those leaves were heavier than those reared on leaves from the shady side of the tree. It is thus possible that the female moths were somehow responding to the nutrients in the leaves rather than to the microclimate.

Some insects show a relatively straightforward relation between host quality and host plant selection; for example, adults of the leaf folding sawfly, *Phyllocolpa* spp., lay their eggs on those *Salix lasiolepis* bushes on which their offspring will grow and develop fastest.[76] In other insects, the relationship between the stimuli attracting the insect to oviposit and the suitability of the host for offspring growth and survival is not so apparent. An excellent example of this type of relationship between oviposition preference and host suitability is shown by the monarch butterfly, *D. plexippus*. As mentioned earlier, adult females are stimulated to oviposit by cardenolides in their milkweed host plants. However, the larvae are adversely affected by very high levels of cardenolides in individual *Asclepias* spp. Females lay their eggs on plants with intermediate cardenolide content, which is also most suitable for the survival rate of their larvae (Figure 6).[240] Many insects appear to be capable of making quite subtle choices of host in relation to plant chemistry. The chrysomelid beetle, *Paropsis atomaria*, for example, has larvae that perform best in terms of growth and survival on *Eucalyptus* hosts that are high in nitrogen, irrespective of the content of essential oils, and the adult beetles choose oviposition sites accordingly.[154]

However, sometimes the ovipositing insect does indeed appear to be unable to distinguish the host plant of higher quality. For example, larvae of the pine beauty moth, *P. flammea*, grow, develop, and survive at higher rates on pine trees that are water stressed.[223] The adult moths, however, although able to distinguish among pine trees of differing suitability because of genetic differences or because they have previously suffered insect attack,[113,128] are unable to distinguish among plants that are of the same genetic material but are in different states of stress.[223] This may, however, be because the host plant stimuli, on which *P. flammea* relies to determine the quality of its host (monoterpenes), are unaffected by water stress. Further work in this area is required. On the other hand, some insects are more adept when faced with this situation. There is, for example, a strong relationship between oviposition preference and larval performance in the shoot galling sawfly, *Euura lasiolepis*.[36] In this case, however, plants that are under water stress are not suitable hosts for the larvae of *E. lasiolepis*, and the adult females are able to distinguish between stressed and unstressed plants to avoid those plants suffering from water stress.[169] This may be the result of the fact that the eggs of *E. lasiolepis* are laid in greater intimacy with the host plant (inside the tissue) than those of *P. flammea*, which are laid on the needles. It is likely that *E. lasiolepis* can use its ovipositor to detect stimuli relating to host plant quality.

Plants may appear suitable for offspring in one respect, such as high nitrogen levels, but may have secondary disadvantages. For example, host quality can be reduced by the activities of other insects. The weevil *Cyrtobagous salviniae* and the moth *Samea multiplicalis* both feed on the aquatic plant *Salvinia molesta*. As with *S. multiplicalis*, *C. salviniae* prefers to oviposit on buds high in nitrogen. However, it avoids plants that have been damaged, especially if they have been damaged by *S. multiplicalis*. The weevils also select their host plants by assessing the number of buds available, so as to maximize the potential food supply of their offspring.[74,200] Other insects also show similar sensitivities. The stored product pest, *Callosobruchus maculatus*, which feeds on cowpea, lays its eggs in the field before harvest. It prefers younger, smaller pods as oviposition sites, and these are indeed more suitable for the larvae.[146] In addition, it is able to determine whether another beetle has already laid its eggs on a pod and to assess the number of eggs already present. The adult females avoid laying eggs on pods with a high egg load thus ensuring that larval competition is reduced. Similarly, the moth *Cactoblastis cactorum* lays its eggs on those *Opuntia* plants that its larvae grow well on, and unlike some insects, lays more eggs on previously attacked plants than on unattacked plants. In this case, signs of a successful previous attack may be an indicator of a high quality host plant.[154a]

The experience that an insect has had also influences its perception of host plant quality. In a choice test, the leaf-mining fly, *Liriomyza trifolii*, preferred tomato plants that had a high foliar nitrogen content, on which their larvae survived and developed better, than plants that had low foliar nitrogen. However, flies with no experience of high quality hosts (those high in nitrogen), showed no preference at first. After exposure to a high quality plant, however, a strong preference for high quality plants was seen. It was postulated that, in nature, *L. trifolii* used plants that maintain or increase in acceptability and dispersed relatively quickly from areas containing plants of low acceptability (and quality). As conditions worsened and host deprivation increased, nutritionally subthreshold plants once again became acceptable as host plants.[147]

2. Host Plant Phenology

Plant phenology also affects insect oviposition strategies. The tephritid flies, *Tephritis bardanae* and *Cerajocera tussilaginis*, time their oviposition so that the establishment and rapid growth of their early instar larvae coincide with the two main periods of physiological activity of their host, the development of the flower heads of the thistle *Arctium minus*. *Tephritis bardanae* lays its eggs in the flower heads at flowering and *C. tussilaginis* at achene maturation. At both these times, nutrient availability is at a maximum, and the young larvae position themselves in those structures where the greatest nutrient flux occurs.[194] Another tephritid, *Eurosta solidaginis*, samples the ramets of its host, *Solidago altissima*, and lays its eggs on those hosts at the phenological stage most suitable for the growth of its larvae.[3]

Leaf miners, whose offspring are commonly confined to one leaf for the whole of their development, are presumably selected for extreme sensitivity to leaf quality, since a poor selection means very high mortality. In addition to the usual details of host selection, they may have to deal with the fact that plants often show a defensive response to attack by leaf miners by shedding leaves earlier than normal.[171,172] For these reasons, as well as nutritional ones, the age of the leaf is very important in oviposition choice. *Lithocolletis quercus*, a miner of oak leaves, selects its leaves in a way that appears to minimize the chance of early death for its larvae.[8] Other insects are constrained in their choice of host plant due to leaf unavailability. Those that feed on deciduous trees in the autumn are in a particularly difficult position. For example, the aphid *Periphyllus californiensis* only colonizes those maple trees with orange-yellow leaves. Trees with red leaves are avoided. Orange-yellow leaves remain on the tree longer, which ensures that the aphids are able to complete their development before leaf fall.[77] The bird cherry aphid, *R. padi*, which colonizes the bird cherry tree, *Prunus padus,* in autumn, shows no preference for different ages of leaves.[108] However, it times the production of the morphs that colonize trees in the autumn in response to the temperature it experiences in July. This cue is, in fact, an excellent predictor of leaf fall.[218] In addition, although these autumn morphs do not feed as adults,[109] they are able to distinguish the relative suitability of their tree hosts for their offspring's offspring. If they are able to leave, they do so, but if they are unable to leave due to wing muscle autolysis, they produce offspring that lay fewer eggs on that particular tree.[114]

The tropical satyrid butterfly, *Mycalesis perseus*, oviposits on plants high in nitrogen but only chooses the youngest leaves on that plant as oviposition sites. It inhabits savannah grassland, which is prone to premature drying, and it thus chooses those grasses that are likely to live longest. This gives its offspring the maximum possible time in which to develop.[149] A similar scenario is seen in the oviposition strategy of the butterfly *Euphydryas chalcedona*. In laboratory trials, the females prefer to lay their eggs on the shrub *Scrophularia californica*. This is nutritionally superior to another shrub, *Diplacus auranticus*, which is also used as a host plant by *E. chalcedona*. However, *D. auranticus* is more drought-tolerant than *S. californica,* and in the wild, *E. chalcedona* will lay a sizable proportion of its eggs on the former shrub. Although *D. auranticus* is nutritionally inferior, its drought-tolerance means that it is more persistent in dry years. Thus, host use reflects a trade-off between nutritional quality and resource availability (persistence).[234]

Once the insect has made the decision to produce its offspring on a particular host plant, then a number of reproductive options are still open to it, and these are greatly dependent on the quality of that host plant.

C. FECUNDITY

Host quality can affect the fecundity of an herbivorous insect in two ways; it can affect the fecundity of the adult arising from the larvae feeding on it, or it can actually affect the fecundity of the adult insect meeting that individual plant for the first time. The latter case is almost certainly just as common as the former.

One aspect of food quality, the level of nutrient nitrogen, is known to be important in several species. For example, the host quality of the larval food plant affects the fecundity of the adult gypsy moth, *Lymantria dispar*. Those adults arising from hosts with low nitrogen levels are less fecund than those reared on high nitrogen level hosts.[93] Hosts plants that are high in nitrogen result in more fecund individuals of the armyworm, *Spodoptera frugiperda*.[163] Spruce budworm larvae that are reared on old, nitrogen-poor foliage develop into adults of reduced fecundity when compared to those arising from larvae reared on young, nitrogen-rich foliage.[23]

When reared on *Holcus lanatus* plants with artificially enhanced nitrogen levels, the leafhoppers *Dicranotropis hamata, Elymana sulphurella, Encelis incisus* and *Zyginidia saitellaris* were all more fecund as adults than those reared on plants with low levels of nitrogen.[168] The collembolan, *Folsomia candida* increases its production of eggs when its host is rich in nitrogen.[206]

The fecundity of the spear-marked black moth, *Rheumaptera hastata*, which feeds on *Pinus nigra*, is affected in two ways by the quality of its host plant. On hosts that have previously been attacked by conspecifics, the sex ratio is skewed in favor of males, and those females that are produced are less fecund than those produced on undamaged host plants.[231] In this case,

FIGURE 7. Relationship between preference index of *Panolis flammea* adults and suitability of the tree for growth of their larvae. (From Leather, S. R., *Ecol. Entomol.*, 10, 213, 1985. With permission.)

damaged plants are of poorer quality than undamaged plants. The fecundity of the pine processionary moth, *Thaumetopoea pityocampa*, is also affected by host plant variation in this way.[13]

The aphid *Sitobion avenae* is more fecund on the ears of cereal plants than on other parts of the plant, and that is where it is most often found.[222] The bird cherry aphid, *R. padi*, is more fecund on young grass and cereal plants than on mature plants, and on mature plants it is most fecund on the stem; this, too, is where it is most often found.[54,122] The cabbage aphid, *Brevicoryne brassicae*, is also strongly affected by host quality, being apparently most fecund on leaves high in nitrogen.[175,210] As herbaceous host plants age, their quality as a food source, as measured by aphid performance in terms of growth and fecundity, shows a decline. This is seen in several species; for example, the asparagus aphid, *Brachycorynella asparagi*, is less fecund on old leaves than it is on young ones.[238] The flower thrips, *Frankliniella occidentalis*, also show a variable fecundity depending on host growth stage.[204]

Plant secondary compounds are themselves also important in some instances. The fecundity of the scale insect *Fiorinia externa* is increased by terpenoid alcohols, as is that of another scale insect, *Nuculaspis tsugae*, both of which feed on species of *Tsuga*.[138] Whereas the fecundity of *F. externa* is negatively correlated with acyclic terpenoids, that of *N. tsugae* is positively correlated with the same compounds.[139] The pine beauty moth, *P. flammea*, is greatly affected by the quality of the pine trees it encounters. Like many species of Lepidoptera, it is able to detect differences among hosts (in this case monoterpene content) and to select those individual plants on which their offspring have a greater fitness in terms of growth and fecundity when they become adults (Figure 7).[127]

The aphid *Metopolophium dirhodum*, in times of nutrient stress, reduces embryo growth and resorbs the youngest embryos, which safeguards reproduction until conditions improve.[81] The reverse process occurs when conditions improve. Host quality greatly affects the fecundity of adult aphids, both in terms of what happens to adults arriving on a novel host plant and in terms of what they were reared upon. An aphid reared on a poor quality host plant but feeding on a good quality host plant will be more fecund than it would have been if it had remained on its original host plant. This has been demonstrated with *R. padi* transferred from oats to *Phleum pratense*.[107,118] This is further complicated by the morph of the mother (whether winged or not). The effect of maternal experience is also seen in Lepidoptera. Larvae of the small ermine moth, *Yponomeuta evonymellus*, do better when transferred as neonate larvae from poor host plants to good host plants, but do not develop as well as those larvae that have been transferred from good quality plants to similarly good quality plants. The reverse is also true; larvae transferred from good quality plants to poor quality plants still develop faster than those larvae reared on poor quality plants for their entire developmental period.[124]

Thus, host quality affects fecundity of adult herbivorous insects both by affecting larval nutrition but also by acting on the physiological and behavioral patterns of the adult reproductive female.

D. CLUTCH SIZE

Clutch size, the number of eggs or offspring produced in a reproductive event, has attracted a great deal of attention from ecologists. The general conclusion among vertebrate zoologists is that large clutches are associated with short-lived, fast developing animals exhibiting little parental care, which leads to increased offspring mortality. On the other hand, small clutches are thought to be characteristic of long-lived, slow maturing animals with a high investment in parental care, which leads to high offspring survival.[203] Clutch size in birds is often closely related to prey–food availability; in years following good food conditions, clutch size tends to be higher.[164] If clutch size is high and the food conditions experienced by the adults are poor, then there will be an increased mortality among the hatchlings since, when more eggs are laid by the bird, each hatchling will be smaller, and its chances of surviving will be decreased.[165] The evolution of insect clutch size has undoubtedly been shaped by similar factors; this assumption is apparent in the literature.

Clutch size in insects represents a trade-off between fitness and dispersal. A detailed review of the costs and benefits of varying clutch sizes in insects is presented by Godfray.[80] However, some of the basic assumptions behind the current models are worth considering.

Clutch size is related to a number of factors: the frequency of ovi- or larviposition, the size of the mother, the number of eggs/offspring produced, and, most importantly, in the context of this review, the quality of the host. Several authors have addressed the question of clutch size, using models,[142,160,161] and their conclusions are in broad agreement with the conclusions reached by the vertebrate zoologists cited above. One difference is that some authors believe that the major factor determining clutch size is not offspring success but the need to achieve maximum egg or larval deposition.[161] This does not contradict the theory that host quality is a major factor in determining clutch size, as it has been shown that host quality affects all the factors listed above.[120]

Offspring production is not a constant over the life of the insect, and the patterns of clutch size variation are, to a certain extent, determined by this fact. In nature, female insects do not generally achieve their maximum life-span,[117] and the offspring produced in the early days of reproductive life are therefore of greater value than those produced later, and thus, in some species, those represented by development of eggs in the upper reproductive tract. It would be expected, then, that most female insects would have a large initial burst of reproduction and that their reproductive rate would then fall away gradually with increasing age. In fact, female insects usually show a pattern of reproduction that starts at a low level as the female first begins to reproduce, builds to a peak (the time taken depending on the probable life-span of the insect), and then drops off to a low level again. This pattern is seen in long-lived Lepidoptera, like the monarch butterfly, *Danaus plexippus*,[239] in relatively short-lived Lepidoptera, such as the pine beauty moth, *P. flammea*,[120] and also in short-lived aphids and bugs, such as *A. fabae* and *Oncopeltus fasciatus*.[63,186,199] Some insects with a high probability of early death lay most of their eggs either in one clutch or on the first night of oviposition. An example of this strategy is seen in *Chilo partellus*, which lays most of its eggs on its first night of oviposition, with clutch size becoming smaller each succeeding night.[17] Some insects with relatively immobile females are constrained to lay their eggs in one large reproductive burst. For example, the gypsy moth, *L. dispar*,[148] and the bird cherry ermine moth, *Y. evonymellus*,[115] both lay their egg clutches in one large reproductive event. Such insects adjust their reproductive effort in other ways and must also show specific adaptations to avoid predation on the large egg masses.[192] Female insects that lay their eggs in such large clusters tend to have a high realized fecundity.[33] This may compensate for the risks entailed in laying all eggs on the same host rather than spreading the risk by moving from one host to another and for the increased probability of death for the mother before oviposition is completed.

How does host quality affect clutch size? If offspring produced earlier in the reproductive life of an insect are of greater value, then, on a good host, it would be expected that peak offspring production would be reached sooner rather than later, as it is to the advantage of the female to deposit its offspring on a good quality host. On a poor quality host, the peak should be achieved later, as the insect must trade-off the risks of waiting for an improvement in host quality for an investment in dispersal to seek out a better quality host. Some insects are not as flexible in their strategies as others. For example, in the flies *Dacus jarvisi* and *D. tryoni*, clutch size is related to food availability. *Dacus jarvisi* normally feeds on large fruit and lays large egg clutches, whereas *D. tryoni* normally feeds on small fruit and consequently lays only a few eggs on each fruit. When presented with the fruit of the opposite size class, they are unable to change their strategies.[73] The pine beauty moth, *P. flammea,* on the other hand, when ovipositing on a good host plant, has a short prereproductive delay, lays a greater proportion of its eggs in the first 3 d of reproduction, and lives longer than when ovipositing on a poor quality host.[127] The pipevine butterfly, *Battus philenor,* is also able to adjust its clutch size in response to differences in host quality, more eggs per clutch being deposited when host quality is good.[166,167]

Thus, as predicted, herbivorous insects tend to lay large clutches early in their reproductive life when host quality is high, which will maximize both female reproductive success and offspring fitness. On poor quality hosts, clutches are small and are produced at a slower rate than on good quality hosts. In general, clutch size is adjusted so as to maximize reproductive effort, which can be done by adjusting the size of the offspring produced.[120]

E. SIZE OF OFFSPRING

One of the ways in which an insect can manipulate its reproductive investment is by having different sizes of offspring. This is generally done in relation to host or habitat quality. Theoretically, the more investment there is in reproductive effort per egg or larva (e.g., by having parental care or large offspring), the greater the likelihood of survival of those offspring.[28] Models show that the more eggs produced per clutch, the smaller those eggs will be.[6] Modelling work on two grasshopper species, *Myrmeleotettix maculatus* and *Chorthippus brunneus*, also shows that egg size should decline with maternal age, as reproductive investment become increasingly risky.[14] Theory indicates that the larger the offspring are at birth, the greater are the advantages to be gained. Large egg size in Lepidoptera and other insects confers an advantage on the offspring, in that they are able to combat poorer conditions.[232] Thus, on a poor quality host, or in situations where the eggs are laid at a distance from the host, insects hatching from large eggs or being deposited as large live young will have greater reserves and be able to establish more successfully than smaller insects.[99] In two grasshopper species, *Chorthippus brunneus* and *C.parallelus*, insects arising from big eggs show faster growth rates.[183] On a good quality host, large size is not as critical, as rapid growth and development are afforded by the host plant. Thus, theoretically, insects depositing offspring on a poor host should deposit fewer but larger offspring than those depositing offspring on a good quality host. The pine beauty moth, *P. flammea,* lays fewer, larger, eggs on Alaskan lodgepole pine, (a poor host), than it does on south coastal lodgepole pine (a good host).[120] The feeding habitats of the insect can also affect the size of eggs laid. For example, in the Japanese skipper butterflies (Lepidoptera: Hesperiidae) where some species are polyphagous and others less so, the more polyphagous species, which are likely to encounter a wider variety of hosts of differing quality, lay larger eggs and are less fecund than the species with the more restricted diets.[155] Although there are many exceptions, there is a general tendency for larger insect species to produce relatively large eggs[48] and for generalist feeders to produce larger eggs than specialists.[221]

The phenonmenon of producing many offspring but of a small size on a good host is seen within as well as between species. Aphids in general produce fewer, larger offspring when host quality is poor.[62] The pine beauty moth, *P. flammea*, provides a very good illustration of the effects that both maternal age and host quality have on egg size. Eggs laid decrease in size as the mother ages.[120] When confined to hosts of differing quality, the female moths respond by delaying the onset of oviposition on some hosts, and when they do begin to oviposit, their eggs are larger on the poor quality hosts than on the better quality hosts. In addition, the greater the clutch size, the smaller the eggs produced.[120] Egg size of the beech leaf-mining weevil, *Rynchaenus fagi*, is decreased when adult nutrition is poor, but is much greater when food quality improves.[9]

Egg size in Lepidoptera is frequently affected by the size of the mother as well as her age. For example, in the butterfly, *Euploea core corrinne*, egg weight decreased with maternal age but was also correlated with maternal weight.[87] The maternal weight is, in turn, an indicator of host quality.[87,88] However, another study on the same species showed no correlation between larval diet and egg weights of resultant females.[87] Egg weight in the butterfly *Pararge aegeria* was also positively correlated with female size and decreased as the mother aged.[98] Further, in the butterflies *D. plexippus*, *Lasiommata petropolitana*, *L. maera*, *L. megera*, *Lopinga achine*, and *Pieris rapae crucivora*, egg weights showed a decrease as the mother aged.[98,103,197] This was seen in a large number of insects in other groups. For example, in the beetle, *Callosobruchus maculatus*, egg size decreased with maternal age and large eggs were more viable than small ones.[220] The same was true in the pentatomid bug, *Graphosoma lineatum*.[106]

The size of eggs of *Pieris rapae* were found in one study to be positively correlated with maternal age and inversely correlated with adult size, which appears to run contrary to the general rule.[97] Large individuals are more fecund but lay smaller eggs. However, the small butterflies that then arise are generally present as adults in poor quality habitats and thus produce large eggs that will produce larvae able to withstand the poor conditions.[97] Another apparent contradiction is seen in the case of the gypsy moth, *L. dispar*. Small eggs are produced in poor host habitat quality forests, where there are large numbers of defoliated trees.[105] This is explained by the fact that small eggs give rise to less mobile larvae, which tend to remain on the hosts on which they were deposited. Suitably foliated host trees being in short supply under these conditions, it is to the advantage of the larvae to remain where they were laid rather than to use energy dispersing to another tree that is just as likely to be of poor host quality as the one it would have left. Large eggs are produced when conditions are good and hosts are in abundance, as is the case when a small insect population is just developing. Under these circumstances, it may be to the advantage of the larvae to disperse to new hosts. Large larvae are more mobile and thus disperse successfully.[105]

There is some dissent among those ecologists working with Lepidoptera as to the relationship between offspring size and fitness. In *P. aegeria*, no fitness function was found in relation to egg size. That is, large larvae did not survive better than small larvae. However, egg size did decrease with maternal age,[232] possibly at least allowing more eggs to be laid. The checkerspot butterfly, *Euphydryas editha*, when fed as an adult, was found in one study to keep its egg weights constant over its entire life span, in contrast to the unfed females, which showed the expected decline in egg weight.[151] However, in other studies, Lepidoptera were also shown to have a decline in egg weight with maternal age even when fed.[120] Clearly, egg weight can alter in response to selection pressure over evolutionary time, but the variability with relation to the host plant within a species or individual is also surprisingly large and apparently adaptive.

VII. CONCLUSIONS

It is apparent that not only are herbivorous insects markedly affected by host quality, but host quality has a number of dimensions that are not always readily perceived. It is in this area that profitable research could be performed, particularly in the cases where interactions between natural enemies, host quality, and herbivore performance are poorly understood. Although much has been done in correlating plant composition with nutrients, and nutrient status with insect performance, further research is still required in this area, particularly in assessing the relationships between attractant and deterrent chemical cues and offspring fitness. It is of paramount importance that more work using artificial diets be initiated and that the results of field studies and whole plant work then be related to the results obtained using artificial diets.

Life cycle evolution in aphids has been clearly linked with host quality; it is apparent that this factor has also affected the evolution of life cycles of many other insect groups.[202] More emphasis on research involving the coevolution of herbivore, host, and natural enemies should be encouraged.

REFERENCES

1. **Acreman, S. J. and Dixon, A. F. G.,** The effects of temperature and host quality on the rate of increase of the grain aphid *Sitobion avenae* on wheat, *Ann. Appl. Biol.,* 115, 3, 1989.
2. **Al Salti, M. N.,** Influence de la temperature pendant la vie imaginale sur les potentialites reproductrices de l'espece Sesamia nonagriodes Lefebvre (Lep., Noctuidae), *Acta Oecol. Oecol. Appl.,* 5, 103, 1984.
3. **Anderson, S. S., McCrea, K. D., Abrahamson, W. G., and Hartzel, L. M.,** Host genotype choice by the ball gallmaker *Eurosta solidagensis* (Diptera: Tephritidae), *Ecology,* 70, 1048, 1989.
4. **Angelo, M. J. and Slansky, F.,** Body building insects: trade-offs in resource allocation with particular reference to migration, *Fla. Entomol.,* 67, 22, 1984.
5. **Ankersmit, G. W., Acreman, T. M., and Dijkmann, H.,** Parasitism of colour forms in Sitobion avenae, *Entomol. Exp. Appl.,* 29, 362, 1981.
6. **Atkinson, D. A. and Begon, M.,** Reproductive variation and adult size in two co-occurring grasshopper species, *Ecol. Entomol.,* 12, 119, 1987.
7. **Auclair, J. L.,** Aphid feeding and nutrition, *Annu. Rev. Entomol.,* 8, 439, 1963.
8. **Auerbach, M. and Simberloff, D.,** Oviposition site preference and larval mortality in a leaf-mining moth, *Ecol. Entomol.,* 14, 131, 1989.
9. **Bale, J. S.,** Bud burst and success of the beech weevil, *Rhychaenus fagi:* feeding and oviposition, *Ecol. Entomol.,* 9, 139, 1984.
10. **Barbosa, P. and Baltensweiler, W.,** Phenotypic plasticity and herbivore outbreaks, in *Insect Outbreaks,* Barbosa, P. and Schultz, J. C., Eds., Academic Press, New York, 1987, 469.
11. **Barbosa, P. and Capinera, J. L.,** Population quality, dispersal and numerical change in the gypsy moth *Lymantria dispar* (L.), *Oecologia,* 36, 203, 1978.
12. **Barbosa, P., Cranshaw, W., and Greenblatt, J. A.,** Influence of food quality and quality on polymorphic dispersal behaviours in the gypsy moth, Lymantria dispar, *Can. J. Zool.,* 59, 293, 1981.
13. **Battisti, A.,** Host–plant relationships and population dynamics of the pine processionary caterpillar *Thaumetopoea pityocampa* (Dennis and Schiffermuller), *J. Appl. Entomol.,* 105, 393, 1988.
14. **Begon, M. and Parker, G. A.,** Should egg size and clutch size decrease with age? *Oikos,* 47, 293, 1988.
15. **Bell, G.,** The costs of reproduction and their consequences, *Am. Nat.,* 116, 45, 1980.
16. **Bennetova, B. and Fraenkel, G.,** What determines the number of ovarioles in a fly ovary? *J. Insect. Physiol.,* 27, 403, 1981.
17. **Berger, A.,** Egg weight, batch size and fecundity of the spotted stalkborer, *Chilo partellus* in relation to weight of females and time of oviposition, *Entomol. Exp. Appl.,* 50, 199, 1989.
17a. **Bernays, E. A. and Chapman, R. F.,** Evolution of plant deterrence to insects, in *Perspectives in Chemoreception and Behavior,* Chapman, R. F., Bernays, E. A., and Stoffolano, J. G., Eds., Springer-Verlag, New York, 1987, 159.

17b. **Bernays, E. A. and Cornelius, M.,** Relationship between deterrence and toxicity of plant secondary compounds for the alfalfa weevil *Hypera brunneipennis, Entomol. Exp. Appl.,* 64, 289.

18. **Bevan, D. and Brown, R. M.,** *Pine Looper Moth,* Forestry Commission Forest Record 119, HMSO, London 1978.

19. **Bink, F. A.,** Acid stress in *Rumex hydrolapathum* (Polygonaceae) and its influence on the phytophage *Lycaena dispar* (Lepidoptera: Lycaenidae), *Oecologia,* 70, 447, 1986.

20. **Bintcliffe, E. J. B. and Wratten, S. D.,** Antibiotic resistance in potato cultivars to the aphid *Myzus persicae, Ann. Appl. Biol.,* 100, 382, 1982.

21. **Blackman, R. L.,** *Aphids,* Ginn and Company, London, 1974, chap. 10.

22. **Blackman, R. L.,** Specificity in aphid/plant genetic interactions with particular reference to the role of the alate colonizer, in *Aphid–Plant Genotype Interactions,* Campbell, R. K. and Eikenbary, R. D., Eds., 1990, Elsevier, Amsterdam, 251.

23. **Blais, J. R.,** Effects of the destruction of the current year's foliage of Balsam fir on the fecundity and habits of flight of the spruce budworm, *Can. Entomol.,* 85, 446, 1953.

24. **Blake, E. A. and Wagner, M. R.,** Foliage age as a factor in food utilization by the western spruce budworm, *Choristoneura occidentalis, Great Basin Nat.,* 46, 169, 1986.

25. **Brakefield, P. M.,** Differential winter mortality and seasonal selection in the polymorphic ladybird *Adalia bipunctata* (L.) in the Netherlands, *Biol. J. Linn. Soc.,* 24, 189, 1985.

26. **Bristow C. M.,** Differential benefits from ant attendance to two species of Homoptera on New York iron weed, *J. Anim. Ecol.,* 53, 715, 1984.

27. **Brower, L. P.,** Chemical defence in butterflies, in *The Biology of Butterflies,* Vane-Wright, R. I. and Ackery, P. R., Eds., Symp. R. Entomol. Soc. Lond., No. 11, Academic Press, 1984, 109.

28. **Calow, P.,** Economics of ontogeny — adaptational aspects, in *Evolutionary Ecology,* Shorrocks, B., Ed., Blackwell Scientific, Oxford, 1984, 81.

29. **Cheng, H. H.,** Oviposition and longevity of the dark-sided cutworm, *Euxoa messoria* (Lepidoptera: Noctuidae), in the laboratory, *Can. Entomol.,* 104, 919, 1972.

30. **Chew, F. S.,** Coevolution of pierid butterflies and their cruciferous food plants. II. The distribution of eggs on potential foodplants, *Evolution,* 31, 568, 1977.

31. **Chew, F. S. and Robbins, R. K.,** Egg laying in butterflies, in *The Biology of Butterflies,* Vane-Wright, R. I. and Ackery, P. R., Eds., Symp. R. Entomol. Soc. Lond., No. 11, Academic Press, London, 1984, 65.

32. **Chroston, J. R.,** Colour Polymorphism in the English Grain Aphid *Sitobion avenae* F., Ph.D. thesis, University of East Anglia, Norwich, England, 1983.

33. **Courtney, S. P.,** The evolution of egg clustering by butterflies and other insects, *Am. Nat.,* 123, 276, 1984.

34. **Courtney, S. P.,** Why insects move between host patches — some comments on risk-spreading, *Oikos,* 47, 112, 1986.

34a. **Courtney, S. P. and Kibota, T. K.,** Mother doesn't know best: selection of hosts by ovipositing insects, in *Insect–Plant Interactions,* Vol. 2, Bernays, E. A., Ed., CRC Press, Boca Raton, FL, 1990, 161.

35. **Craig, T. P., Itami, J. K., and Price, P. W.,** A strong relationship between oviposition preference and larval performance in a shoot-galling sawfly, *Ecology,* 70, 1691, 1989.

36. **Craig, T. P., Price, P. W., and Itami, J. K.,** Resource regulation by a stem-galling sawfly on the arroyo willow, *Ecology,* 67, 419, 1986.

37. **Crawley, M. J.,** *Herbivory,* Blackwell Scientific, Oxford, 1985, chap. 3.

38. **Cull, D. C. and van Emden, H. F.,** The effect on *Aphis fabae* of diel changes in their food quality, *Physiol. Entomol.,* 2, 109, 1977.

39. **Danthanarayana, W.,** Factors determining variation in fecundity of the light brown apple moth, *Epiphyas postvittana* (Walker) (Tortricidae), *Aust. J. Zool.,* 23, 439, 1975.

40. **Danthanarayana, W.,** Environmentally cued size variation in light-brown apple moth, *Epiphyas postvittana* (Walk.) (Tortricidae), and its adaptive value in dispersal, *Oecologia,* 26, 121, 1976.

41. **Day, K. R.,** Phenology, polymorphism and insect–plant relationships of the larch budmoth, *Zeiraphera diniana* (Guenee) (Lepidoptera: Tortricidae) on alternative conifer hosts in Britain, *Bull. Entomol. Res.,* 74, 47, 1984.

42. **Dennis, R. L. H. and Shreeve, T. G.,** Host plant–habitat structure and the evolution of butterfly mate-locating behaviour, *Zool. J. Linn. Soc.,* 94, 301, 1988.

43. **Denno, R. F. and Grissell, E. E.,** The adaptiveness of wing-dimorphism in the salt marsh inhabiting planthopper *Prokelisia marginata* (Homoptera: Delphacidae), *Ecology,* 60, 221, 1979.

44. **Denno, R. F. and McCloud, E. S.,** Predicting fecundity from body size in the planthopper, *Prokelisia marginata* (Homoptera: Delphacidae)., *Environ. Entomol.,* 14, 846, 1985.

45. **Denno, R. F., Douglas, L. W., and Jacobs, D.,** Effects of crowding and host plant nutrition: environmental determinants of wing form in the migratory planthopper, *Prokelisia marginata, Ecology,* 66, 1588, 1985.

46. **Denno, R. F., Larsson, S., and Olmstead, K. L.,** Role of enemy-free space and plant quality in host-plant selection by willow beetles, *Ecology,* 7, 124, 1990.

47. **Denno, R. F., Olmstead, K. L., and McCloud, E. S.,** Reproductive cost of flight capability: a comparison of life history traits in wing dimorphic planthoppers, *Ecol. Entomol.,* 14, 31, 1989.

48. **Derr, J. A., Alden, B., and Dingle, H.,** Insect life histories in relation to migration, body size and host plant assay: a comparative study of *Dysdercus, J. Anim. Ecol.,* 50, 181 1981.
49. **Dewar, A. M.,** Assessment of methods for testing varietal resistance to aphids in cereals, *Ann. Appl. Biol.,* 87, 183, 1977.
50. **Dixon, A. F. G.,** Reproductive activity of the sycamore aphid, *Drepanosiphum platanoides* (Schr.) (Hemiptera, Aphididae), *J. Anim. Ecol.,* 32, 33, 1963.
51. **Dixon, A. F. G.,** The effect of population density and nutritive status of the host on the summer reproductive activity of the sycamore aphid, *Drepanosiphum platanoides* (Schr.), *J. Anim. Ecol.,* 35, 105, 1966.
52. **Dixon, A. F. G.,** Population dynamics of the sycamore aphid *Drepanosiphum platanoides* (Schr.) (Hemiptera: Aphididae): migratory and trivial flight activity, *J. Anim. Ecol.,* 38, 585, 1969.
53. **Dixon, A. F. G.,** Quality and availability of food for a sycamore aphid population, in *Animal Populations in Relation to their Food Resources,* Watson, A., Ed., Blackwell Scientific, Oxford, 1970, 277.
54. **Dixon, A. F. G.,** The life cycle and host preferences of the bird cherry-oat aphid, *Rhopalosiphum padi* (L.), and their bearing on the theories of host alternation in aphids, *Ann. Appl. Biol.,* 68, 135, 1971.
55. **Dixon, A. F. G.,** Control and significance of the seasonal development of colour forms in the sycamore aphid, *Drepanosiphum platanoides* (Schr.), *J. Anim. Ecol.,* 41, 689, 1972.
55a. **Dixon, A. F. G.,** Fecundity of brachypterous and macropterus alatae in *Drepanosiphum dixoni* (Callaphididae, Aphididae), *Entomol. Exp. Appl.,* 15, 335, 1972.
56. **Dixon, A. F. G.,** Effect of population density and food quality on autumnal reproductive activity in the sycamore aphid, *Drepanosiphum platanoides* (Schr.), *J. Anim. Ecol.,* 33, 297, 1975.
57. **Dixon, A. F. G.,** Reproductive strategies of the altate morphs of the bird cherry-oat aphid *Rhopalosiphum padi* (L.), *J. Anim. Ecol.,* 45, 1976.
58. **Dixon, A. F. G.,** Aphid ecology: life cycles, polymorphism and population regulation, *Annu. Rev. Ecol. Syst.,* 8, 329, 1977.
59. **Dixon, A. F. G.,** *Aphid Ecology,* Blackie, London, 1985, chap. 6.
60. **Dixon, A. F. G. and Dharma, T. R.,** Number of ovarioles and fecundity in the black bean aphid, *Aphis fabae, Entomol. Exp. Appl.,* 28, 1, 1980.
61. **Dixon, A. F. G. and Glen, D. M.,** Morph determination in the bird cherry-oat aphid, *Rhopalosiphum padi* (L.), *Ann. Appl. Biol.,* 68, 11, 1971.
62. **Dixon, A. F. G. and Wellings, P. W.,** Seasonality and reproduction in aphids, *Int. J. Inv. Rep.,* 5, 83, 1982.
63. **Dixon, A. F. G. and Wratten, S. D.,** Laboratory studies on aggregation size and fecundity in the black bean aphid, *Aphis fabae* (Scop.), *Bull. Entomol. Res.,* 61, 97, 1971.
64. **Dunlap-Pianka, H. L.,** Ovarian dynamics in *Heliconius* butterflies: correlations among daily oviposition rates, egg weights, and qualitative aspects of oogenesis, *J. Insect. Physiol.,* 25, 741, 1979.
65. **Dunlap-Pianka, H. L., Boggs, C. L., and Gilbert, L. E.,** Ovarian dynamics in Heliconiine butterflies: programmed senescence versus eternal youth, *Science,* 197, 487, 1977.
66. **Elliott, W. M.,** A method of predicting short term population trends of the green peach aphid, *Myzus persicae* (Homoptera: Aphididae), on potatoes, *Can. Entomol.,* 105, 11, 1973.
67. **Fairbairn, D. J.,** Adaptive significance of wing dimorphism in the absence of dispersal: a comparative study of wing morphs in the waterstrider, *Gerris remigis, Ecol. Entomol.,* 13, 273, 1988.
68. **Fatzinger, C. W. and Merkel, E. P.,** Oviposition and feeding preferences of the southern pine coneworm (Lepidoptera, Pyralidae) for different host-plant materials and observations on monoterpenes as an oviposition stimulant, *J. Chem. Ecol.,* 11, 689, 1985.
69. **Feeny, P. P.,** Seasonal changes in oak leaf tannins and nutrient as a cause of spring feeding by winter moth caterpillars, *Ecology,* 51, 565, 1970.
70. **Firempong, S. and Zalucki, M. P.,** Host plant selection by *Helicoverpa armigera* (Hubner) (Lepidoptera: Noctuidae); the role of certain plant attributes, *Aust. J. Zool.,* 37, 675, 1990.
71. **Fisher, R. A.,** *The Genetical Theory of Natural Selection,* Oxford University Press, Oxford, 1930, chap. 2.
72. **Fitt, G. P.,** Variation in ovariole number and egg size of species of *Dacus* (Diptera: Tephritidae) and their relation to host specialization, *Ecol. Entomol.,* 15, 255, 1990a.
73. **Fitt, G. P.,** Comparative fecundity, clutch size, ovariole number and egg size of *Dacus tryoni* and *D. jarvisi,* and their relationship to body size, *Entomol. Exp. Appl.,* 55, 11, 1990b.
74. **Forno, I. W. and Bourne, A. S.,** Oviposition by the weevil *Cyrtobagous salviniae* Calder and Sands when its host plant, *Salvinia molesta,* is damaged, *J. Appl. Entomol.,* 106, 85, 1988.
75. **Fowler, S. V. and Lawton, J. H.,** Foliage preferences of birch herbivores: a field manipulation experiment, *Oikos,* 42, 239, 1984b.
76. **Fritz, R. S. and Nobel, J.,** Host plant variation in mortality of the leaf-folding sawfly on the arroyo willow, *Ecol. Entomol.,* 15, 25, 1990.
77. **Furuta, K.,** Host preferences and population dynamics in an autumnal population of the maple aphid, *Periphyllus californiensis* Shinji (Homoptera: Aphididae), *Z. Ang. Entomol.,* 102, 93, 1986.
78. **Gilbert, L. E.,** The biology of butterfly communities, in *The Biology of Butterflies,* Vane-Wright, R. I. and Ackery, P. R., Eds., Symp. R. Entomol. Soc. Lond., No. 11, Academic Press, 1984, 41.

79. **Gilchrist, G. W. and Rutowski, R. L.,** Adaptive and incidental consequences of the Alba polymorphism in an agricultural population of *Colias* butterflies — female size, fecundity and differential dispersion, *Oecologia,* 68, 235, 1986.

80. **Godfray, H. C. J.,** The evolution of clutch size in invertebrates, *Oxford Surv. Evol. Biol.,* 4, 117, 1987.

81. **Gruber, K. and Dixon, A. F. G.,** The effect of nutrient stress on development and reproduction in an aphid, *Entomol. Exp. Appl.,* 47, 23, 1988.

82. **Gunn, A., Gatehouse, A. G., and Woodrow, K. P.,** Trade-off between flight and reproduction in the African armyworm moth, *Spodoptera exempta, Physiol. Entomol.,* 14, 419, 1989.

83. **Harrison, S. and Karban, R.,** Effects of an early-season folivorous moth on the success of a later-season species, mediated by a change in the quality of the shared host, *Lupinus arboreus* Sims, *Oecologia,* 69, 354, 1986.

84. **Haslam, E.,** Plant polyphenols (syn. vegetable tannins) and chemical defense — a reappraisal, *J. Chem. Ecol.,* 14, 1789, 1988.

85. **Haukioja, E.,** Induced defences in plants against insects, *Annu. Rev. Entomol.,* 36, 25, 1991.

86. **Heliovaara, K. and Vaisanen, R.,** Changes in population dynamics of pine insects induced by air pollution, in *Population Dynamics of Forest Insects,* Watt, A. D., Leather, S. R., Hunter, M. D., and Kidd, N. A. C., Eds., Intercept Press, Andover, MA, 1990, 209.

87. **Hill, C. J.,** The effect of adult diet on the biology of butterflies 2. The common crow butterfly, *Euploea core corinna, Oecologia,* 81, 258, 1989.

88. **Hill, C. J. and Pierce, N. E.,** The effect of adult diet on the biology of butterflies. 1. The common imperial blue, *Jalmenus evagoras, Oecologia,* 81, 249, 1989.

89. **Hillyer, R. J. and Thorsteinson, A. J.,** The influence of the host plant or males on ovarian development or oviposition in the diamondback moth *Plutella maculipennis* (Curt.), *Can. J. Zool.,* 47, 805, 1969.

90. **Honek, A.,** Ecophysiological differences between brachypterous and macropterus morphs in *Pyrrhocoris apterus* (Heteroptera, Pyrrhocoridae), *Acta Entomol. Bohem.,* 82, 347, 1985.

91. **Honek, A.,** Effect of plant quality and microclimate on population growth and maximum abundances of cereal aphids, *Metopolophium dirhodum* (Walker) and *Sitobion avenae* (F.) (Hom. Aphididae), *J. Appl. Entomol.,* 104, 304, 1987.

92. **Honek, A.,** Host plant energy allocation to and within ears, and abundance of cereal aphids, *J. Appl. Entomol.,* 110, 68, 1990.

93. **Hough, J. A. and Pimentel, D.,** Influence of host foliage on development, survival and fecundity of the gypsy moth, *Envir. Entomol.,* 7, 97, 1978.

94. **Howlett, R. J. and Majerus, M. E. N.,** The understanding of industrial melanism in the peppered moth (*Biston betularia*) (Lepidoptera: Geometridae), *Biol. J. Linn. Soc.,* 30, 31, 1987.

95. **Imms, A. D.,** *A General Textbook of Entomology,* 9th ed., Chapman and Hall, London, 1957.

96. **James, W. O.,** *An Introduction to Plant Physiology,* Oxford University Press, Oxford, 1973, chap. 7.

97. **Jones, R. E., Hart, J. R., and Bull, G. D.,** Temperature, size and egg production in the cabbage butterfly, *Pieris rapae* L., *Aust. J. Zool.,* 30, 223, 1982.

98. **Karlsson, B.,** Variation in egg weight, oviposition rate and reproductive reserves with female age in a natural population of the speckled wood butterfly, *Pararge aegeria, Ecol. Entomol.,* 12, 473, 1987.

99. **Karlsson, B. and Wiklund, C.,** Egg weight variation in relation to egg mortality and starvation endurance of newly hatched larvae in some satyrid butterflies, *Ecol. Entomol.,* 10, 205, 1985.

100. **Keese, M. C. and Wood, T. K.,** Host-plant mediated geographic variation in the life history of *Platycotis vittata* (Homoptera: Membracidae), *Ecol. Entomol.,* 16, 63, 1991.

101. **Kidd, N. A. C. and Tozer, D. J.,** On the significance of post-reproductive life in aphids, *Ecol. Entomol.,* 10, 357, 1985.

102. **Kimmerer, T. W. and Potter, D. A.,** Nutritional quality of specific leaf tissues and selective feeding by a specialist leaf miner, *Oecologia,* 71, 548, 1987.

103. **Kimura, K. and Tsubaki, Y.,** Egg weight variation associated with female age in *Pieris rapae crucivora* Boisduval (Lepidoptera; Pieridae), *Appl. Entomol. Zool.,* 20, 500, 1985.

104. **Kukal, O. and Dawson, T. E.,** Temperature and food quality influence feeding behviour, assimilation efficiency and growth rate of Arctic wooly bear caterpillars, *Oecologia,* 79, 526, 1989.

105. **Lance, D. R.,** Host-seeking behaviour of the gypsy moth: the influence of polyphagy and highly apparent host plants, in *Herbivorous Insects — Host Seeking Behaviour and Mechanisms,* Ahmad, S., Ed., Academic Press, New York, 1983, 201.

106. **Larsson, F. K.,** Female longevity and body size as predictors of fecundity and egg length in *Graphosoma lineatum* L., *Deutsche Entomol. Zeitschr.,* 36, 329, 1989.

107. **Leather, S. R.,** Aspects of the Ecology of the Bird Cherry-oat Aphid, Rhopalosiphum padi (L.), Ph.D. thesis, University of East Anglia, Norwich, England, 1980.

108. **Leather, S. R.,** Reproduction and survival: a field study of gynoparae of the bird cherry-oat aphid, *Rhopalosiphum padi* (Homoptera; Aphididae), on its primary host *Prunus padus, Ann. Entomol. Fenn.,* 47, 131, 1981.

109. **Leather, S. R.,** Do gynoparae and males need to feed? An attempt to allocate resources in the bird cherry oat aphid, *Rhopalosiphum padi, Entomol. Exp Appl.,* 31, 386, 1982.
110. **Leather, S. R.,** Evidence of ovulation after adult moult in the bird cherry-oat aphid, *Rhopalosiphum padi, Entomol. Exp. Appl.,* 33, 348, 1983.
111. **Leather, S. R.,** The effect of adult feeding on the fecundity, weight loss and survival of the pine beauty moth, *Panolis flammea* (D & S), *Oecologia,* 65, 70, 1984.
112. **Leather, S. R.,** Factors affecting pupal survival and eclosion in the pine beauty moth, *Panolis flammea* (D & S), *Oecologia,* 63, 75, 1984.
113. **Leather, S. R.,** Oviposition preferences in relation to larval growth rates and survival in the pine beauty moth, *Panolis flammea, Ecol. Entomol.,* 10, 213, 1985.
114. **Leather, S. R.,** Host monitoring by aphid migrants: do gynoparae maximise offspring fitness? *Oecologia,* 68, 367, 1986.
115. **Leather, S. R.,** Insects on bird cherry. I. The bird cherry ermine moth, *Yponomeuta evonymellus* (L.) (Lepidoptera: Yponmeutidae), *Entomol. Gaz.,* 37, 209, 1986.
116. **Leather, S. R.,** Pine monoterpenes stimulate oviposition in the pine beauty moth, *Panolis flammea, Entomol. Exp. Appl.,* 43, 295, 1987.
117. **Leather, S. R.,** Size, reproductive potential and fecundity in insects. Things aren't as simple as they seem, *Oikos,* 51, 386, 1988b.
118. **Leather, S. R.,** Do alate aphids produce fitter offspring? The influence of maternal rearing history and morph on life-history parameters of *Rhopalosiphum padi* (L.), *Funct. Ecol.,* 3, 237, 1989.
119. **Leather, S. R.,** The role of host quality, natural enemies, competition and weather in the regulation of autumn and winter populations of the bird cherry aphid, in *Population Dynamics of Forest Insects,* Watt, A. D., Leather, S. R., Hunter, M. D., and Kidd, N. A. C., Eds., Intercept Press, Andover, MA, 1990, 35.
120. **Leather, S. R. and Burnand, A. C.,** Factors affecting life-history parameters of the pine beauty moth, *Panolis flammea* (D & S): the hidden costs of reproduction, *Funct. Ecol.,* 1, 331, 1987.
121. **Leather, S. R. and Dixon, A. F. G.,** The effect of cereal growth stage and feeding site on the reproductive activity of the bird-cherry aphid, *Rhopalosiphum padi, Ann. Appl. Biol.,* 97, 135, 1981.
122. **Leather, S. R. and Dixon, A. F. G.,** Growth, survival and reproduction of the bird-cherry aphid, *Rhopalosiphum padi,* on its primary host, *Ann. Appl. Biol.,* 99, 115, 1981.
123. **Leather, S. R. and Dixon, A. F. G.,** Secondary host preferences and reproductive activity of the bird cherry-oat aphid, *Rhopalosiphum padi, Ann. Appl. Biol.,* 101, 219, 1982.
124. **Leather, S. R. and Mackenzie, G. A.,** Factors affecting the population development of the bird cherry ermine moth, *Yponomeuta evonymellus* (L.), *Entomologist,* in press, 1993.
125. **Leather, S. R. and Wellings, P. W.,** Ovariole number and fecundity in aphids, *Entomol. Exp. Appl.,* 30, 128, 1981.
126. **Leather, S. R., Ward, S. A., and Dixon, A. F. G.,** The effect of nutrient stress on life history parameters of the black bean aphid, *Aphis fabae* Scop., *Oecologia,* 57, 156, 1983.
127. **Leather, S. R., Watt, A. D., and Barbour, D. A.,** The effect of host plant and delayed mating on the fecundity and lifespan of the pine beauty moth, *Panolis flammea* (Dennis & Schiffermuller) (Lepidoptera: Noctuidae): their influence on population dynamics and relevance to pest management, *Bull. Entomol. Res.,* 75, 641, 1985.
128. **Leather, S. R., Watt, A. D., and Forrest, G. I.,** Insect-induced changes in young lodgepole pine (*Pinus contorta*): the effect of previous defoliation on oviposition, growth and survival of the pine beauty moth, *Panolis flammea, Ecol. Entomol.,* 12, 275, 1987.
129. **Leather, S. R., Wellings, P. W., and Dixon, A. F. G.,** Habitat quality and the reproductive strategies of the migratory morphs of the bird cherry-oat aphid, *Rhopalosiphum padi* (L.), colonizing secondary host plants. *Oecologia,* 59, 302, 1983.
130. **Leather, S. R., Wellings, P. W., and Walters, K. F. A.,** Variation in ovariole number within the Aphidoidea, *J. Nat. Hist.,* 22, 381, 1988.
131. **Lees, A. D.,** Action spectra for the photoperiodic control of polymorphism in the aphid *Megoura viciae, J. Insect Physiol.,* 27, 761, 1981.
132. **Lees, A. D.,** Parturition and alate morph determination in the aphid *Megoura viciae, Entomol. Exp. Appl.,* 35, 93, 1984.
133. **Lees, A. D.,** The photoperiodic responses and phenology of the pea aphid *Acyrthosiphon pisum, Ecol. Entomol.,* 14, 69, 1989.
134. **Loxdale, H. D. and Brookes, C. P.,** Electrophoretic study of enzymes from cereal aphid populations. V. Spatial and temporal genetic similarity of holocyclic populations of the bird cherry-oay aphid, *Bull. Entomol. Res.,* 78, 241, 1988.
135. **Loxdale, H. D. and Brookes, C. P.,** Temporal genetic stability within and restricted migration (gene flow) between local populations of the blackberry-grain aphid *Sitobion fragariae* in south-east England, *J. Anim. Ecol.,* 59, 497, 1990.
136. **Lukefahr, M. J. and Martin, D. F.,** The effect of various larval and adult diets on the fecundity and longevity of the bollworm, tobacco budworm and the cotton leafworm, *J. Econ. Entomol.,* 57, 233, 1964.

137. **Mackay, D. A. and Jones, R. E.,** Leaf shape and the host-finding behaviour of two ovipositing monophagous butterfly species, *Ecol. Entomol.,* 14, 423, 1989.

138. **McClure, M. S.,** Reproduction and adaptation of exotic hemlock scales (Homoptera: Diapsidae) on their new and native hosts, *Environ. Entomol.,* 12, 1811, 1983.

139. **McClure, M. S. and Hare, J. D.,** Foliar terpenoids in *Tsuga* species and the fecundity of scale insects, *Oecologia,* 63, 185, 1984.

140. **Maelzer, D. A.,** The biology and main causes of changes in numbers of the rose aphid, *Macrosiphum rosae* (L.) on cultivated roses in South Australia, *Aust. J. Zool.,* 25, 269, 1977.

141. **Mejerus, M. E. N.,** Genetic control of two melanic forms of *Panolis flammea* (Lepidoptera: Noctuidae), *Heredity,* 49, 171, 1982.

142. **Mangel, M.,** Oviposition site selection and clutch size in insects, *J. Math. Biol.,* 25, 1, 1987.

143. **Mansingh, A.,** Developmental response of *Antheraea pernyi* to seasonal changes in oak leaves from two localities, *J. Insect Physiol.,* 18, 1395, 1972.

144. **Markkula, M.,** Studies on the pea aphid, *Acyrthosiphon pisum* Harris (Hom., Aphididae), with special reference to the differences in the biology of the green and red forms, *Ann. Agric. Fenn.,* 2 (Suppl. 1), 1, 1963.

145. **Matsumoto, K. and Tsuji, H.,** Occurrence of two colour types in the green peach aphid *Myzus persicae,* and their susceptibility to insecticides. (in Japanese), *Jap. J. Appl. Entomol. Zool.,* 20, 92, 1979.

146. **Messina, F. J.,** Influence of cowpea pod maturity on the oviposition choices and larval survival of a bruchid beetle *Callosobruchus maculatus, Entomol. Exp. Appl.,* 35, 241, 1984.

147. **Minkenberg, O. P. and Fredrix, M. J. J.,** Preference and performance of an herbivorous fly, *Liriomyza trifolii* (Diptera: Agromyzidae), on tomato plants differing in leaf nitrogen, *Ann. Entomol. Soc. Am.,* 82, 350, 1989.

148. **Montgomery, M. E. and Wallner, W. E.,** The gypsy moth — a westward migrant, in *Dynamics of Forest Insect Populations: Patterns, Causes, Implications,* Berryman, A. A., Ed., Plenum Press, New York, 1988, 353.

149. **Moore, G. J.,** Host plant discrimination in tropical satyrine butterflies, *Oecologia,* 70, 592, 1986.

150. **Moore, L. V., Myers, J. H., and Eng, R.,** Western tent caterpillars prefer the sunny side of the tree, but why? *Oikos,* 51, 321, 1988.

151. **Moore, R. A. and Singer, M. C.,** Effects of maternal age and adult diet on egg weight in the butterfly *Euphydryas editha, Ecol. Entomol.,* 12, 401, 1987.

152. **Moran, N. A.,** Seasonal shifts in host usage in *Uroleucon gravicorne* (Homoptera: Aphididae) and implications for the evolution of host alternation in aphids, *Ecol. Entomol.,* 8, 371, 1983.

153. **Morris, R. F.,** Influence of parental food quality on the survival of *Hyphantria cunea, Can. Entomol.,* 99, 24, 1967.

154. **Morrow, P. A. and Fox, L. R.,** Effects of variation in Eucalyptus essential oil yield on insect growth and grazing damage, *Oecologia,* 45, 209, 1980.

154. **Myers, J. H., Monro, J., and Murray, N.,** Egg clumping, host plant selection and population regulation in *Cactoblastis cactorum* (Lepidoptera), *Oecologia,* 51, 1981.

155. **Nakasuji, F.,** Egg size of skippers (Lepidoptera: Hesperiidae) in relation to their host specificity and to leaf toughness of host plants, *Ecol. Res.,* 2, 175, 1987.

156. **Newton, C. and Dixon, A. F. G.,** Methods of hatching the eggs and rearing the fundatrices of the English grain aphid, *Sitobion avenae, Entomol. Exp. Appl.,* 45, 277, 1987.

157. **Newton, C. and Dixon, A. F. G.,** Embryonic growth rate and birth weight of the offspring of apterous and alate aphids: a cost of dispersal, *Entomol. Exp. Appl.,* 55, 223, 1990.

158. **Noguchi, H.,** Mating frequency, fecundity, and egg hatchability of the smaller tea tortrix moth, *Adoxyphyes* sp. (Lepidoptera: Tortricidae), *Jpn. J. Appl. Entomol. Zool.,* 25, 259, 1981.

159. **Oyeyele, S. O. and Zalucki, M. P.,** Cardiac glycosides and oviposition by *Danaus plexippus* on *Asclepias fruiticosa* in south-east Queensland (Australia), with particular notes on the effect of plant nitrogen content, *Ecol. Entomol.,* 15, 177, 1990.

160. **Parker, G. A. and Begon, M.,** Optimal egg size and clutch size: effects of environment and maternal phenotype, *Am. Nat.,* 128, 573, 1986.

161. **Parker, G. A. and Courtney, S. P.,** Models of clutch size in insect oviposition, *Theor. Pop. Biol.,* 26, 27, 1984.

162. **Parry, W. H. and Spires, S.,** Studies on the factors affecting the population levels of the *Adelges cooleyi* (Gillette) on Douglas fir, 4. Polmorphism in progredientes, *Z. Ang. Entomol.,* 94, 253, 1982.

163. **Pencoe, N. L. and Martin, P. B.,** Fall armyworm (Lepidoptera: Noctuidae) larval development and adult fecundity on five grass hosts, *Environ. Entomol.,* 11, 720, 1982.

164. **Petty, S. J.,** Fur and feather — tawny owls in 1988, *Entopath. News,* 92, 7, 1989.

165. **Pianka, E. R.,** *Evolutionary Ecology,* 2nd ed., Harper and Row, New York, 1978, chap. 5.

166. **Pilson, D. and Rausher, M. D.,** Clutch size adjustment by a swallowtail butterfly, *Nature,* 333, 361, 1988.

167. **Pilson, D. and Rausher, M. D.,** In response to Tabor, *Oikos,* 55, 136, 1989.
168. **Prestidge, R. A.,** Instar duration, adult consumption, oviposition and nitrogen utilization efficiencies of leafhoppers feeding on different quality food (Auchenorrhyncha: Homoptera), *Ecol. Entomol.,* 7, 91, 1982.
169. **Preszler, R. W. and Price, P. W.,** Host quality and sawfly populations: a new approach to life table analysis, *Ecology,* 69, 2012, 1988.
170. **Price, P. W.,** *Insect Ecology,* John Wiley & Sons, London, 1975, chap. 8.
171. **Pritchard, I. M. and James, R.,** Leaf miners: their effect on leaf longevity, *Oecologia,* 64, 132, 1984.
172. **Pritchard, I. M. and James, R.,** Leaf fall as a source of leafminer mortality, *Oecologia,* 64, 140, 1984b.
173. **Puttick, G. M. and Bowers, M. D.,** Effect of qualitative and quantitative variation in allelochemicals on a generalist insect: iridoid glycosides and the southern armyworm, *J. Chem. Ecol.,* 14, 335, 1988.
174. **Rausher, M. D.,** Larval habitat suitability and oviposition preference in three related butterflies, *Ecology,* 60, 503, 1979.
175. **Raworth, D. A., McFarlane, S., Gilbert, N., and Frazer, B. D.,** Population dynamics of the cabbage aphid, *Brevicoryne brassicae* (Homoptera: Aphididae) at Vancouver, British Columbia III Development, fecundity and morph determination vs. aphid density and plant quality, *Can. Entomol.,* 116, 879, 1984.
176. **Renwick, J. A. A.,** Chemical ecology of oviposition in phytophagous insects, *Experientia,* 45, 223, 1989.
177. **Richards, L. J. and Myers, J. H.,** Maternal influences on size and emergence time of the cinnabar moth, *Can. J. Zool.,* 58, 1452, 1980.
178. **Roff, D. A.,** The cost of being able to fly: a study of wing polymorphism in two species of cricket, *Oecologia,* 63, 30, 1984.
179. **Schroeder, L. A.,** Changes in tree leaf quality and growth performance of lepidopteran larvae, *Ecology,* 67, 1628, 1986.
180. **Schwartz, C. C. and Hobbs, N. T.,** Forage and range evaluation, in *Bioenergetics of Wild Herbivores,* Hudson, R. J. and White, R. G., Eds., CRC Press, Boca Raton, FL, 1985, 25.
181. **Scriber, J. M.,** Host-plant suitability, in *Chemical Ecology of Insects,* Bell, W. J. and Carde, R. T., Eds., Chapman and Hall, London, 1984, 159.
182. **Scriber, J. M. and Slansky, F.,** The nutritional ecology of immature insects, *Annu. Rev. Entomol.,* 26, 183, 1981.
183. **Sibly, R. and Monk, K.,** A theory of grasshopper life cycles, *Oikos,* 48, 186, 1987.
184. **Simmonds, M. S. J. and Blaney, W. M.,** Effects of rearing density on development and feeding behaviour in larvae of *Spodoptera exempta, J. Insect. Physiol.,* 32, 1043, 1986.
185. **Singer, M. C.,** Butterfly-hostplant relationships: host quality, adult choice and larval success, in *The Biology of Butterflies,* Vane-Wright, R. I. and Ackery, P. R., Eds., Symp. R. Entomol. Soc. Lond., Academic Press, 1984, 81.
186. **Slansky, F.,** Food consumption and reproduction as affected by tethered flight in female milkweed bugs (*Oncopeltus fasciatus*), *Entomol. Exp. Appl.,* 28, 1980.
187. **Slansky, F.,** Insect nutrition: an adaptationist's perspective, *Fla. Entomol.,* 65, 46, 1982.
188. **Sopp, P. I., Sunderland, K. D., and Coombes, D. S.,** Observations on the number of cereal aphids on the soil in relation to aphid density in winter wheat, *Ann. Appl. Biol.,* 111, 53, 1987.
189. **Speight, M. R. and Wainhouse, D.,** *Ecology and Management of Forest Insects,* Oxford Science Publications, Oxford, 1989, chap. 3.
190. **Springer, P. and Boggs, C. L.,** Resource allocation to oocytes: heritable variation with altitude in *Colias philodice eriophyle* (Lepidoptera), *Am. Nat.,* 127, 252, 1986.
191. **Stadler, E.,** Host plant stimuli affecting oviposition behavior of the eastern spruce budworm, *Entomol. Exp. Appl.,* 17, 176, 1974.
192. **Stamp, N. E.,** Egg deposition patterns in butterflies: why do some species cluster their eggs rather than deposit them singly? *Am. Nat.,* 115, 367, 1980.
193. **Stern, V. M. and Smith, R. F.,** Factors affecting egg production and oviposition in populations of *Colias philodice eurytheme* Boisduval (Lepidoptera: Pieridae), *Hilgardia,* 29, 411, 1960.
194. **Straw, N. A.,** The timing of oviposition and larval growth by two tephritid fly species in relation to host-plant development, *Ecol. Entomol.,* 14, 443, 1989.
195. **Strong, D. R., Lawton, J. H., and Southwood, T. R. E.,** *Insects on Plants — Community Patterns and Mechanisms,* Blackwell Scientific Publications, Oxford, 1984, chap. 2.
196. **Sutton, R. D.,** The effect of host plant flowering on the distribution and growth of hawthorn psyllids (Homoptera: Psylloidea), *J. Anim. Ecol.,* 53, 37, 1984.
197. **Svard, L. and Wiklund, C.,** Fecundity, egg weight and longevity in relation to multiple matings in females of the monarch butterfly, *Behav. Ecol. Sociobiol.,* 23, 39, 1988.
198. **Tauber, M. J., Tauber, C. A., and Masaki, S.,** *Seasonal Adaptations of Insects,* Oxford University Press, New York, 1986, chap. 3.
199. **Taylor, L. R.,** Longevity, fecundity and size; control of reproductive potential in a polymorphic migrant, *Aphis fabae* Scop., *J. Anim. Ecol.,* 44, 135, 1975.

200. **Taylor, M. F. J. and Forno, I. W.,** Oviposition preferences of the Salvinia moth *Samea multiplicalis* Guenee (Lep., Pyralidae) in relation to host plant quality and damage, *J. Appl. Entomol.*, 104, 73, 1987.

200a. **Thomas, C. D.,** Butterfly larvae reduce host plant survival in vicinity of alternative host plants, *Oecologia*, 70, 113, 1986.

201. **Thompson, J. N.,** Evolutionary ecology of the relationship between oviposition preference and performance of offspring in phytophagous insects, *Entomol. Exp. Appl.*, 47, 3, 1988.

202. **Thompson, J. N.,** Coevolution and the evolutionary genetics of interactions among plants and insects and pathogens, in *Pests, Pathogens and Plant Communities*, Burdon, J. J. and Leather, S. R., Eds., Blackwell Scientific Publications, Oxford, 1990, 249.

203. **Tinkle, D. W., Wilbur, H. M., and Tilley, S. G.,** Evolutionary strategies in lizard reproduction, *Evolution*, 24, 55, 1969.

204. **Trichilo, P. J. and Leigh, T. F.,** Influence of resource quality on the reproductive fitness of flower thrips (Thysanoptera: Thripidae), *Ann. Entomol. Soc. Am.*, 81, 64, 1988.

205. **Turner, J. R. G.,** Darwin's coffin and Doctor Pangloss — do adaptationist models explain mimicry? in *Evolutionary Ecology*, Shorrocks, B., Ed., Blackwell Scientific Publications, Oxford, 1984, 313.

206. **van Amelsvoort, P. A. M. and Usher, M. B.,** A method for assessing the palatability of senesced leaf litter using *Folsomia candida* (Collembola: Isotomidae), *Pedobiologia*, 33, 193, 1989.

207. **van der Meijden, E., van Zoolen, A. M., and Soldaat, L. L.,** Oviposition by the cinnabar moth, *Tyria jacobaeae*, in relation to nitrogen, sugars and alkaloids of ragwort, *Senecio jacobaea*, *Oikos*, 54, 337, 1989.

208. **van der Meijden, E., van Bemmelen, M., Kooi, R., and Post, B. J.,** Nutritional quality and chemical defence in the ragwort–cinnabar moth interaction, *J. Anim. Ecol.*, 53, 443, 1984.

209. **van Emden, H. F.,** Aphids as phytochemists, in *Phytochemical Ecology*, Harborne, J. B., Ed., Academic Press, London, 1972, 26.

210. **van Emden, H. F. and Bashford, M. A.,** A comparison of the reproduction of *Brevicoryne brassicae* and *Myzus persicae* in relation to soluble nitrogen concentration and leaf age (leaf position) in the Brussels sprout plant, *Entomol. Exp. Appl.*, 12, 351, 1969.

211. **van Emden, H. F. and Bashford, M. A.,** The performance of *Brevicoryne brassicae* and *Myzus persicae* in relation to plant age and leaf amino acids, *Entomol. Exp. Appl.*, 14, 349, 1971.

212. **Vepsalainen, K.,** The life cycle and wing lengths of Finnish *Gerris* Fabr. species (Heteroptera, Gerridae), *Acta Zool. Fenn.*, 141, 1, 1974.

213. **Volney, W. J. A., Waters, W. E., Askers, R. P., and Liebhold, A. M.,** Variation in spring emergence patterns among western *Choristoneura* spp. (Lepidoptera: Tortricidae) populations in southern Oregon, *Can. Entomol.*, 115, 199, 1983.

214. **Walters, K. F. A. and Dixon, A. F. G.,** The effect of host quality and crowding on the settling and take-off of cereal aphids, *Ann. Appl. Biol.*, 101, 211, 1982.

215. **Walters, K. F. A. and Dixon, A. F. G.,** Migratory urge and reproductive investment in aphids: variation within clones, *Oecologia*, 58, 70, 1983.

216. **Walters, K. F. A., Brough, C., and Dixon, A. F. G.,** Habitat quality and reproductive investment in aphids, *Ecol. Entomol.*, 13, 337, 1988.

217. **Ward, S. A. and Dixon, A. F. G.,** Selective resorption of aphid embryos and habitat changes relative to life span, *J. Anim. Ecol.*, 51, 859, 1982.

218. **Ward, S. A., Leather, S. R., and Dixon, A. F. G.,** Temperature prediction and the timing of sex in aphids, *Oecologia*, 62, 230, 1983.

219. **Ward, S. A., Wellings, P. W., and Dixon, A. F. G.,** The effect of reproductive investment on pre-reproductive mortality in aphids, *J. Anim. Ecol.*, 52, 305, 1983.

220. **Wassermann, S. S. and Asami, T.,** The effect of maternal age upon fitness of progeny in the southern cowpea weevil, *Callosobruchus maculatus*, *Oikos*, 45, 191, 1985.

221. **Wassermann, S. S. and Mitter, C.,** The relationship of body size to breadth of diet in some Lepidoptera, *Ecol. Entomol.*, 3, 115, 1978.

222. **Watt, A. D.,** The effect of cereal growth stages on the reproductive activity of *Sitobion avenae* and *Metopolophium dirhodum*, *Ann. Appl. Biol.*, 91, 147, 1979.

223. **Watt, A. D.,** The performance of the pine beauty moth on water-stressed lodgepole pine plants: a laboratory experiment, *Oecologia*, 70, 578, 1986.

224. **Watt, A. D.,** The effect of shoot growth stage of *Pinus contorta* and *Pinus sylvestris* on the growth and survival of *Panolis flammea* larvae, *Oecologia*, 72, 429, 1987.

224a. **Watt, A. D. and Leather, S. R.,** The pine beauty in Scottish lodgepole pine plantations, in *Dynamics of Forest Insect Populations*, Berryman, A. A., Ed., Plenum Press, New York, 1988, 243.

225. **Webb, J. W. and Moran, V. C.,** The influence of the host plant on the population dynamics of *Acizzia russellae* (Homoptera: Psyllidae), *Ecol. Entomol.*, 3, 313, 1978.

226. **Wellings, P. W. and Dixon, A. F. G.,** Physiological constraints on the reproductive activity of the sycamore aphid: the effect of developmental experience, *Entomol. Exp. Appl.*, 34, 227, 1983.

227. **Wellings, P. W., Leather, S. R., and Dixon, A. F. G.,** Seasonal variation in reproductive potential: a programmed feature of aphid life cycles, *J. Anim. Ecol.,* 49, 975, 1980.
228. **Wellington, W. G.,** Atmospheric circulation processes and insect ecology, *Can. Entomol.,* 86, 312, 1954.
229. **Wellington, W. G.,** Qualitative changes in populations in unstable environments, *Can. Entomol.,* 96, 436, 1964.
230. **Wellington, W. G. and Maelzer, D. A.,** Effects of farneysl methyl ether on the reproduction of the western tent caterpillar, *Malacosma pluviale:* some physiological, ecological and practical implications, *Can. Entomol.,* 99, 249, 1967.
231. **Werner, R. A.,** Influence of host foliage on development, survival, fecundity and oviposition of the spear-marked black moth, *Rheumaptera hastata* (Lepidoptera: Geometridae), *Can. Entomol.,* 111, 317, 1979.
232. **Wiklund, C. and Persson, A.,** Fecundity, and the relation of egg weight variation to offspring fitness in the speckled wood butterfly *Pararge aegeria,* or why don't female butterflies lay more eggs? *Oikos,* 40, 53, 1983.
233. **Wiktelius, S.,** Wind dispersal of insects, *Grana,* 20, 205, 1981.
234. **Williams, K. S.,** The coevolution of Euphydryas chalcedona butterflies and their larval host plants III. Oviposition behaviour and host plant quality, *Oecologia,* 56, 336, 1983.
235. **Wipking, W. and Neumann, D.,** Polymorphism in the larval hibernation strategy of the burnet moth, *Zygaena trifolii,* in *The Evolution of Insect Life Cycles,* Taylor, F. and Karban, R., Eds., Springer-Verlag, Berlin, 1986, 125.
236. **Wratten, S. D.,** Aggregation in the birch aphid, *Euceraphis punctipennis* (Zett.) in relation to food quality, *J. Anim. Ecol.,* 43, 191, 1974.
237. **Wratten, S. D., Edwards, P. J., and Winder, L.,** Insect herbivory in relation to changes in host plant quality, *Biol. J. Linn. Soc.,* 35, 339, 1988.
238. **Wright, L. C. and Cone, W. W.,** Population statistics for the asparagus aphid, *Brachycorynella asparagi* (Homoptera: Aphididae) on different ages of asparagus foliage, *Environ. Entomol.,* 17, 699, 1988.
239. **Zalucki, M. P.,** The effects of age and weather on egg laying in *Danaus plexippus* L. (Lepidoptera: Danaidae), *Res. Popul. Ecol.,* 23, 318, 1981.
240. **Zalucki, M. P., Brower, L. P., and Malcolm, S. B.,** Oviposition by *Danaus plexippus* in relation to cardenolide content of three *Asclepias* species in the southeastern U.S.A., *Ecol. Entomol.,* 15, 231, 1990.

Chapter 7

THE CHEWING HERBIVORE GUT LUMEN: PHYSICOCHEMICAL CONDITIONS AND THEIR IMPACT ON PLANT NUTRIENTS, ALLELOCHEMICALS, AND INSECT PATHOGENS

Heidi M. Appel

TABLE OF CONTENTS

0-8493-4125-6/94/$0.00+$.50

I. INTRODUCTION

Foliage comprises potential toxins, pathogens, and nutrients refractory to digestion. As a result, herbivorous insects have evolved a diverse array of physiological strategies to reduce toxicity, limit infection, and enhance nutrient extraction. These physiological strategies influence insect performance and host preference and influence our ability to manipulate pest species by natural and human-made toxins.

Nutrient extraction and most detoxification of food occurs in the gut. The gut is an irregular cylinder comprised of a single layer of epithelium surrounding a food-filled lumen. Food is digested in the lumen until small enough to be absorbed by the epithelium and is then transferred, with some repackaging, to the hemolymph. Until recently, the epithelium received most of the experimental attention, with studies of ultrastructure, ion-nutrient transport, digestive enzyme secretion, and detoxification. As a result, we are beginning to understand the events occurring *during* and *after* absorption of nutrients and toxins by the gut. What have received far less attention, however, are events occurring *prior* to absorption. With the exception of studies of digestive enzyme activity, we know very little about the physicochemical environment of the lumen and the biochemical transformations which occur there.

We do know, however, that food must undergo dramatic transformation in the gut lumen, such that it may bear little biochemical resemblance to what was in the leaf. While this is normally viewed as a consequence of insect digestive enzyme activity, it also results from the action of foliar enzymes and the physicochemical conditions of the gut lumen. Mastication, by destroying cellular compartmentalization, increases the activity of some plant enzymes, especially peroxidases and phenol oxidases. Many of these foliar enzymes are active in the gut lumen, resulting in unexpected transformations of food constituents. Similarly, some herbivores generate extreme physicochemical conditions of pH and redox potential in the lumen, which drive nonenzymatic transformations of food constituents. Thus, the availability of nutrients and the activity of allelochemicals and pathogens depend on the combined activity of insect enzymes, plant enzymes, and the physicochemical conditions in the gut lumen.

In this paper, the physicochemical environment of the gut lumen is described, recent work on nutrient extraction and the activity of allelochemicals and pathogens in the lumen is summarized, and additional biotransformations based on physicochemical conditions are predicted. Almost exclusively folivorous herbivores, that is, grasshoppers and katydids (Orthoptera), walking sticks (Phasmatodea), caterpillars (Lepidoptera), beetles (Coleoptera), and sawflies (Symphyta), are discussed although much of what is said is applicable to herbivores feeding on other plant parts and to detritivores. Three major conclusions arise from this review: 1) physicochemical conditions of the gut lumen vary dramatically among herbivores, 2) plant constituents, especially enzymes and oxidants, can be highly active in the gut lumen, and 3) gut physicochemical conditions and foliage constituents can cause significant biotransformations of nutrients, toxins, and pathogens prior to their digestion, absorption, or infection.

II. THE HERBIVORE GUT LUMEN

A. MORPHOLOGY

Plant tissue is low in protein and high in allelochemicals compared with animal tissue. As a result, the digestive tracts of herbivores are designed to maximize nutrient extraction and detoxification. The gut is large to process large volumes of food and occupies most of the body cavity in nonreproductive life stages. Gross morphology varies to some extent among herbivorous Orthoptera, Lepidoptera, Coleoptera, and Symphyta, and there is extensive specialization in ultrastructural and physiological traits important to the digestion and absorption of nutrients and the detoxification of allelochemicals. There are excellent reviews on the evolution of digestive systems,[77] midgut function,[20] and hindgut function.[61]

The gut is an irregular cylinder comprised of a single layer of epithelium surrounding a lumen filled with food. Herbivorous insects have three major gut regions, the foregut, midgut, and hindgut. The cuticle-lined foregut, consisting of the pharynx, esophagus, crop, and proventriculus, has no secretory cells but is often the initial site of food digestion by salivary or regurgitated midgut enzymes. While most herbivores have a well-developed foregut, some caterpillars have a poorly developed one in which the crop is lacking. At the posterior end of the foregut is the proventriculus, a valve which also functions as a grinding organ in some Orthoptera and Coleoptera.

The midgut is the major site of nutrient digestion and absorption.[20] The midgut lacks a cuticle and is lined by a peritrophic membrane which functions, in part, to protect the epithelium from the abrasions of ingested food. The multilayered peritrophic membrane, composed of chitin and protein, is produced by the midgut epithelium. It is delaminated from the microvilli of epithelial cells and is impermeable to molecules with a radius larger than 7 to 8 nm or to proteins larger than about 100,000 kDa,[68,92] although there is a report of a pore size of 200 nm in one species.[1] As a consequence, digestion is compartmentalized; large macromolecules are hydrolyzed within the space defined by the peritrophic membrane (endoperitrophic space) until they are small enough to diffuse into the ectoperitrophic space for further hydrolysis and absorption. In Orthoptera, gastric caecae around the anterior midgut participate in nutrient absorption. Caeca along the midgut of Coleoptera may function similarly. Lepidoptera and Symphyta lack gastric caecae.

The midgut epithelium is extremely active metabolically and is the best-studied portion of the gut, particularly in Lepidoptera. It is the site of active transport of nutrients from lumen to hemolymph as well as digestive enzyme synthesis and secretion. Midgut epithelium has extremely high rates of ion transport,[19] making it the "frog skin" of the invertebrate world. With a surface area greatly exceeding that of the external skeleton, the midgut must also function as a major barrier to allelochemicals and infectious microbes. As a consequence, the midgut epithelium produces many detoxification enzymes, including polysubstrate mono-oxygenases, reductases, glycosidases, esterases, epoxide hydrolases, glutathione transferases, and the antioxidant enzymes superoxide dismutase, catalase, and ascorbate reductase.[34,46]

The hindgut is a major site of resorption of water and ions from the food bolus and from urine secreted by the Malpighian tubules, which empty at the anterior end of the hindgut.[61] Most of our information about hindgut function comes from studies of Orthoptera. The hindgut is cuticle-lined and impermeable to molecules with a radius larger than about 0.6 nm. The hindgut may also be a site of digestion by midgut enzymes or by micro-organisms in hindgut caeca of Coleoptera. The midgut and hindgut epithelia are bathed on their outer surfaces by hemolymph, in which absorbed nutrients are transported to the rest of the body.

There is little physical mixing of the particulate contents of the gut lumen. As a consequence, masticated food is plant-like in the foregut but increasingly digested and transformed in the midgut and hindgut. In contrast, there can be substantial mixing of the fluid contents of the lumen. Although most of the radial movement of digestive enzymes and their products is accomplished by diffusion, there is posterior-to-anterior movement of gut fluid in the ectoperitrophic space of some herbivores. In grasshoppers, gut fluid in the ectoperitrophic space moves anteriorly in the midgut, driven by nutrient absorption in anterior gastric caecae and water secretion in the posterior midgut.[18] Although this countercurrent model has also been proposed for caterpillars,[66,67] the rapid movement of food through the gut, the absence of caecae, and the lack of evidence for water secretion by the posterior midgut make this unlikely.[20]

B. LUMEN MICROBIOLOGY

The gut microbiology of folivores is poorly characterized because it has been assumed that microbes play only a minor, if any, role in the digestion and nutrition of this group. This assumption is based on the absence of specialized structures along the gut to harbor resident microbes in caterpillars, the presumed absence of resident microbes in Orthoptera, the lack of studies of microbes in folivorous beetles, and the rapid throughput of food in many folivores, providing little time for metabolic contributions from facultative species. Anaerobic bacterial fermentation has been demonstrated in the hindguts of wood-feeding and detritivorous beetles, roaches, and termites but has not been examined in folivores.

There are reports of facultative and epimurally-attached bacteria whose function is unknown in folivores. There are facultative bacteria and yeast in the gut lumen of the tobacco hornworm[80] and facultative bacteria in the gut lumen of velvetworm caterpillars, whose composition is influenced by host plant.[45] There are bacteria attached epimurally to the foregut and hindgut epithelium and incorporated into the peritrophic membrane of migratory grasshoppers[54] and bacteria attached to rectal setae in mole crickets.[58] Resident or facultative microbes may produce antimicrobial substances that improve disease resistance. In the desert locust, antifungal phenolics are often present in the gut lumen but are absent in bacteria-free individuals,[17] although a microbial origin for these phenolics has not been established clearly.

C. LUMEN PHYSICOCHEMISTRY

Folivores have evolved multiple solutions to the problem of extracting foliar nutrients made less digestible by their association with allelochemicals. As a consequence, there are large differences among folivores in the physicochemical conditions of the digestive tract. These physicochemical conditions are ones that influence the redox reactions of digestion and detoxification and the infectivity of pathogenic microbes. They include the availability of hydrogen ions (pH) and electrons (Eh). Measured by pH and redox electrodes, pH and Eh reflect the net availability of protons and electrons in biochemical reactions in the lumen. Although they do not describe the dynamics of specific redox reactions, they indicate the likelihood of overall proton and electron gain and loss and thus the likelihood of individual redox reactions.[5]

Most caterpillars, all sawflies examined, and at least one species of walking stick maintain extreme physicochemical conditions in the digestive tract and high rates of food passage through the gut. They maintain alkaline midguts ranging from pH 8 to 12.[3,5,6,10] Alkaline pHs increase the solubility of foliar protein[33] and may reduce the activity of some allelochemicals (see Section IV). Although the midgut lumens of caterpillars, sawflies, and walking sticks are alkaline, they differ widely in midgut redox potentials, ranging from highly oxidizing (+400 mV) to highly reducing (–200 mV).[3,5] The impact of midgut redox conditions on nutrient extraction and the activity of allelochemicals is unknown (see next section). Food retention times average 0.5 to 2.0 h on preferred host plants during the active feeding stages of development.

Grasshoppers and katydids maintain pH neutral, mildly oxidizing gut conditions and moderate rates of food passage. They have midgut lumens with pH ranging from 5 to 8, and with Eh from 0 to +300mV.[3] Food retention times often exceed 8 h on preferred host plants during the active feeding stages of development.[8,73] Beetles maintain moderate to long rates of food passage and lumens with neutral to acid pH, although a few species have mildly alkaline midguts.[14] Gut redox conditions have not been measured in folivorous beetles.

Physicochemical conditions of the gut lumen are primarily insect-driven in caterpillars. Midgut lumen pH is closely regulated by the insect, and although macerated foliage varies in pH from acid to neutral, it is rapidly titrated to the alkaline pH of the midgut.[3,71] This has not been examined for sawflies, walking sticks, and beetles. Alkaline pHs are generated in caterpillars by a powerful potassium pump,[17] with significant cost to the insect; Dow has estimated that maintenance of highly alkaline gut pHs in caterpillars may consume 10% of their ATP.[20] The particular costs depend on the initial leaf pH and its buffering capacity, both of which vary among plant species.[3,71] Thus, plants may differ in the energy costs of pH titration they impose on folivores.

Midgut lumen Eh is also regulated by caterpillars. Although macerated foliage varies in Eh from highly to mildly oxidizing, it is rapidly titrated to the average Eh of the folivore gut lumen.[3] The mechanism of Eh regulation is unknown. In the keratin-feeding webbing clothes moth, the reducing conditions of the gut lumen are maintained by sulfides released from cysteine by the action of cysteine lyases.[94] Bismuth-sulfide accumulates in the posterior midgut epithelium of tobacco hornworm caterpillars fed bismuth subnitrate, suggesting that sulfur may also play an important role in maintenance of reducing conditions in folivores.[57] The metabolic and nutritional costs to the insect of maintaining a reducing gut lumen have not been determined. If reducing equivalents are sulfur-based, they would impose significant sulfur demands on larval nutrition and may influence the evolution of host choice (see Section VI).

Although lumen pH and Eh are regulated by the folivore, foliage constituents can influence the rate of particular redox reactions. Foliage contains a large number of oxidants and reductants, including tocopherols, flavonoids, chromenes, benzofurans, furanocoumarins, organic peroxides, and metal cations.[42] In addition, foliage contains oxidizing and reducing enzymes,[42] some of which retain their activity in the insect gut.[31] As a consequence, foliage constituents other than nutrients and allelochemicals may influence nutrient availability and allelochemical toxicity.

A major consequence of differences among folivores in gut physicochemical conditions is that foliage may undergo dramatically different biochemical transformations in the guts of different herbivores. Even among herbivores feeding on the same plant, the availability of nutrients, the activity of allelochemicals, and the infectivity of pathogenic microbes may differ widely.

III. DIGESTION AND ABSORPTION OF NUTRIENTS

A. PROTEINS

The growth of folivores is often nitrogen-limited.[52] The digestion of foliar protein requires the lysis of plant cell walls and membranes and the solubilization of cellular protein. Although some maceration of plant cells occurs during ingestion, most cell lysis occurs in the gut by the action of phospholipases and carbohydrases, and, in the case of caterpillars, sawflies, and walking sticks, by alkaline hydrolysis. Once solubilized, the lysis of foliar protein is accomplished by digestive proteinases.

In caterpillars, trypsin- and/or chymotrypsin-like serine proteinases hydrolyze internal peptide bonds of polypeptides, and carboxypeptidases and aminopeptidases hydrolyze terminal amino acids from the polypeptide fragments.[40,44] Produced by the midgut epithelium, serine proteinases are thought to be transported to the lumen in vesicles produced by the microvilli, where they are released by alkaline solubilization and/or exocytosis.[24,67] They may also be incorporated in the peritrophic membrane as it is delaminated from the microvilli.[25,68] Carboxy- and amino-peptidases are found in the midgut epithelium and the gut lumen, although they may be restricted to the ectoperitrophic space by their comparatively larger size.[68] Amino acids are transported from the lumen across the midgut epithelium via a potassium cotransport system, in which the electrochemical gradient established by potassium transport drives amino acid transport.[35] Amino acids can be transported to the hemolymph from the cells by a passive, potassium-independent amino acid exchange mechanism.[59]

Little is known about protein digestion in grasshoppers, and digestive proteinases have not been characterized. Amino acids are absorbed (probably passively) by the midgut caecae along with water.[56,83] In walking sticks, amino acid absorption appears similar to that of caterpillars.[64]

In beetles, much of protein digestion results from the action of cysteine proteinases, rather than the trypsin-like serine proteinases of caterpillars.[57,93] These tend to have pH optima in the acid to neutral range, unlike the highly alkaline optima of lepidopteran proteinases. Although the biochemistry and morphological organization of protein digestion are assumed to be similar to other herbivores, they have not been investigated in folivorous beetles.

Lumen physicochemistry is likely to have a major impact on amino acid availability through its influence on the efficiency of extraction. The extraction of protein from foliage will be greater under highly alkaline or acid conditions or under highly reducing or oxidizing conditions. The solubility of foliar protein has been shown to increase at higher pHs, increasing the availability of essential amino acids to caterpillars.[33] Differences in gut redox potential are also likely to influence protein solubility but have not been examined. Allelochemicals, especially phenolics and protease inhibitors, can have a major impact on amino acid availability by processes that are pH and Eh dependent (see Section IV).

B. CARBOHYDRATES

Folivores digest most leaf carbohydrates except cellulose and are not usually considered carbohydrate-limited on natural diets. Caterpillars and grasshoppers produce amylases, glucosidases, galactosidases, and sometimes pectinases and fructosidases.[55,78] In caterpillars, carbohydrases are produced by the midgut epithelium and transported in vesicles to the lumen.[67] In grasshoppers, carbohydrate digestion begins in the foregut with amylases produced by salivary glands.[2,26] Digestion may also occur in the foreguts of caterpillars by enzymes in regurgitated midgut fluid.

The absorption of carbohydrates has not been examined in the same detail as amino acid and lipid absorption but is assumed to be passive. In grasshoppers, absorption is greatest in the anterior midgut and caeca.[81,82] Absorption is unstudied in caterpillars and folivorous beetles.

Lumen physicochemistry is likely to influence the availability of carbohydrates by its effect on the efficiency of extraction. As with protein, the extraction of carbohydrates from foliage will be greater under highly alkaline or acid conditions or under highly reducing or oxidizing conditions. The interactions of carbohydrates with other foliage constituents, such as metals and phenolics, are also likely to be influenced by lumen conditions.

C. LIPIDS

Lipids are important to folivorous insects as nutrients and as gut surfactants, which facilitate the absorption of nutrients. The digestion of lipids has been studied in caterpillars and, to some extent, grasshoppers but is uncharacterized in other folivores. Leaf glycero- and phospholipids (primarily galactosyl diglycerides, phosphatidylcholines, phosphatidylethanolamines, and phosphatidylglycerols) are enzymatically hydrolyzed in the gut lumen of caterpillars to free fatty acids, sugar residues, glycerol, phosphatidic acid, choline, and ethanolamine.[79,89-91] Little is known about the enzymes responsible. Individual lipid components are absorbed by the midgut epithelium, resynthesized to di- and triglycerides, and released as diacylglycerols to lipophorins, the hemolymph transport proteins.[85,89] In one species of caterpillar, *Heliothis zea* (Lepidoptera: Noctuidae), hydrolyzed lipids are resynthesized in the lumen and then rehydrolyzed for absorption.[87] There is morphological specialization in lipid absorption in some species, in which absorption is greatest in the anterior two-thirds of the midgut epithelium.[88] Absorption, metabolism, and/or transport of fatty acids in the midgut epithelium may involve fatty acid binding proteins in the midgut cytosol of some species.[75]

Midgut surfactants solubilize lipophilic compounds by reducing the surface tension of midgut fluid. Lysolecithins, primarily lysophosphatidylcholine and lysophosphatidylethanolamine, are synthesized in caterpillars from ingested lysolecithins and the enzymatic hydrolysis of ingested dietary lecithin.[86] Not all gut surfactants may be of dietary origin. Surfactant-producing micro-organisms have been isolated from the midgut of a sawfly, although their contribution *in vivo* has not been demonstrated.[60]

Lipids and surfactants can interact with a variety of other compounds. Gut surfactants have been shown to prevent certain types of protein-phenolic binding, to increase the solubility of foliar proteins, and to inhibit the activity of some foliar enzymes in the midgut lumens of caterpillars.[33,51] Gut surfactants can be precipitated by phenolics from gut fluid *in vitro*, indicating that phenolics could impose lipid deficiency and interfere with normal digestion and absorption.[16] Gut lipids can be substrates for foliar lipoxygenases, generating radicals like hydroperoxides, carbonyls, and epoxides in the lumen.[72] Phospholipids may also synergize the antioxidant properties of some phenolics,[38] protecting midgut membranes against damage from dietary oxidants. Although interactions among dietary lipids, surfactants, and other food constituents are rarely considered in studies of insect digestion, they warrant further examination.

Lumen physicochemistry is likely to have a major influence on the interactions of lipids and surfactants with other compounds. The type of bond and the participants involved depend on the redox state of the gut. Micellar interactions and the formation of hydrophobic, hydrogen, or covalent bonds by the fatty acid, glycerol, or base moieties of lipids will be sensitive to the pH and redox conditions of the lumen. These effects may extend to interactions at the midgut epithelium, including membrane transport and cell recognition phenomena.

D. MINERALS

Although it is often assumed that herbivorous insects are not mineral limited, there are no quantitative studies of dietary mineral requirements or availability.[53] The numerous *in vitro* studies of ion transport across gut epithelia indicate that minerals are probably absorbed as ions by active and passive transport mechanisms. Their absorption may be coupled with that of other nutrients, such as the potassium–amino acid cotransport system in the midgut epithelium of caterpillars. The impact of lumen pH and Eh on mineral availability has not been examined, but mineral oxidation state is sensitive to pH and Eh. For example, manganese, iron, and sulfur are oxides at neutral to alkaline pHs under oxidizing conditions. Furthermore, under alkaline, oxidizing conditions, metals may be complexed to phenolics, lipids, or carbohydrates. Thus, minerals may be in different forms in the lumen of herbivores with different gut conditions, and be differentially available as nutrients.

IV. ACTIVITY OF ALLELOCHEMICALS

Although nutrient–allelochemical interactions other than protein–tannin interactions have been emphasized by several investigators,[39,74] it is only recently that their significance has been widely appreciated. We now know that there are complex interactions between a variety of foliage constituents that influence the availability of nutrients and the activity of allelochemicals. Foliar phenolics, proteinase inhibitors, alkaloids, saponins, and hydroxamic acids interact with nutrients in the gut lumen to reduce herbivore growth. Some of these interactions are driven by foliar enzymes, which have been shown to be active in the gut lumen.[23,24] Gut physicochemical conditions may inhibit or stimulate interactions by their impact on the redox state and stability of allelochemicals.

A. PHENOLICS

Plant phenolics influence the food preference and performance of a large diversity of insect herbivores by acting as feeding deterrents, binding agents, and apparent toxins.[12,69] Their action as binding agents and toxins occurs in the gut lumen and is discussed here. Phenolics, including tannins, were initially assumed to reduce digestion by forming hydrogen bonds with dietary protein or digestive enzymes in herbivore guts,[27,28] but this mode of action is unlikely in many herbivores.[4] There are no demonstrations of hydrogen bond formation in the insect gut. Some authors have attributed the absence of hydrogen binding to gut surfactants,[50,51] and presumably hydrogen bonding could occur when surfactants are depleted. There is, however, evidence that covalent phenolic binding inhibits digestion in the gut of caterpillars. In tomato, the monomeric phenolic chlorogenic acid forms covalent bonds with amino acids that inhibit their digestion.[31,33] Binding depends on the oxidation of chlorogenic acid to chloroquinone by foliar polyphenol oxidases that retain half of their activity in the alkaline (pH 8 to 9) gut.[31] Phenolics may also impair lipid digestion by covalent binding in the lumen, as phenolic precipitation of gut fluid surfactants is higher at alkaline pHs conducive to oxidation.[16]

Phenolic toxicity has been invoked to explain reductions in growth on phenolic-containing diets when there is no digestion inhibition. Oxygen radicals, generated during phenolic oxidation, are proposed to disrupt membrane integrity and metabolism in the gut epithelium and may be responsible for the gut lesions observed in herbivores exposed to novel tannins.[11,62,76] Oxidized phenolics can also inhibit antioxidant enzyme systems of herbivores,[43] thereby enhancing the effect of oxygen radicals they generate. Herbivores have several antioxidant enzyme systems, which scavenge oxygen radicals generated by phenolic oxidation. These consist of the superoxide dismutase, catalase, glutathione peroxidase, and glutathione reductase complex[46] and the ascorbate-free radical reductase, dehydroascorbate reductase, and glutathione reductase complex.[34] The value of an antioxidant enzyme system depends on its location relative to where the oxygen radicals are generated. If radicals are generated in the gut lumen, then a system in the gut epithelium may be of little benefit in preventing membrane damage or enzyme inactivation. Unfortunately, investigators routinely assay enzyme activity of whole insects or specific tissues and not of the gut lumen. Luminal activities can be expected to be high in herbivores, except when reducing redox conditions prevent oxygen radical formation.

Lumen physicochemistry is critical to the action of phenolics because it influences phenolic oxidation state. Phenolics only participate in covalent bonds with nutrients or generate oxygen radicals when they are oxidized to quinones. Phenolic oxidation may occur nonenzymatically by autoxidation at pHs greater than 9, the pKa of the phenolic hydroxyl, or enzymatically by phenol oxidases and peroxidases in foliage.[4,23] In general, alkaline, oxidizing conditions promote autoxidation, and reducing conditions promote phenolic reduction. Thus, phenolics are more likely to be active in the oxidizing midguts of some herbivores and less likely to be active in the reducing midguts of other herbivores.

Lumen physicochemistry may also be important to the activity of phenolics because it influences hydrolysis. The degree to which phenolics are hydrolyzed in the gut depends on gut conditions, phenolic structure, and enzyme activity. In general, monomers are more readily hydrolyzed than polymers, and hydrolyzable tannins more readily than condensed tannins. Hydrolyzable tannins (gallotannins and ellagitannins) are readily hydrolyzed nonenzymatically under acidic and basic conditions,[36] such as those encountered in many folivore guts. However, tannic acid remains intact in the digestive tract of some caterpillars.[9] Condensed tannins are only oxidatively degraded under strongly acid conditions in the presence of metal catalysts,[36] such as those encountered in microbially-mediated fermentation.

If phenolic activity in herbivores depends on phenolic oxidation, then it would at first appear maladaptive to maintain an alkaline, oxidizing gut. Yet large numbers of caterpillars, sawflies, and some walking sticks do so, even on host plants with high phenolic concentrations. There may be advantages in maintaining alkaline, oxidizing guts that outweigh disadvantages of phenolic oxidation, such as enhanced nutrient extraction and inactivation of foliar enzymes.[33] In addition, oxidized phenolics protect some caterpillars from microbial pathogens, resulting in enhanced survivorship at the expense of slowed growth (see Section V).[6,70] Furthermore, some folivores may inactivate ingested phenolics by oxidative polymerization and/or binding with other compounds.[3,63] Thus, conditions promoting phenolic oxidation may result in a net benefit to the folivore when all aspects of life history are considered.

B. OTHER ALLELOCHEMICALS

Several other classes of allelochemicals are known to be active in the gut lumen and are likely to be influenced by lumen physicochemistry, including proteinase inhibitors, alkaloids, terpene aldehydes, saponins, and hydroxamic acids. Proteinase inhibitors are produced by a number of plant taxa in response to insect damage, and they inhibit the digestive proteinases of some herbivores.[65] In caterpillars, proteinase inhibitors reduce growth by the depletion of sulfur-containing amino acid reserves resulting from the hyperproduction of serine proteinases.[13] Their activity is influenced by other foliage constituents in the lumen, in much the same way as are dietary proteins. Chlorogenoquinone, generated from chlorogenic acid by foliar polyphenol oxidases in the gut lumen, binds and inactivates proteinase inhibitors in tomato, reducing their activity in caterpillars.[29] Thus, the activity of proteinase inhibitors will depend not only on their concentration but also on the concentration of phenolics and their oxidation by phenol oxidases. Furthermore, differences among folivores in lumen physicochemistry may result in different rates of phenolic oxidation and proteinase inhibitor inactivation on similar foliage.

Alkaloids are usually considered to have their major effects on herbivores after they are absorbed from the digestive tract. However, there are numerous examples of alkaloid inhibition of carbohydrases *in vitro*,[21] and their potential activity in the gut lumen of herbivores warrants investigation. Lumen physicochemistry is likely to have a major influence on the absorption of alkaloids by the gut epithelium. The tertiary amine of alkaloids is amphipathic, remaining largely protonated at acid pHs and unprotonated at alkaline pHs. As a consequence, alkaloids are more hydrophilic and less readily absorbed at acid pHs and more lipophilic and readily absorbed at alkaline pHs.

Plant allelochemicals other than phenolics may inhibit digestion by binding to proteins in the gut lumen. The terpene aldehyde gossypol binds to gut enzymes and to cottonseed protein, reducing amino acid availability in mammals, as it may also do in insects.[74] Alkylisothiocyanates, produced by damage-induced hydrolysis in some crucifers, may bind and reduce the digestibility of dietary protein.[22] Saponins have been shown to inhibit the growth of some insect herbivores. The mechanism by which they do so has not been determined but probably involves inhibition of proteases and/or lipid and sterol metabolism in the gut lumen, as addition of protein, cholesterol, or β-sitosterol alleviates growth inhibition.[37]

DIMBOA, a hydroxamic acid produced by cereal grains, reduces growth of some caterpillars by inhibition of serine proteinases in the gut lumen.[15] Lumen physicochemistry is likely to have a major influence on the activity of DIMBOA, because conversion to the active form is a pH-dependent process. The inhibition is accomplished by the aldol tautomer, which is formed from DIMBOA in aqueous solution at a rate that is accelerated at alkaline pHs.[15]

V. TRITROPHIC LEVEL INTERACTIONS

Many viruses and bacteria, including microbial pesticides, are ingested on foliage and infect insects through the midgut epithelium. As a consequence, plant allelochemicals and lumen physicochemistry can have a major impact on pathogen activity. Several classes of allelochemicals have been shown to influence pathogens, including phenolics, proteinase inhibitors, nonprotein amino acids, and alkaloids. Their effects can be inhibitory or stimulatory, depending on the allelochemical and system examined. The viability of pathogens can be reduced by digestion of microbial proteins by insect enzymes and by denaturing by gut lumen conditions that are highly acid or alkaline or highly oxidizing or reducing. The majority of work in this area has focused on pathogens used in microbial pesticides, although lumen events are likely to influence other pathogens as well.

A. BACULOVIRUSES

Plant phenolics inhibit infection of some caterpillars by naturally occurring baculoviruses. The simple phenolic chlorogenic acid inhibits infection of several noctuid caterpillars by a baculovirus.[32] Chlorogenic acid, oxidized by tomato phenol oxidases in the gut lumen, inhibits infection by formation of covalent bonds with the virus. Polymeric oak phenolics (tannins) inhibit viral infection of gypsy moth caterpillars[70] also by a mechanism requiring polyphenol oxidase activity.[3] Foliar constituents other than phenolics and phenol oxidases may influence baculovirus activity but have not been examined.

B. *BACILLUS THURINGIENSIS*

Although not a pathogen in the strict sense because it does not reproduce in insects, the soil bacterium *Bacillus thuringiensis* produces a toxin that kills insects by disrupting midgut epithelium cells. The toxicity of the protein endotoxin is modified by several classes of plant allelochemicals, and the effect is stimulatory or inhibitory depending on the allelochemical examined. For example, proteinase inhibitors increase the toxicity of endotoxin to caterpillars and colorado potato beetles,[49] and nonprotein amino acids increase endotoxin toxicity to tobacco hornworm caterpillars.[30] However, the alkaloid nicotine reduces toxicity to tobacco hornworm caterpillars.[41] Phenolics appear to influence the activity of endotoxin in a size-dependent manner: polymeric phenolics (tannins) reduce toxicity, whereas monomeric phenolics enhance it. Tannins reduce the toxicity of endotoxin to gypsy moth caterpillars[3] and to cabbage butterfly caterpillars.[48] In contrast, the simple phenolic chlorogenic acid, oxidatively activated by tomato polyphenol oxidases, increases the toxicity of endotoxin to corn earworm caterpillars.[47] Similarly, phenolic glycosides of aspen increase toxicity of endotoxin to gypsy moth caterpillars.[7]

There are several systems in which plant allelochemicals have been shown to have no influence on endotoxin activity. Linear furanocoumarins in celery have no effect on endotoxin toxicity to armyworm caterpillars,[84] and the flavonoid phenolic rutin does not influence endotoxin activity to tobacco hornworm caterpillars.[41]

C. DESIGN OF MICROBIAL PESTICIDES

Events occurring in the gut lumen that influence activity of microbial pathogens provide targets for improvements in microbial pesticides. Plant allelochemicals that enhance the activity of microbial pathogens are potentially useful adjuvants in commercial formulations. Allelochemicals that reduce the activity of microbial pathogens may be prevented from doing so by appropriate adjuvants.

Insect enzymes and physicochemical conditions of the gut lumen are also a barrier to pathogen infection and thus to microbial pesticides. To the extent that gut conditions vary among target insects, efficacy may also vary. Gut barriers may be reduced by improvements in the stability of pathogens in the gut lumen to proteinases and extremes of pH and redox potential. Although this stability may arise from structural changes in the viral or bacterial proteins, it may also be achieved by appropriate adjuvants. These adjuvants could include inhibitors of insect proteinases, already shown to be effective in noncommerical formulations (see above), or compounds that alter gut physicochemical conditions like pH and Eh.

Microbial pesticides are a rapidly expanding economic market. Of major importance as the site of infection, the gut is also rapidly attracting interest. However, efficient selection of potential adjuvants to reduce host plant resistance requires knowledge of the mode of action of allelochemicals on microbial pesticides. Similarly, it requires understanding of the mode of action of gut physicochemistry on pathogen viability. As a consequence, ecological and physiological studies will become increasingly important, in order to identify sources and mechanisms of resistance important to formulation improvement.

VI. GUT CONDITIONS AND EVOLUTION OF HOST CHOICE

An herbivore's ability to avoid toxicity, disease, and predation while obtaining nutrients determines its host plant preference in evolutionary time. The gut lumen, as a major site of digestion, allelochemical activity, and pathogen infection, represents a major arena for the interaction of these factors. A failure to appreciate their interdependence results in an inability to explain the variation we observe in any one factor over ecological time. A failure to appreciate their interdependence also results in an inability to appreciate the constraints imposed by other factors in evolutionary time. Many of the controversies among physiological and chemical ecologists working in plant-insect interactions have resulted from such a single factor approach.

In this review, two major sources of variation in herbivore performance are highlighted that are often overlooked: constitutive differences among insects and facultative differences among host plants in the biochemical background against which nutrient acquisition, allelochemical activity, and pathogen infection occur. Their importance to herbivore performance has significant consequences for plant and insect evolution. To plants, variation among herbivores in gut physicochemical conditions represents a suite of highly divergent selection pressures from an herbivore community. To insects, variation among host plants in the biochemical background they produce in the gut, against which nutrient acquisition, allelochemical activity, and pathogen infection occur, requires a complex accounting of chemical traits of hosts that may vary in space and time.

The widely divergent gut conditions observed in herbivorous insects suggest that there are multiple strategies for dealing with these tradeoffs. Evolution of a particular gut condition probably results from a combination of host plant traits (allelochemical type, structure, and concentration, availability of specific nutrients), pathogen traits (mode of infection, sensitivity to host plant allelochemicals), and phylogenetic constraint. For example, classes of allelochemicals will differ in their sensitivity to gut redox conditions. The activity of phenolics, in particular, will be highly dependent on gut redox state. Within a class, structural differences will influence the rate of hydrolysis under highly alkaline or acidic gut conditions or the ease of oxidation and reduction. Even the concentration of allelochemical commonly encountered by the herbivore may set limits on the usefulness of certain gut conditions. The availability of particular nutrients may also constrain the evolution of gut conditions. Reducing guts may be possible only when host plants provide an adequate supply of reducing equivalents, like sulfur-based amino acids. Similarly, the pathogens of herbivores may influence gut conditions if they infect via the gut and are sensitive to host plant allelochemical and/ or gut physicochemical conditions. Selection may favor gut conditions inhibitory to pathogens, even at the expense of growth.

VII. CONCLUSIONS

The gut is the primary arena in which the variable constituents of food, including nutrients, allelochemicals, and pathogens, are transformed for absorption and use. As a consequence, events occurring there have a major impact on herbivore performance and evolution. Although the gut is known as the site of digestion and absorption, most attention has focused on the gut epithelium, with the underlying assumption that events occurring in the lumen were either unimportant or too complex to explore experimentally. There is now sufficient evidence, however, to suggest that events in the lumen are extremely important to the availability of nutrients and the activity of toxins and pathogens in insect herbivores. Physicochemical conditions of the gut lumen vary dramatically among herbivores, suggesting differential adaptation to host plants and pathogens. Plant constituents other than nutrients and allelochemicals, especially enzymes and oxidants, can be highly active in the gut lumen. Gut physicochemical conditions and foliage constituents can cause significant biotransformations of nutrients, toxins, and pathogens prior to their digestion, absorption, or infection.

To scientists working on plant–insect interactions, the dependence of nutrients, allelochemicals, and pathogens on biochemical events occurring in the gut lumen has four important consequences. First, what is measured in the leaf may not predict the response of the herbivore. If nutrients, allelochemicals, and pathogens undergo significant transformations in the lumen by other foliar constituents or gut conditions, then their availability or activity may not be correlated with foliar levels or dose. Second, the response of one herbivore may not predict the response of others if gut conditions vary among them. If nutrients, allelochemicals, or pathogens undergo significant transformations in the lumen as a result of physicochemical conditions, then we also need to know how herbivores differ in these traits. Third, artificial diet experiments may not predict what happens with foliage if they lack the additional foliar constituents responsible for transformations in the lumen. Fourth, *in vitro* experiments may provide little information about what happens *in vivo* if they are not designed to mimic the physicochemical conditions of the lumen.

REFERENCES

1. **Adang, M. J. and Spence, K. D.**, Permeability of the peritrophic membrane of the douglas fir tussock moth (*Orgyia pseudotsugata*), *Comp. Biochem. Physiol.*, 75a, 233, 1983.
2. **Anstee, J. H. and Charnley, A. K.**, Effects of frontal ganglion removal and starvation on activity and distribution of six gut enzymes in *Locusta*, *J. Insect Physiol.*, 23, 965, 1977.
3. **Appel, H. M.**, unpublished data.
4. **Appel, H. M.**, Phenolics in ecological interactions: the importance of oxidation, *J. Chem. Ecol.*, 19, 1521, 1993.
5. **Appel, H. M. and Martin, M. M.**, Gut redox conditions in herbivorous lepidopteran larvae, *J. Chem. Ecol.*, 16, 3277, 1990.
6. **Appel, H. M. and Schultz, J. C.**, Activity of phenolics in insects may require oxidation, in *Plant Polyphenols: Biogenesis, Chemical Properties, and Significance*, Hemingway, R. W., Ed., Plenum Press, 1992, 609.
7. **Arteel, G. E. and Lindroth, R. L.**, Effects of aspen phenolic glycosides on gypsy moth (Lepidoptera: Lymantriidae) susceptibility to *Bacillus thuringiensis*, *Great Lakes Entomol.*, 25, 239, 1992.
8. **Baines, D. M., Bernays, E. A., and Leather, E. M.**, Movement of food through the gut of fifth-instar males of *Locusta migratoria* migratorioides, *Acrida*, 2, 319, 1973.
9. **Barbehenn, R. V. and Martin, M. M.**, The protective role of the peritrophic membrane in the tannin-tolerant larva of *Orgyia leucostigma* (Lepidoptera), *J. Chem Ecol.*, in press.
10. **Berenbaum, M.**, Adaptive significance of midgut pH in larval Lepidoptera, *Am. Nat.* 115, 138, 1980.
11. **Bernays, E. A.**, Tannins: an alternative viewpoint, *Entomol. Exp. Appl.* 34, 245, 1978.
12. **Bernays, E. A., Cooper Driver, G., and Bilgener, M.**, Herbivores and plant tannins, *Adv. Ecol. Res.* 19, 263, 1989.
13. **Broadway, R. M. and Duffey, S. S.**, Plant proteinase inhibitors: mechanism of action and effect on the growth and digestive physiology of larval *Heliothis zea* and *Spodoptera exigua*, *J. Insect Physiol.*, 32, 827, 1986.
14. **Crowson, R. A.**, *The Biology of the Coleoptera*, Academic Press, N. Y., 1981.
15. **Cuevas, L., Niemeyer, H. M., and Perez, F. J.**, Reaction of DIMBOA, a resistance factor from cereals, with alpha-chymotrypsin, *Phytochemistry*, 29, 1429, 1990.
16. **De Veau, E. J. I. and Schultz, J. C.**, Reassessment of the interaction between gut detergents and phenolics in Lepidoptera and significance for gypsy moth larvae, *J. Chem. Ecol.*, 18, 1437, 1992.
17. **Dillon, R. J. and Charnley, A. K.**, Inhibition of *Metarhizium anisopliae* by the gut bacterial flora of the desert locust: characterisation of antifungal toxins, *Can. J. Microbiol.*, 34, 1075, 1988.
18. **Dow, J. A. T.**, Countercurrent flows, water movements and nutrient absorption in the locust midgut, *J. Insect Physiol.*, 27, 579, 1981.
19. **Dow, J. A. T.**, Extremely high pH in biological systems: a model for carbonate transport, *Am. J. Physiol.*, 246, R633, 1984.

20. **Dow, J. A. T.**, Insect midgut function, *Adv. Insect Physiol.*, 19, 187, 1986.
21. **Dreyer, D. L., Jones, K. C., and Molyneux, R. J.**, Feeding deterrency of some pyrrolizidine indolizidine and quinolizidine alkaloids towards pea aphid (*Acyrthosiphon pisum*) and evidence for phloem transport of indolizidine alkaloid swainsonine, *J. Chem. Ecol.*, 11, 1045, 1985.
22. **Duffey, S. S. and Felton, G. W.**, Plant enzymes in resistance to insects, in *Biocatalysis in Agricultural Biotechnology*, Whitaker, J. R. and Sonnett, P. E., Eds., Am. Chem. Soc. Symp. Series 389, Washington, D.C., 1989, 166.
23. **Duffey, S. S. and Felton, G. W.**, Enzymatic antinutritive defenses of the tomato plant against insects, in *Naturally Occuring Pest Bioregulators*, Hedin, P. A., Ed., Am. Chem. Soc. Symp. Series 449, Washington, D.C., 1991, 166.
24. **Eguchi, M. and Arai, M.**, Relationship between alkaline proteases from the midgut lumen and epithelia of the silkworm: solubilization and activation of epithelial protease (6B3), *Comp. Biochem. Physiol.*, 75B, 589, 1983.
25. **Eguchi, M., Iwamoto, A., and Yamauchi, K.**, Interrelation of proteases from the midgut lumen, epithelium, and peritrophic membrane of the silkworm, *Bombyx mori* L., *Comp. Biochem. Physiol.*, 72A, 359, 1982.
26. **Evans, W. A. L. and Payne, D. W.**, Carbohydrases aof the alimentary tract of the desert locust, *Schistocerca gregaria* Forsk., *J. Insect Physiol.*, 10, 657-674.
27. **Feeny, P.**, Inhibitory effect of oak leaf tannins on the hydrolysis of proteins by trypsin, *Phytochemistry*, 8, 2119, 1969.
28. **Feeny, P.**, Seasonal changes in oak leaf tannins and nutrients as a cause of spring feeding by winter moth caterpillars, *Ecology*, 51, 565, 1970.
29. **Felton, G. W., Broadway, R. M., and Duffey, S. S.**, Inactivation of protease inhibitor activity by plant-derived quinones: complications for host-plant resistance against noctuid herbivores, *J. Insect Physiol.*, 35, 981, 1989.
30. **Felton, G. W. and Dahlman, D. L.**, Allelochemical induced stress: effects of L-canavanine on the pathogenicity of *Bacillus thuringiensis* in *Manduca sexta*, *J. Invert. Pathol.*, 44, 187, 1984.
31. **Felton, G. W., Donato, K., Del Vecchio, R. J., and Duffey, S. S.**, Activation of plant foliar oxidases by insect feeding reduces nutritive quality of foliage for noctuid herbivores, *J. Chem. Ecol.*, 12, 2667, 1989.
32. **Felton, G. W. and Duffey, S. S.**, Inactivation of baculovirus by quinones formed in insect-damaged plant tissues, *J. Chem. Ecol.*, 16, 1221, 1990.
33. **Felton, G. W. and Duffey, S. S.**, Reassessment of the role of gut alkalinity and detergency in insect herbivory, *J. Chem. Ecol.*, 17, 1821, 1991.
34. **Felton, G. W. and Duffey, S. S.**, Ascorbate oxidation reduction in *Helicoverpa zea* as a scavenging system against dietary oxidants, *Arch. Insect Biochem. Physiol.*, 19, 27, 1992.
35. **Giordana, B., Sacchi, F. V., and Hanozet, G. M.**, Intestinal amino acid absorption in lepidopteran larvae, *Biochim. Biophys. Acta*, 692, 81, 1982.
36. **Hagerman, A. E. and Butler, L. G.**, Tannins and lignins, in *Herbivores: Their Interactions with Secondary Plant Metabolites*, 2nd Ed., Vol. 1. *The Chemical Participants*, Rosenthal, G. A. and Berenbaum, M. R., Eds., Academic Press, San Diego, CA, 1991, 355.
37. **Harmatha, J., Mauchamp, B., Arnault, C., and Slama, K.**, Identification of a spirostane-type saponin in the flowers of leek with inhibitory effects on growth of leek-moth larvae, *Biochem. Syst. Ecol.*, 15, 113, 1987.
38. **Hudson, B. J. F. and Lewis, J. I.**, Polyhydroxy flavonoid antioxidants for edible oils. Phospholipids as synergists, *Food Chem.* 10, 111, 1983.
39. **Ishaaya, I.**, Nutritional and allelochemical insect–plant interactions relating to digestion and food intake, in *Insect–Plant Interactions*, Miller, J. R. and Miller, T. A., Eds., Springer-Verlag, NY, 1986, 191.
40. **Johnston, K. A., Lee, M. J., Gatehouse, J. A., and Anstee, J. H.**, The partial purification and characterisation of serine protease activity in midgut of larval *Helicoverpa armigera*, *Insect Biochem.*, 21, 389, 1991.
41. **Krischik, V. A., Barbosa, P., and Reichelderfer, C. F.**, Three trophic level interactions: allelochemicals, *Manduca sexta* (L.), and *Bacillus thuringiensis* var. *kurstaki* Berliner, *Environ. Entomol.*, 17, 476, 1988.
42. **Larson, R. A.**, The antioxidants of higher plants, *Phytochemistry*, 27, 969, 1988.
43. **Lee, K.**, Glutathione S-transferase activities in phytophagous insects: induction and inhibition by plant phototoxins and phenols, *Insect Biochem.*, 21, 353, 1991.
44. **Lenz, C. J., Kang, J., Rice, W. C., McIntosh, A. H., Chippendale, G. M., and Schubert, K. R.**, Digestive proteinases of larvae of the corn earworm, *Heliothis zea*: characterization, distribution, and dietary relationships, *Arch. Insect Biochem. Physiol.*, 16, 201, 1991.
45. **Lighthart, B.**, Some changes in gut bacterial flora of field-grown *Peridroma saucia* (Lepidoptera: Noctuidae) when brought into the laboratory, *Appl. Environ. Microbiol.*, 54, 1896, 1988.
46. **Lindroth, R. L.**, Differential toxicity of plant allelochemicals to insects: roles of enzymatic detoxication systems, in *Insect–Plant Interactions*, Bernays, E., Ed., CRC Press, Boca Raton, FL, 1991, 1.
47. **Ludlum, C. T., Felton, G. W., and Duffey, S. S.**, Plant defenses: chlorogenic acid and polyphenol oxidase enhance toxicity of *Bacillus thuringiensis* subsp. *Kurstaki* to *Heliothis zea*, *J. Chem. Ecol.*, 17, 217, 1991.

48. **Luthy, P., Hofmann, C., and Jaquet, F.,** Inactivation of delta-endotoxin of *Bacillus thuringiensis* by tannin, *FEMS Microbiol. Lett.,* 28, 31, 1985.
49. **MacIntosh, S. C., Kishore, G. M., Perlak, F. J., Marrone, P. G., Stone, T. B., Sims, S. R., and Fuchs, R. L.,** Potentiation of *Bacillus thuringiensis* insecticidal activity by serine protease inhibitors, *J. Agric. Food Chem.,* 38, 1145, 1990.
50. **Martin, M. M., and Martin, J. S.,** Surfactants: their role in preventing the precipitation of proteins in insect guts, *Oecologia,* 61, 342, 1984.
51. **Martin, J. S., Martin, M. M., and Bernays, E. A.,** Failure of tannic acid to inhibit digestion or reduce digestibility of plant protein in gut fluids of insect herbivores: implication for theories of plant defense, *J. Chem. Ecol.,* 13, 605, 1987.
52. **Mattson, W. J.,** Herbivory in relation to plant nitrogen content, *Annu. Rev. Ecol. Syst.,* 11, 503, 1980.
53. **Mattson, W. J. and Scriber, J. M.,** Nutritional ecology of insect folivores of woody plants: nitrogen, water, fiber, and mineral considerations, in *Nutritional Ecology of Insects, Mites, and Spiders,* Slansky, F., Jr. and Rodriguez, J. G., Eds., John Wiley & Sons, NY, 1987, 105.
54. **Mead, L. J., Khachatourians, G. G., and Jones, G. A.,** Microbial ecology of the gut in laboratory stocks of the migratory grasshopper, *Melanoplus sanguinipes* (Fab.) (Orthoptera: Acrididae), *Appl. Environ. Microbiol.,* 54, 1174, 1988.
55. **Morgan, M. R. J.,** Gut carbohydrases in locusts and grasshoppers, *Acrida,* 5, 45, 1976.
56. **Murdock, L. L. and Koidl, B.,** Blood metabolites after intestinal absorption of amino acids in locusts, *J. Exp. Biol.,* 56, 795, 1972.
57. **Murdock, L. L., Brookhart, G., Dunn, P. E., Foard, D. E., Kelley, S., Kitch, L., Shade, R. E., Shukle, R. H., and Wolfson, J. L.,** Cysteine digestive proteinases in Coleoptera, *Comp. Biochem. Physiol.,* 87B, 783, 1987.
58. **Nation, J. L.,** Specialization in the alimentary canal of some mole crickets (Orthoptera: Gryllotalpidae), *Int. J. Insect Morphol. Embryol.,* 12, 2, 1983.
59. **Nedergaard, S.,** Amino acid exchange mechanism in the basolateral membrane of the midgut epithelium from the larva of *Hyalophora cecropia, J. Membrane Biol.,* 58, 175, 1981.
60. **Ohmart, C. P., Thomas, J. R., and Bubela, B.,** Surfactant-producing microorganisms isolated from the gut of a *Eucalyptus*-feeding sawfly, *Perga affinis affinis, Oecologia,* 77, 140, 1988.
61. **Phillips, J. E., Hanrahan, J., Chamberlin, M., and Thomson, B.,** Mechanism and control of reabsorption in insect hindgut, *Adv. Insect Physiol.,* 19, 329, 1986.
62. **Raubenheimer, D.,** Tannic acid, protein, and digestible carbohydrate: dietary imbalance and nutritional compensation in locusts, *Ecology,* 73, 1012, 1992.
63. **Rhoades, D. F.,** The antiherbivore chemistry of Larrea, in *Creosote Bush: Biology and Chemistry of Larrea in the New World Deserts,* Mabry, T. J., Hunziker, J. H., and DiFeo, D. R., Eds. Dowden, Hutchinson & Ross, Stroundburg, PA, 1977, 135.
64. **von Rutschke, E., Gerhardt, W., and Herrmann, V.,** Studies on amino acid transport by the intestine of the stick insect *Carausius morosus, Br. Zool. Jb. Physiol.,* 80, 24, 1976.
65. **Ryan, C. A.,** Proteinase inhibitor gene families: strategies for transformation to improve plant defenses against herbivores, *BioEssays,* 10, 20, 1989.
66. **Santos, C. D., Ferreira, C., and Terra, W. R.,** Consumption of food and spatial organization of digestion in the cassava hornworm, *Erinnyis ello, J. Insect Physiol.,* 29, 707, 1983.
67. **Santos, C. D., Ribeiro, A. F., Ferreira, C., and Terra, W. R.,** The larval midgut of the cassava hornworm (*Erinnyis ello*), *Cell Tissue Res.,* 237, 565, 1984.
68. **Santos, C. D. and Terra, W. R.,** Distribution and characterization of oligomeric digestive enzymes from *Erinnyis ello* larvae and inferences concerning secretory mechanisms and the permeability of the peritrophic membrane, *Insect Biochem.,* 16, 691, 1986.
69. **Schultz, J. C.,** Tannin–insect interactions, in *Chemistry and Significance of Condensed Tannins,* Hemingway, R. W. and Karchesy, J. J., Eds., Plenum Press, N.Y., 1989, 417.
70. **Schultz, J. C. and Keating, S. T.,** Host-plant-mediated interactions between the gypsy moth and a baculovirus, in *Microbial Mediation of Plant–Herbivore Interactions,* Barbosa, P., Krischik, V. A., and Jones, C. G., Eds., John Wiley & Sons, New York, 1991, 489.
71. **Schultz, J. C. and Lechowicz, M. J.,** Hostplant, larval age, and feeding behavior influence midgut pH in the gypsy moth (*Lymantria dispar*), *Oecologia,* 71, 133, 1986.
72. **Shukle, R. H. and Murdock, L. L.,** Lipoxygenase, trypsin inhibitor, and lectin from soybeans: effects on larval growth of *Manduca sexta* (Lepidoptera: Sphingidae), *Environ. Entomol.,* 12, 787, 1983.
73. **Simpson, S. J.,** The pattern of feeding, in *Biology of Grasshoppers,* Chapman, R. F. and Joern, A., Eds., John Wiley & Sons, New York, 1990, 73.
74. **Slansky, F., Jr.,** Allelochemical–nutrient interactions in herbivore nutritional ecology, in *Herbivores: Their Interactions with Secondary Plant Metabolites,* 2nd Ed., Vol.2. *Ecological and Evolutionary Processes,* Rosenthal, G. A. and Berenbaum, M. R., Eds., Academic Press, San Diego, CA, 1992, 135.

75. **Smith, A. F., Tsuchida, K., Hanneman, E., Suzuki, T. C., and Wells, M. A.**, Isolation, characterization, and cDNA sequence of two fatty acid-binding proteins from the midgut of *Manduca sexta* larvae, *J. Biol. Chem.*, 267, 380, 1992.
76. **Steinly, B. A. and Berenbaum, M. R.**, Histopathological effects of tannins on the midgut epithelium of *Papilio polyxenes* and *Papilio glaucus*, *Entomol. Exp. Appl.*, 39, 3, 1985.
77. **Terra, W. R.**, Evolution of digestive systems of insects, *Annu. Rev. Entomol.*, 35, 181, 1990.
78. **Terra, W. R., Valentin, A., and Santos, C. D.**, Utilization of sugars, hemicellulose, starch, protein, fat, and minerals by *Erinnyis ello* larvae and the digestive role of their midgut hydrolases, *Insect Biochem.*, 17, 1143, 1987.
79. **Thomas, K. K.**, Studies on the absorption of lipid from the gut of desert locust, *Schistocerca gregaria*, *Comp. Biochem. Physiol.*, 77A, 707, 1984.
80. **Toth-Prestia, C. and Hirshfield, I. N.**, Isolation of plasmid-harboring *Serratia plymuthica* from facultative gut microflora of the tobacco hornworm, *Manduca sexta*, *Appl. Environ. Microbiol.*, 54, 1855, 1988.
81. **Treherne, J. E.**, The absorption of glucose from the alimentary canal of the locust, *Schistocerca gregaria* (Forsk.), *J. Exp Biol.*, 35, 297, 1959.
82. **Treherne, J. E.**, The absorption and metabolism of some sugars in the locust, *Schistocerca gregaria* (Forsk.), *J. Exp Biol.*, 35, 611, 1959.
83. **Treherne, J. E.**, Amino acid absorption in the locust (*Schistocerca gregaria* Forsk.), *J. Exp. Biol.*, 36, 533, 1959.
84. **Trumble, J. T., Moar, W. J., Brewer, M. J., and Carson, W. G.**, Impact of UV radiation on activity of linear furanocoumarins and *Bacillus thuringiensis* var. *kurstaki* against *Spodoptera exigua*: implications for tritrophic interactions, *J. Chem. Ecol.*, 17, 973, 1991.
85. **Tsuchida, K. and Wells, M. A.**, Digestion, absorption, transport and storage of fat during the last larval stadium of *Manduca sexta*. Changes in the role of lipophorin in the delivery of dietary lipid to the fat body, *Insect Biochem.*, 18, 263, 1988.
86. **Turunen, S.**, Absorption and utilization of essential fatty acids in lepidopterous larvae: metabolic implications, in *Metabolic Aspects of Lipid Nutrition in Insects*, Mittler, T. E. and Dadd, R. H., Eds., Westview, Boulder, CO, 1983, 57.
87. **Turunen, S.**, Uptake of dietary lipids: a novel pathway in *Pieris brassicae*, *Insect Biochem.*, 18, 499, 1988.
88. **Turunen, S.**, Absorption of choline, myo-inositol, and oleic acid in the midgut of *Pieris brassicae*: sectional differentiation and uptake into the haemolymph, *J. Insect Physiol.*, 36, 737, 1990.
89. **Turunen, S. and Chippendale, G. M.**, Relationship between dietary lipids, midgut lipids, and lipid absorption in eight species of Lepidoptera reared on artificial and natural diets, *J. Insect Physiol.*, 35, 627, 1989.
90. **Weintraub, H. and Tietz, A.**, Triglyceride digestion and absorption in the locust, *Locusta migratoria*, *Biochem. Biophys. Acta*, 306, 31, 1973.
91. **Weintraub, H. and Tietz, A.**, Lipid absorption by isolated intestinal preparations, *Insect Biochem.*, 8, 267, 1978.
92. **Wolfersberger, M. G., Spaeth, D. D., and Dow, J. A. T.**, Permeability of the peritrophic membrane of tobacco hornworm larval midgut, *Am. Zool.* 26, 74A, 1986.
93. **Wolfson, J. L. and Murdock, L. L.**, Diversity of digestive proteinase activity among insects, *J. Chem. Ecol.*, 16, 1089, 1990.
94. **Yoshimura, T., Tabata, H., Nishiho, M., Ide, E., Yamaoko, R., and Kayashiya, K.**, l-Cystein lyase of the webbing clothes moth, *Tineola bisselliella*, *Insect Biochem.*, 18, 771, 1988.

SUBJECT INDEX

Scientific names of plants and insects are listed alphabetically in the Taxonomic Index, page 235.

A

Active defense response (ADR) theory, 7–8
Alkaloids, 3, 5–6, 9–10, 109, 112, 141–142, 217–218, *see also* Ergot alkaloids
  age-associated concentration of, 7
  biosynthesis references, 152
  catabolism of, 146
  damage-induced increase in, 10
  defoliation and, 4
  in endophyte–plant associations, 86
  in floral volatiles, 64–65
  Herivore-deterrent activity of, 11–12
  loline alkaloids, 86, 90, 95
  in nitrogen transport, 12
  as pheromone precursors, 64–65
  production costs of, 149
  "sunscreen" role of, 14
  synergistic activity of, 92
  tissue allocation of, 94
  turnover in, 13
  UV-induced increase in, 13–14
  variable spatial/temporal production of, 96
  vascular uptake of, 133
Allelochemicals, 210, 214
  activity of, 215–217, 219
  and microbial pesticides, 218
Antennal odor receptors, 59, *see also* Electro-antennogram
Antiherbivore defense compounds
  in cell vacuoles, 127–131
  in glandular trichomes, 69, 127, 129
  in resin ducts, 127
  in secretory cavities, 127–128
Antiherbivore defense costs, 106–107, 149–150
  of autotoxicity avoidance adaptations, 129–132, 149
  of biosynthesis, 149, *see also* Biosynthesis
  biosynthetic machinery, 116–126
  raw materials, 107–116
  and carbon–nutrient balance, 148–149
  of defense compound modifications, 129
  of defense compound reduction, 143–148
  of defense compound storage, 126–127, 129–131, 149
  of defense compound storage site construction, 127–128
  of maintenance, 149
  of metabolic turnover, 137–142
  of transport, 132–133, 149
    intercellular, 134–137
    intracellular, 133–134
  of volatilization/leaching/exudation, 143
Antiherbivore volatiles, *see* Floral volatiles
Antioxidants, in photosystem protection, 14
Antiparasite responses, 11–12

Ants, 54
Aphids (Aphididae), 89–90, 97, 178, 180–181, 183, 194, 196, 199
  adelgid (*Gilletteella cooleyii*), 182
  asparagus aphid (*Brachycorynella asparagi*), 195
  birch aphid (*Euceraphis punctipennis*), 185–186
  bird cherry aphid (*Rhopalosiphum padi*), 194–195
  bird cherry-oat aphid (*Rhopalosiphum padi*), 185–186, 188
  black bean aphid (*Aphis fabae*), 188, 196
  cabbage aphid (*Brevicoryne brassicae*), 185, 195
  grain aphid (*Sitobion avenae*), 176, 182
  larval (nymphal) form, 178
  (*Myzus persicae*), 91
  nettle aphid (*Microlophium carnosum*), 182
  oak tree hopper (*Platycotis vitata*), 186
  pea aphid (*Acyrthosiphon pisum*), 181–182
  peach potato aphid (*Myzus persicae*), 182
  planthopper (*Prokelisia marginata*), 180
  rose aphid (*Macrosiphum rosae*), 182
  sapsuckers (*Prokelisia marginata*), 186–187
  seasonal adaptations of, 188, 190
  sycamore aphid (*Drepanosiphum platanoidis*), 180, 182, 185, 189
  winged morphs differences, 188, 190
Aphids (aphididae)
  Russian wheat aphid, 92
Arthropods (noninsect), floral volatiles and, 68–69

B

Baculoviruses, 218
Bees (Apidae)
  behavioral training of, 60–62
  bumble bee, 51, 70
  carpenter bee (*Xylocopa varipuncta*), 54
  euglossine, 62
  honey bee (*Apis mellifera*), 51, 60–61, 69
  oriental honey bee (*Apis cerana japonica*), 64
  pollen volatiles and, 70
Beetles (Coleoptera), 210, 212
  Argentine stem weevil (*Listronotus bonariensis*), 88–89, 91–92, 95–97
  beech leaf-mining weevil (*Rynchaenus fagi*), 198
  billbug, 91, 97
  black beetle, 90
  bluegrass billbugs (*Sphenophorus parvulus*), 90
  cabbage stem flea beetle, 91
  chrysomelid beetle (*Paropsis atomaria*), 192
  Colorado potato beetle, 218
  corn flea beetle (*Chaetocnema pulicaria*), 90
  cowpea beetle (*Callosobruchus maculatus*), 193
  elm bark beetle (*Scolytus multistriatus*), 93
  flour beetle (*Tribolium castaneum*), 90

TAXONOMIC INDEX

Common names of plants and insects are found in the Subject Index, page 225.

A

235

Milton Keynes UK
Ingram Content Group UK Ltd.
UKHW051933141024
449569UK00027B/1480